URBAN DYNAMICS AND SPATIAL CHOICE BEHAVIOUR

THEORY AND DECISION LIBRARY

AN INTERNATIONAL SERIES
IN THE PHILOSOPHY AND METHODOLOGY OF THE
SOCIAL AND BEHAVIORAL SCIENCES

VOLUME 49

URBAN DYNAMICS AND SPATIAL CHOICE BEHAVIOUR

Edited by

JOOST HAUER
Department of Geography, University of Utrecht, The Netherlands

HARRY TIMMERMANS
Department of Urban Planning, Eindhoven University of Technology, The Netherlands

and

NEIL WRIGLEY
Department of Town Planning, University of Wales, Cardiff, United Kingdom

KLUWER ACADEMIC PUBLISHERS
DORDRECHT / BOSTON / LONDON

1989

ISBN 0-7923-0391-1

Published by Kluwer Academic Publishers,
P.O. Box 17, 3300 AA Dordrecht, The Netherlands.

Kluwer Academic Publishers incorporates
the publishing programmes of
D. Reidel, Martinus Nijhoff, Dr W. Junk and MTP Press.

Sold and distributed in the U.S.A. and Canada
by Kluwer Academic Publishers,
101 Philip Drive, Norwell, MA 02061, U.S.A.

In all other countries, sold and distributed
by Kluwer Academic Publishers Group,
P.O. Box 322, 3300 AH Dordrecht, The Netherlands.

Printed in the Netherlands

PREFACE

This book arises from The Fourth European Colloquium on Theoretical and Quantitative Geography which was held in Veldhoven, The Netherlands in September 1985. It contains a series of papers on spatial choice dynamics and dynamical spatial systems which were presented at the colloquium, together with a few other sollicited ones. The book is intended primarily as a state-of-the art review of mainly European research on these two fastly growing problem areas.

As a consequence of this decision, the book contains a selection of papers that differs in terms of focus, level of sophistication and conceptual background. Evidently, the dissimination of ideas and computer software is a time-related phenomenon, which in the European context is amplified by differences in language, the profile of geography and the formal training of geographers. The book reflects such differences.

It would have been impossible to produce this book without the support of the various European study groups on theoretical and quantitative geography. Without their help the meetings from which this volumes originates would not have been held in the first place. We are also indebted to the Royal Dutch Academy of Science for partly funding the colloquium, and to SISWO and TNO/PSC for providing general support in the organisation of the conference.

Special thanks go out to Mrs. Irene Borgers - de Jong for typing the manuscripts and to Aloys Borgers for implementing the necessary word-processing software in the Eindhoven computer environment.

Finally, we owe a lot to our families who have been inspirational in their understanding, patience and encouragement during the many hours that were devoted to this project in a very hectic period of our lives due to important changes in our professional careers.

November, 1988

Joost Hauer
Harry Timmermans
Neil Wrigley

TABLE OF CONTENTS

JOOST HAUER, HARRY TIMMERMANS AND NEIL WRIGLEY

INTRODUCTION

The development of quantitative geography has been characterised by a number of phases in which particular types of methods or models have received major attention. Once the potential advantages and limitations of such methods or models were sufficiently understood they were incorporated into the geographer's apparatus in studying spatial phenomena. The last two decades have witnessed special interest in such topics as spatial statistics, multivariate statistical analysis such as factor analysis and cluster analysis, gravity and entropy - maximizing models, categorical data analysis and discrete choice / decompositional multiattribute preference analysis.

The latest fields of special interest, at least in European quantitative geography, can be subsumed under the general heading of spatial dynamics. Conventional choice models which predict the probability that an individual will choose a single choice alternatives from the set of available alternatives have been extended to predict the dynamics of choice. Likewise static urban models were gradually replaced by dynamic models with a concern for the evolution of spatial phenomena, bifurcations etc.

This book is intended to give an overview of typical European studies in the field of spatial dynamics. Purposefully, contributions of the various European countries have been selected. This may disrupt the consistency of this volume, but at the same time reflects the different theories and methodologies as adopted by scholars from different countries who may work in different traditions.

This general purpose is reflected in two main sections. The first section is on dynamic choice models. In the context of this volume, both multipurpose trips / trip chaining and inter-temporal repeated choices are considered to be examples of dynamic choice behaviour. Much of the modelling effort in this area is an extension of traditional random utility models. The first chapter in the volume, written by Timmermans and Borgers, provides an overview of different modelling approaches concerned with dynamic choice behaviour. It is included in this volume to sketch the context, not only with respect to those approaches that are included in this volume, but also with a respect to approaches that are not. It should give the reader an overview of the different approaches and enable him to place the following more specific chapters into proper perspective.

The first of these, a higly empirical chapter, is by Thill and Thomas. They study the problem of trip chaining. Previous research has suggested spatial behaviour to be influenced by temporal variability and urban spatial structure. The authors wish to complement such findings by analyzing whether parameters of spatial choice behaviour are dependent upon the types of services that are patronized. In particular, the spatial choice behaviour to pharmacies versus post offices is compared. Their findings suggest that some clear differences exist between travel patterns associated with the choice of these two services. This is an important result in the light of how to model trip chaining behaviour, because their results suggests that one should include purpose-specific parameters.

Uncles, in his chapter, addresses the problem of repeated choices of consumers. The use of beta-logistic and Dirichlet models in studying travel choices is discussed. Several examples are presented. Special attention is paid to the testing of the goodness-of-fit and the temporal independence of these models.

Davies addresses the problem of misspecification in models of dynamic choice behaviour. Attention is focused on quasi-likelihood as a method which may provide insight into the misspecifion of Poisson and binomial models. Empirical examples concerning store choice frequencies and voting behaviour are presented to show how quasi-likelihood models can be used to correct for misspecification and how misspecification models can result in misleading conclusions.

In a second paper, co-written with Pickles, the problem of inference from cross-sectional data on dynamic choice processes are discussed. Limitations on the scope for inferences using cross-sectional data due to the presence of omitted variables, heterogeneity and endogeneity, are clearly illustrated in the context of housing mobility. The authors argue that longitudinal models have more to offer, although some issues need careful consideration. In particular, they discuss the problems of identification, initial conditions and generalisation.

The estimation of such models requires the availability of longitudinal data. Such data have their own specific problems. One of these is attrition: the fact that some respondents drop out as the panel proceeds. Attrition might be a source of bias if there exists some covariance between the phenomenon under investigation and the influences on attrition. The assessment of attrition in panel data is addressed by Hensher. He relates the problem to correction procedures of sample weighting. The existence of attrition bias is assessed in a study on household vehicle use and possession. Although no evidence of attrition bias is found, the author concludes that a model specification which accomodates a correlated error structure between time periods is required.

The final chapter of the first section is a paper written by Nijkamp and Reggiani who explore the extent to which conventional spatial interaction models are in agreement with random utility theory. In particular, they show that an analytical correspondence between the conventional Alonso model and a discrete choice model can be established. They continue by demonstrating that dynamic spatial interaction models models can be interpreted in terms of a dynamic disaggregate choice model by using optimal control theory. Also in this dynamic framework, an analytic correspondence between discrete choice models and spatial interaction models can be found under certain conditions.

The second part of this volume is concerned with dynamic urban models. Dynamic models already entered the scene in the 1960's but due to the ideas on self-organising systems of Prigogine, the interest in dynamic urban models was renewed in the early 1980's. The isssue whether we are dealing with new wine or new bottles is addressed in a state-of-the-art paper by Pumain. She briefly discusses catastrophe theory, Volterra-Lotka models, the master equation approach, the Leeds approach and some simulation models. In addition, comments are made regarding conceptualisation, data availability, calibration and interpretation problems related to these approaches.

Many of these modelling approaches are presented in the subsequent chapters by Birkin & Wilson, Lombardo & Rabino, Haag & Frankhauser and Durand & LeBerre. The models presented by Birkin & Wilson are a dynamic version of the familiar spatial interaction model of spatial choice behaviour. A previous dynamic model is extended to include prices and rents. A series of dynamic models results and numerical experiments are conducted to explore the properties of these models.

Lombardo & Rabino present a closely related model. They focus on the range of stability of spatial structure, the speed of convergence to the equilibria solutions of the models, the shape of the market areas and the pattern of trips to services. In the second part of their paper, they argue that dynamic spatial interaction models can be used as a clustering technique and two examples are provided.

An example of the master equation approach is provided by Haag & Frankhauser. They attempt to combine dynamic modelling approaches which focus on the macro states of the system and the interactions of aggregate variables and modelling approaches on individual behaviour at the micro scale. The dynamics of individual decision processes are formulated in terms of a master equation. This leads to a macro configuration of the urban system. This idea is incorporated into a dynamic model of the urban service sector. The stationary patterns of the expenditure flows and the scale of provision are investigated by numerical simulations.

The final model in this volume discussed by Durand & LeBerre is an integrated dynamic model. It is a more or less traditional approach based on the simulation language DYNAMO. It includes the elements of population, social relations, jobs, economic factors, housing and space.

The model was developed in a planning context. Several simulation results are presented. In addition, the authors discuss problems of representation of space, data requirements, spatial transferability and spatial heterogeneity and their implication for dynamic modelling.

The volume is concluded by Longley and Batty's chapter on measuring and simulating the structure and form of cartographic lines. At first glance, this chapter does not fit well into the general themes of this volume. However, we have decided to include it because the methods outlined in this chapter may prove useful to the description of the dynamics of urban form and the mapping of model forecasts.

This collection of papers illustrates the increasing interest in spatial choice dynamics. It is hoped that this volume will illuminate the kind of modelling approaches that are currently used by (mainly) European scholars. So far, most analyses have been rather theoretical. Future comparative empirical research should learn whether the potential theoretical advantages of dynamic approaches justify their application given the inevitable higher costs of data collection and calibration associated with such approaches.

PART I

DYNAMIC CHOICE MODELS

HARRY TIMMERMANS AND ALOYS BORGERS

DYNAMIC MODELS OF CHOICE BEHAVIOUR: SOME FUNDAMENTALS AND TRENDS

1. INTRODUCTION

In this chapter we will discuss some models of dynamic decision-making and choice behaviour. The term dynamic indicates that we are interested in choice behaviour over time, i.e. in possible changes in choices. The discussion will be restricted to three general types of modelling approaches. First, stochastic models of buying behaviour will be discussed. This is followed by a summary of newly developed variety-seeking models. Finally a brief summary is given of some of the latest developments in the field of dynamic discrete choice models. It should be noted that these models constitute only a small part of this rapidly growing field of research and that the structure of this section is rather arbitrary. Other reviews can be found in Halperin and Gale (1984), Halperin (1985) and Hensher and Wrigley (1984).

2. STOCHASTIC MODELS OF BUYING BEHAVIOUR

The class of stochastic models of buying behaviour consists of brand choice models and purchase incidence models. Brand choice models predict which choice alternative will be chosen given that a choice is made at a particular point in time. Purchase incidence models predict when an alternative will be chosen or how many alternatives will be chosen in a specified interval of time.

Several models belonging to the class of stochastic buying behaviour will be discussed in the following subsections. For a more detailed survey, we refer to Massy, Montgomery and Morrison (1970).

2.1. *Bernoulli Models*

Perhaps the most simple model of dynamic choice behaviour is the Bernoulli model. It is based on the assumption that the probability of choosing alternative i is constant over time. It implies that the past history of the process has no effect on the choice probabilities. Hence, the model may be expressed as:

$$p_t = p \tag{1}$$

where p_t is the probability that a particular alternative is chosen at time t;
p is the initial probability of choosing the alternative.

Another limiting property of the Bernoulli model is that it assumes homogeneity. That is, it is assumed that the choice probabilities apply to all individuals. Nevertheless the model has been frequently

J. Hauer et al. (eds.), Urban Dynamics and Spatial Choice Behaviour, 3–26.

used with success, especially in a marketing context. Burnett (1975) applied the Bernoulli model in a study of spatial shopping behaviour using panel data. The model did not perform very well though.

2.2. *The Compound Beta Bernoulli Model*

Perhaps the most important limitation of the standard Bernoulli model is its homogeneity assumption. The compound beta Bernoulli model has been developed to relax this assumption. The model still assumes that every individual in the population has a constant probability p of choosing an alternative and a probability of $(1-p)$ of choosing another alternative, but the homogeneity assumption is replaced by assuming that p has a beta distribution over individuals in the population. This beta distribution has the form:

$$b(p) = \begin{cases} \dfrac{\Gamma(\alpha + \beta)}{\Gamma(\alpha)\,\Gamma(\beta)}\, p^{\alpha-1}(1-p)^{\beta-1}, & 0 < p < 1 \\ 0, \text{ otherwise} & \end{cases} \tag{2}$$

where $\Gamma(\cdot)$ is the gamma function;
α and β are parameters to be estimated (α, $\beta > 0$).

2.3. *Markov Models*

Markov models have been used to study the dynamics of choice behaviour. Especially the first-order models have been applied frequently. These models are typically based on deriving a transition probability matrix which expresses the probability that alternative j will be chosen at time t+1 given that alternative i has been chosen at time t. Consequently, these transition or switching probabilities are independent from the choices at times t-1, t-2, ... Moreover, in the conventional models homogeneity and stationarity is assumed: the transition probabilities apply to all individuals in the population and the transition probability matrix is independent of t. Given these assumptions, choice behaviour and market shares at some future point at time t^* can be calculated easily by raising the transition probability matrix to the power t^* and multiplying this matrix by an initial state vector. Likewise, the vector of steady state probabilities can be calculated in a straightforward manner.

This standard Markov model has been very popular in marketing science. A geographical example is provided in Burnett (1974, 1978) and Crouchley, Davies and Pickles (1982), although it should be noted that Markov chain analysis as such is a well-known technique in geography which has been used for a variety of purposes.

It is evident that the first-order Markov chain is based on some rigorous assumptions: first order, homogeneity and stationarity. Hence, several authors have attempted to develop more sophisticated Markov models which relax such assumptions. Higher-order Markov models have been developed to incorporate the effect of choice behaviour at times t-1, t-2, etc. on choice behaviour at time t+1. Non-stationary models have been proposed to make the transition probability matrix dependent upon the time period. Heterogeneity has been introduced by segmentation of the population into groups. Some of these developments are well known in geographical research. In particular, such sophisticated Markov models have been used in studying migration (e.g. Ginsberg, 1971, 1972, 1973, 1978, 1979).

2.4. *Brand Loyal Models*

Brand loyal models have been developed by marketing scientists, but in a geographic context they have been called place loyal models. Basically, this model is a compound Markov model. It is based on the following assumptions: each individual's choice behaviour can be described in terms of a first-order process; the parameter describing this process has a probability distribution $b(p)$ among the individuals in the population. In particular, it is assumed that the first-order 0-1 process has the transition matrix:

	1	0
1	p	$1-p$
0	kp	$1-kp$

(3)

where k is a constant for each individual ($0 < k < 1$).

$b(p)$ is beta distributed, although any arbitrary probability density is allowed.

2.5. *Last Purchase Loyal Models*

An alternative to the brand loyal model is the last purchase loyal model which has the following transition matrix:

	1	0
1	p	$1-p$
0	$1-kp$	kp

(4)

Apart from the structure of the transition matrix, this model is based on the same assumptions as the brand loyal model.

2.6. *Linear Learning Models*

Linear learning models were originally developed in psychology for the purpose of describing data from laboratory experiments on adaptive behaviour. However, they have also played an important role in the study of brand choice behaviour. In a geographical context, the model has hardly been used, an exception being the study by Burnett (1977) on spatial shopping behaviour. Actually, she found that the linear learning model outperformed both the Bernoulli model and the Markov model.

The basic assumption of the linear learning model is that choice probabilities at time $t+1$ are a linear function of the choice probabilities at time t. Further, the model assumes quasi-stationarity in the sense that the parameters of the model do not change over short periods of time and that all individuals exhibit adaptive behaviour that can be described by a single set of parameters. The model may be expressed as follows:

$$p_{t+1} = \begin{cases} \alpha + \beta + \lambda p_t & \text{if the choice alternative is chosen at time } t \\ \alpha + \lambda p_t & \text{if the choice alternative is not chosen at time } t \end{cases} \tag{5}$$

where α, β are parameters denoting the intercept of the rejection operator and the difference in intercept between the acceptance and the rejection operator;
λ is the slope of both operators.

The expectation of p_t for $t > 0$ is equal to:

$$E(p_t | p_0) = \left[\alpha + \alpha(\beta + \lambda) + \alpha(\beta + \lambda)^2 + \ldots + \alpha(\beta + \lambda)^{t-1}\right] + (\beta + \lambda)^t \; p_0 \tag{6}$$

This equation can be used to predict dynamic choice behaviour. The linear learning model can be considered as a generalization of both the zero order Bernoulli and the first-order Markov models. If $\lambda = 0$, the linear learning model collapses into a Markov model; if $\alpha = \beta = 0$ and $\lambda = 1$, we have a Bernoulli model.

2.7. *Purchase Incidence Models*

Again, these models were originally developed by marketing scientists in the fifties. The main differences between previously discussed dynamic models and purchase incidence models is that the latter are primarily concerned with the problem of *when* an alternative will be chosen or, equivalently, how many choices will be made in a specified interval of time, whereas the former are concerned with predicting which alternative will be chosen given that a choice is made at a particular point in time.

A well-known model in this class is Ehrenberg's negative binomial model. It is based on the following assumptions: the average number of purchases of a particular alternative is constant over time periods; the purchases of an individual over successive equal-length time periods can be described by the Poisson distribution:

$$\Pr\{N_t = k | \mu\} = f(N_t | \mu) = \exp(-\mu)\mu^k \,/\, k!, \qquad k \geq 0 \tag{7}$$

where N_t is the total number of purchases during a fixed interval of time;
μ is the average purchase rate,

and, finally, that these average purchase rates are distributed over the individuals according to a gamma distribution with parameters α and β:

$$f(\mu) = \beta \exp(-\beta\mu)(\beta\mu)^{\alpha-1} / \Gamma(\alpha), \qquad \mu > 0 \tag{8}$$

Given these assumptions, the aggregate distribution of purchase events follows the following negative binomial model:

$$\begin{aligned} f(N_t) &= \int_0^\infty f(N_t|\mu)f(\mu)d\mu \\ &= \left[\frac{\beta}{1+\beta}\right]^\alpha \left[\frac{(N_t + \alpha)}{N_t\ \Gamma(\alpha)}\right] \left[\frac{1}{1+\beta}\right]^{N_t} \end{aligned} \tag{9}$$

Other interesting measures, such as the incidence of repeat buying, average purchase frequency and market penetration can be derived in a straightforward manner (see Ehrenberg, 1972 for details).

The basic model has been extended in a number of important ways to relax its assumptions. Chatfield, Ehrenberg and Goodhardt (1966) have shown that the distribution of the total number of purchases in a fixed time interval for individuals who purchase the item at least once during the interval can be approximated by the logarithmic series distribution:

$$f(N_t|N_t>0) = -q^{N_t} / N_t \ \ln(1-q) \tag{10}$$

where q is the parameter of the distribution.

Other distributions, such as the zero-truncated negative binomial (Zufryden, 1977) and the geometric model have been used.

The original model is based on the assumption that the number of purchases of a particular brand by a single individual in equal-length successive time periods are independent and follow a Poisson distribution, implying that inter-purchase times are exponentially distributed. Some authors have argued that such inter-purchase times are more regular and they have introduced the Erlang 2 distribution to describe inter-purchase times (Herniter, 1971; Jeuland *et al.*, 1980). Empirical evidence however has not substantiated this claim: only a small improvement in fit has been obtained by using the resulting negative binomial distribution. In a geographical context, Dunn, Reader and Wrigley (1983) also found that the negative binomial model holds for the majority of their respondents.

The negative binomial model has been applied successfully in marketing science (see Ehrenberg, 1968, 1972) for brand choice behaviour. Wrigley (1980) was the first geographer to use this model in a study of purchasing patterns at particular store types. In a follow up study, the model was successfully applied to purchasing at individual stores in Cardiff (see Wrigley and Dunn, 1984a; Dunn, Reader and Wrigley, 1983).

Another important development is the generalization of the model to more brands or stores. This extension is the Dirichlet model which specifies probabilistically how many purchases each consumer makes in a time-period and which brand/store is chosen on each occasion (Goodhardt, Ehrenberg and Chatfield, 1984), The model combines aspects of purchase incidence and choice aspects and is based on the following assumptions:

- individuals' choice probabilities are constant over time and independent over successive purchases, implying that the number of purchases of each alternative an individual makes in a sequence of purchases can be modelled by a multinomial model;
- these choice probabilities vary across individuals according to a Dirichlet distribution;
- successive purchases of an individual are independent with a constant mean rate, implying that the number of purchases made in each of a succession of equal-length time periods follows a Poisson distribution;
- the mean purchasing rates vary among individuals according to a gamma distribution;
- the choice probabilities and average purchase frequencies of different individuals are distributed independently over the population.

The model itself is then obtained by mixing these multinomial, Dirichlet, Poisson and gamma distributions. The model has worked remarkably well in marketing science (Goodhardt *et al.*, 1984) but also for predicting brand purchases within store groups (Kau and Ehrenberg, 1984) and multistore purchase patterns within individual stores (Wrigley and Dunn, 1984b, 1984c).

A similar model was developed by Jeuland, Bass and Wright (1980) but they replaced the Poisson distribution by the Erlang 2 distribution, which results in a multiple hyper-geometric model. This model worked well in a study of purchases of cooking oil. Yet another model has been advanced by Zufryden (1977). His model is also based on the Erlang and gamma distribution of purchase incidence and the assumption of independence between choice behaviour and purchase incidence behaviour, but he uses a linear model with purchase probabilities varying over the population according to a beta distribution rather than a multinomial distribution. His model also performed extremely well (Zufryden, 1978).

All of the above models do not incorporate explanatory variables. Consequently, the effects of managerial or planning decisions on behaviour cannot be assessed. In order to circumvent this disadvantage a number of authors have proposed models which basically add explanatory variables to the models discussed in this section. For example, Jones and Zufryden (1980) proposed a logit model to explain brand choice probability as a function of purchase explanatory variables. Following Ehrenberg, they assume the negative binomial distribution to describe the product class purchase distribution over the population. The probability of choosing a particular alternative given that the class purchase is being made is then modelled by a logit model assuming heterogeneity among individuals with respect to these choice probabilities (beta distribution), independence from past purchase outcomes and time invariance. Examples of applications of this approach can be found in Jones and Zufryden (1980, 1982). In a similar vein, Paull (1978) used a polytomous logit regression approach to predict discrete purchase quantities in a generalized negative binomial distribution. Broom and Wrigley (1983) and Wrigley and Dunn (1985) have proposed to incorporate explanatory variables into negative binomial and Dirichlet models using loglinear forms. Such extensions clearly are an important step forward in building policy-relevant models of dynamic choice behaviour.

3. MODELS OF VARIETY-SEEKING BEHAVIOUR

Two types of variety-seeking behaviour may be distinguished: structural variety-seeking behaviour and temporal variety-seeking behaviour. Structural variety is the variety that is present within a set of objects whereas temporal variety is the variety that is implied by a sequence of choices (Pessemier, 1985). In the context of spatial choice behaviour temporal variety-seeking is perhaps most important. However, since the basic ideas underlying the models directed at these two types of variety are very similar, we will discuss both approaches in this section.

An initial distinction will be made between two types of models: inventory-based models and non-inventory-based models. The specification of the former type of model is explicitly based on the assumption that the attributes of the chosen alternatives are accumulated in attribute inventories, whereas the latter type of model is not explicitly parameterized in this respect.

3.1. *Inventory-based Variety-seeking Models*

One of the first models of variety-seeking behaviour, originally developed for the case of structural variety, was proposed by McAlister (1979). This so-called model of attribute satiation was built on two basic assumptions:

- attributes are cumulative, implying that the total amount of a particular attribute inherent in a group can be calculated by summing the attribute values across the alternatives belonging to that group;
- the marginal utility for each attribute is a decreasing function.

McAlister selected a quadratic utility function to represent the second assumption. More specifically, it was assumed that the square of the difference between the summed attribute values and an individual's ideal point is an appropriate functional form. Hence, it was assumed that a combination of alternatives will be chosen if:

$$U(g) > U(h) \qquad \forall\, h \neq g \tag{11}$$

where $U(g) = - \sum_{k=1}^{K} w_k (X_{g.k} - \hat{X}_k)^2$ (12)

$U(g)$ is the utility of group g of choice alternatives;
$X_{g.k}$ is the sum of the values for attribute k across the choice alternatives in group g;
$\hat{X}_k$ is the ideal (most preferred) level for the k-th attribute;
w_k is the importance weight of the k-th attribute;
K is the total number of attributes.

In the same study, a more sophisticated version based on Farquhar and Rao's balance model (Farquhar and Rao, 1976) was tested. They divided the attributes in four types. First, they made a distinction between desirable and undesirable attributes. Their relationship to preference is assumed to be reflected in a linearly increasing respectively decreasing function. Second, they assume that preference for a collection of choice alternatives is also influenced by the diversity within the collection. If diversity increases preference, the attribute is called 'counterbalancing'. If, in contrast, preference decreases with increasing diversity, the attribute is termed 'equibalancing'. For both these types of attributes, Farquhar and Rao posit a linear relationship with preference for the collection of choice alternatives.

In a follow-up paper, McAlister (1982) extended the attribute satiation model to the case of temporal variety-seeking. This dynamic attribute satiation model (DAS) differs from its predecessor in that a time-related additional assumption is built in. More specifically it is assumed that a consumption history may be converted into an inventory by a time-related inventory retention factor λ_k. The model may be expressed as follows:

$$U_i^{(T)} > U_j^{(T)} \tag{13}$$

The utility of alternative i at time T is defined as:

$$U_i^{(T)} = - \sum_{k=1}^{K} w_k \left[(I_k^{(T)} + X_{ik}) - \hat{X}_k \right]^2 \tag{14}$$

where $I_k^{(T)}$ is the inventory of attribute k at time T;
X_{ik} is the amount of alternative i on attribute k.

The inventory of attribute k at time T is defined as:

$$I_k^{(T)} = \sum_{t=1}^{T-1} \lambda_k^{T-t} \; X_k^{(t)} \qquad (15)$$

where $X_k^{(t)}$ is the amount of the alternative chosen at time t on attribute k;
λ_k is an inventory retention factor for the k-th attribute, $0 \leq \lambda_k \leq 1$.

Note that if λ_k is 1 for all k, the above model is equal to the attribute satiation model. If $\lambda_k \neq 1$, a consumption history is converted into an inventory by an inverse function of the inventory retention factor.

In order to calibrate these models, the researcher should collect the following data. Subjects' perceptions of the degree in which a choice alternative possesses the selected attributes should be measured. In addition, subjects should be asked to rank the (combinations of) choice alternatives from most likely to least likely to choose at a particular point in time. In case of the DAS model, individuals' consumption histories should also be collected. The ranking task may then be exploited to derive paired comparisons. The parameters of the model, the ideal point and importance weight for each attribute, may then be derived by using linear programming techniques. In case of the DAS model, the inventory retention factors should preferably be calibrated as well, but McAlister assumed these factors all equalled 0.5.

McAlister and Pessemier (1982) extended the DAS model by a term which represents the stimulation contribution to preference. This additional term may be expressed as:

$$w_{K+1}(D_i^{(T)} - \hat{X}_{K+1})^2 \qquad (16)$$

where w_{K+1} represents the importance of the stimulation contribution to preference;
$D_i^{(T)}$ is the total stimulation that will result from choosing alternative i at time T;
$\hat{X}_{K+1}$ is the ideal point for stimulation.

The total amount of stimulation is recursively calculated as:

$$D_i^{(T)} = \lambda_{K+1} D_i^{(T-1)} + \sum_{t=1}^{T-1} \lambda_{K+1}^{T-t} \left[\sum_{k=1}^{K} w_k (X_{ik} - X_k^{(t)})^2 \right] \qquad (17)$$

where λ_{K+1} is a stimulation retention factor;

$D_i^{(0)} = 0$.

Pessemier (1985) has presented a still more sophisticated model of variety-seeking behaviour. He assumed that change in utility results from each attribute of a choice alternative and from interpersonal and intrapersonal variety which the object conveys. Especially, the notion of interpersonal variety is new and it represents an individual's need for group affiliation and personal identity. The model can be best appreciated if it is worked through backwards. An individual's utility for a choice alternative is assumed to be a linear function of the squared distance between the individual's ideal point and the inventory position of that choice alternative in a space of $K+2$ dimensions. In formula:

$$U_i^{(T)} = a + b\, d_{i,T+1}^2 \tag{18}$$

where
$$d_{i,T}^2 = \sum_{k=1}^{K+2} w_k (I_{ik}^{(T)} - \hat{X}_k)^2 \tag{19}$$

$I_{ik}^{(T)}$ is the inventory of the k-th attribute of choice alternative i at time T;

$\hat{X}_k$ is an individual's ideal point for the k-th attribute;

w_k is the importance or salience of the k-th attribute;

a,b are regression coefficients.

The space can be divided into K dimensions associated with the attributes, 1 dimension associated with intrapersonal variety and 1 dimension with interpersonal variety. The definition of the inventories depends on the three kinds of dimensions. The individual inventory level maintained for a particular attribute is assumed to be a function of the times at which increments of the attributes were required, the size of the increments and the consumption rate. The inventory level at time T for attribute k is defined as:

$$I_{ik}^{(T)} = \alpha_k \sum_{t=1}^{T-1} \left[X_k^{(t)} (1 + r_k)^{-t} \right] \qquad k=1,2,\ldots,K \tag{20}$$

where $X_k^{(t)}$ is the amount of attribute k for the choice alternative chosen at time t;

r_k is a time discount rate for attribute k ($r_k \leq 0$);

α_k is a scaling factor for attribute k.

All the r_k's and α_k's are determined by nonlinear least squares methods. The individual ideal points are the dependent observations.

The inventory level of intrapersonal varied experiences is defined by:

$$I_{i,K+1}^{(T)} = \alpha_{K+1} \sum_{t=1}^{T-1} \left[d^*_{t,t-1} (1 + r_{K+1})^{-t} \right] \tag{21}$$

d^* measures the dissimilarity of the alternative chosen at time t to the alternative chosen at t-1. It is defined as the Euclidean distance from alternative i to alternative j, modified by counting only noticeable differences:

$$d^*_{ij} = \left[\sum_{k=1}^{K} X_{ijk}^2 \right]^{0.5} \tag{22}$$

$$\text{where } X_{ijk} = \begin{cases} |X_{ik} - X_{jk}| \text{ if } |X_{ik} - X_{jk}| > c_k X_{min,k} \\ 0. \text{ otherwise} \end{cases} \tag{23}$$

c_k is a simple fraction.

Finally, the interpersonal inventory level is defined by:

$$I_{i,K+2}^{(T)} = \alpha_{K+2} \sum_{t=1}^{T-1} \left[\delta_t (1 + r_{K+2})^{-t} \right] + \alpha_{K+3} \sum_{t=1}^{T-1} \left[\beta_t (1 + r_{K+3})^{-t} \right] \tag{24}$$

where δ_t measures the similarity of an individual ideal point to the mean of the ideal points for friends ($\bar{X}_k^{(t)}$), associates and role models;
β_t measures the individuality of each choice.

More specifically:

$$\delta_t = \left[\sum_{k=1}^{K} (\bar{X}_k{}^{(t)} - X_k{}^{(t)})^2 \right]^{0.5} \tag{25}$$

$$\beta_t = -\ln \gamma_t \tag{26}$$

where γ_t is the share of valued peer's choices going to the alternative chosen by the individual during the period ending at t.

The ideal points of the attributes are derived from a joint space analysis. Object ratings or paired similarity ratings are used to develop the object space.

3.2. *Non-inventory-based Variety-seeking Models*

The models described in the previous section all are explicitly based in some way on the attributes of the choice alternatives. The models to be discussed in this section are not. Most of these models are based on the concept of first-order Markov chains. In fact, they attempt to predict switching probabilities from concepts of variety-seeking.

A somewhat different model has been suggested by Jeuland (1978). He assumed that the utility of a given choice alternative is a function of the past experience of an individual with that alternative and its unique characteristics. Thus:

$$U_i^{(T)} = f(u_i, E_i^{(T)}) \tag{27}$$

where $E_i^{(T)}$ represents the amount of experience with choice alternative i at time T;
u_i accounts for the unique characteristics of choice alternative i.

He assumed that a choice alternative will be chosen if its utility exceeds that of other alternatives by at least a positive constant or threshold Δ; that is:

$$U_i^{(T)} > U_j^{(T)} + \Delta, \qquad \forall\ j \neq i \tag{28}$$

Jeuland postulated the following experience function:

$$E_i^{(T)} = E_i^{(o)} \exp(-\lambda T) + \delta_i^{(T-1)} \tag{29}$$

where $E_i^{(o)}$ is the amount of experience with alternative i at the previous time it was chosen;
λ is a parameter which accounts for the declining over time of the experience function in the absence of choices of the alternative;
$\delta_i^{(T)}$ is equal to zero if a choice alternative other than i is chosen at time T, else $\delta_i^{(T)}$ is equal to 1.

The utility expression $U_i^{(T)}$ itself was defined as:

$$U_i^{(T)} = u_i \; / \; [1 + \theta E_i^{(T)}] \tag{30}$$

where θ is a parameter.

Unfortunately, Jeuland only provided simulation results for this model. It was not estimated on real-world data, hence the predictive properties of this model remain unclear.

Givon (1984) proposed a first-order Markov model which is based on the assumption that a variety-seeker evaluates change positively regardless of the alternative previously chosen. His model assumes that the probability of choosing alternative i given that alternative j was chosen on a previous occasion is a function of the preference for choice alternative i and of preference for switching. Individuals may have a negative switching preference. Givon also extended this model to the situation in which an individual partitions choice alternatives according to some underlying attribute and seeks variety by switching among partitions. In this case, the probability of choosing alternative i given that alternative j was chosen on the previous occasion is a function of preference for alternative i and of preference for all alternatives in the partition with alternative j.

This model was further extended by McAlister (1984) and Lattin and McAlister (1985). Following Tversky's (1977) ideas on similarity, they assumed that similarity between choice alternatives is a function of the features they share. The probability $p_{i|j}$ of choosing alternative i given that alternative j was chosen last time then equals:

$$p_{i|j} = \frac{\Pi_i - vS_{ij}}{1 - (v \sum_{i'=1} S_{i'j})} \tag{31}$$

where Π_i is a parameter which reflects the sum of features except the universal ones of alternative i ($\Pi_i \geq 0, \; \sum_i \Pi_i = 1$);

S_{ij} is a parameter which reflects the features shared by alternatives i and j ($0 \leq S_{ij} \leq \min(\Pi_i, \Pi_j)$);

v is a variety-intensity parameter ($0 \leq v \leq 1$).

The model is estimated by solving a constrained optimization problem, which minimizes the sum of squared differences between the observed switching probabilities and the predicted probabilities. More specifically, the estimation task can be written as:

$$\text{Minimize} \sum_i \sum_j \left[p_{i|j} - \frac{\Pi_i - vS_{ij}}{1 - (v \sum_{i'=1} S_{i'j})} \right]^2 \tag{32}$$

4. DYNAMIC DISCRETE CHOICE MODELS

Over the past few years, standard discrete choice theory as applied to static choice behaviour has been extended to the case of dynamic choice behaviour. Seminal work in this area has been conducted by Heckmann (1981), whose major concern was to distinguish between true state dependence and spurious state dependence, resulting from serial correlation. He developed a general framework for analyzing dynamic choice. He assumes a random sample for which data on the presence or absence of an event in each of T equispaced time intervals exists. He also assumes that the event occurs in period t for individual i if and only if a continuous latent variable $Y_{i,t}$ crosses a threshold. This random variable is supposed to consist of two components: a function of exogenous, predetermined and measured endogenous variables that affect current choices and a stochastic disturbance component. The disturbance component may take on various specifications, but in the present paper Heckmann assumed that the disturbances are jointly normally distributed, similar to the multinomial probit model.

Given these assumptions, the general model of dynamic choice may be written as:

$$Y_i^{(t)} = \mathbf{X}_i^{(t)} \boldsymbol{\beta} + \sum_{j=1}^{\infty} \gamma_{t-j,t} d_{i,t-j} + \sum_{j=1}^{\infty} \lambda_{j,t-j} \prod_{\ell=1}^{j} d_{i,t-\ell} + G(L) Y_i^{(t)} + \varepsilon_i^{(t)} \tag{33}$$

where $G(L)$ is a general lag operator of order K, ($G(0) = 0$);
$d_{i,t}$ is a dummy representing whether the event has occurred ($Y_i^{(t)} \geq 0$ or $Y_i^{(t)} < 0$);
$\mathbf{X}_i^{(t)}$ represents a set of exogenous variables;
$\boldsymbol{\beta}$ is a parameter vector;
γ and λ are parameters;
$\varepsilon_i^{(t)}$ is an error term.

The first term on the right-hand side of the equation represents the effects of exogenous variables on utilities at time t. The second term represents the effect of the entire past history on choice behaviour at time t. The third term represents the cumulative effect of the most recent continuous experience in a state and the fourth term captures the effect of habit persistence. Heckmann then shows that several models such as Bernoulli models, models with structural state dependence, renewal models, models with general correlations in the error term and habit persistence models can be accommodated in this general model.

Another important publication stems from Tardiff (1979). He suggested the following utility function for the dynamic case:

$$U_{qi}^{(t)} = \mathbf{X}_{qi}^{\prime(t)} \boldsymbol{\beta} + \sum_j \beta^*_{ij} C_{jq}^{(t-1)} + \tilde{\varepsilon}_{qi} + \varepsilon^{*(t)}_{qi} \tag{34}$$

where $U_{qi}^{(t)}$ is the utility of choice alternative i to individual q at time t;

$C_{jq}^{(t-1)}$ is 1 if individual q chooses alternative j at time $(t\text{-}1)$ and 0 otherwise;

β and β^* are parameters;

$\tilde{\varepsilon}_{qi}$ is an error term that varies among individuals but not among time periods;

$\varepsilon^{*(t)}_{qi}$ is an error term that varies among both individuals and time periods.

By setting various components of the above utility function at zero, some special cases arise. For example in the cases that $\beta^*_{ij} = 0 \ \forall \ i,j$ and $\varepsilon^{*(t)}_{qi} = 0 \ \forall \ q,i$, the usual type of discrete choice models can be applied directly to the dynamic problem. However, in the case in which the error terms are assumed to be correlated over time, standard estimation procedures are no longer valid. In this case one should either adopt a fixed effects approach, in which the $\tilde{\varepsilon}_{qi}$-terms are explicitly identified and standard discrete choice models are applied directly, or a random effects approach, in which the error variance structure is dealt with directly.

Daganzo and Sheffi (1982) showed that the choice of a state dependence model, a serial correlation model or a hybrid thereof is simply a specification issue, implying that existing computer codes can be used to estimate such models of dynamic choice behaviour. The choice model is specified as:

$$U_i^{(t)} = \boldsymbol{\beta}^{\prime(t)} \mathbf{X}_i^{(t)} \tag{35}$$

where $U_i^{(t)}$ is the utility of alternative i at time t;

$\mathbf{X}_i^{(t)}$ is a vector of attribute values for alternative i at time t;

$\boldsymbol{\beta}^{(t)}$ is a vector of parameters at time t.

The vector $\boldsymbol{\beta}'$, $\boldsymbol{\beta}' = (\boldsymbol{\beta}^{(1)}, \ldots, \boldsymbol{\beta}^{(t)}, \ldots, \boldsymbol{\beta}^{(T)})$ is assumed to be multivariate normal distributed:

$$\boldsymbol{\beta}' \sim \text{MVN}(\bar{\boldsymbol{\beta}}', \Sigma_\beta) \tag{36}$$

If we let c_t denote the choice in period t, the probability of a particular sequence of choices, given conventional discrete choice theory is equal to:

$$p(c_1,\ldots,c_T) = \Pr\left\{U_{c_1}^{(1)} > U_j^{(1)}, \quad \forall\, j \neq c_1; \right.$$
$$\left. \text{and} \ldots \text{and } U_{cT}^{(T)} > U_j^{(T)}, \quad \forall\, j \neq c_T\right\} \tag{37}$$

Daganzo and Sheffi introduce an auxiliary model which reduces the number of alternatives from N^T, where N is the total number of choice alternatives, to $((N-1)T + 1)$. In particular, they define a $N(N-1)$ matrix $\mathbf{\Delta}_i$ such that:

$$\mathbf{X}^{(t)}\mathbf{\Delta}_i = \left[\mathbf{X}_1^{(t)}-\mathbf{X}_i^{(t)},\ldots,\mathbf{X}_{i-1}^{(t)}-\mathbf{X}_i^{(t)},\ldots,\right.$$
$$\left.\mathbf{X}_{i+1}^{(t)}-\mathbf{X}_i^{(t)},\ldots,\mathbf{X}_N^{(t)}-\mathbf{X}_i^{(t)}\right] \quad \forall\, i,t \tag{38}$$

implying that the probability equation may be rewritten as:

$$p(c_1,\ldots,c_T) = \Pr\{\boldsymbol{\beta}'\mathbf{X\Delta} \leq (0,\ldots,0)\} \tag{39}$$

where $\mathbf{X}$ is a $(KT\times NT)$ block-diagonal data-attribute matrix for K attributes;
$\mathbf{\Delta}$ is a $(NT\times(N-1)T)$ block diagonal matrix.

This equation corresponds to a multinomial probit function for a $((N-1)T+1)$-dimensional probit model with the following specification:

$$u_0 = 0, \quad (u_1,u_2,\ldots,u(N-1)T) = \boldsymbol{\beta}'\mathbf{X\Delta} \tag{40}$$

For the state dependence problem, Daganzo and Sheffi assume the following autoregressive process on the utilities for each time period:

$$\mathbf{U}'^{(t)} = \rho\mathbf{U}'^{(t-1)} + \boldsymbol{\beta}'^{(t)}\mathbf{X}^{(t)} \tag{41}$$

$$\mathbf{U}'^{(t-1)} = \rho\mathbf{U}'^{(t-2)} + \boldsymbol{\beta}'^{(t-1)}\mathbf{X}^{(t-1)} \tag{42}$$

$$\vdots$$

$$\mathbf{U}'^{(1)} = \boldsymbol{\beta}'^{(1)}\mathbf{X}^{(1)} \tag{43}$$

To estimate this model, matrix $\mathbf{X}$ should have the following specification:

$$\mathbf{X} = \begin{bmatrix} \mathbf{X}_1 & \rho\mathbf{X}^{(1)} & \rho^2\mathbf{X}^{(1)} & \dots & \rho^{T-1}\mathbf{X}^{(1)} \\ 0 & \mathbf{X}^{(2)} & \rho\mathbf{X}^{(2)} & \dots & \rho^{T-2}\mathbf{X}^{(2)} \\ \vdots & \vdots & \vdots & & \vdots \\ \vdots & \vdots & \vdots & & \vdots \\ \vdots & \vdots & \vdots & & \vdots \\ 0 & 0 & 0 & \dots & \mathbf{X}^{(T)} \end{bmatrix} \quad (44)$$

An application of their approach to two-period panel data can also be found in Johnson and Hensher (1982).

Another dynamic disaggregate choice model has been advanced by Krishnan and Beckmann (1979). Basically their dynamic model is an extension of the Krishnan's static logit model for binary choices which incorporates threshold effects, δ. Introducing time, the basic model can be written as:

$$p(c_t=i) = p(Y^{(t)}=i) + p(c_t=i\,|Y^{(t)}=3)\; p(Y^{(t)}=3), \quad i=1,2 \quad (45)$$

where $p(c_t=i)$ denotes the probability that choice alternative i is chosen at time t;

$p(Y^{(t)}=i)$ is the probability that choice alternative i is preferred at time t;

$p(Y^{(t)}=3)$ is the probability that the individual is indifferent.

The probabilities $p(Y^{(t)}=1)$ and $p(Y^{(t)}=2)$ are defined as:

$$p(Y^{(t)}=1) = 1 \;/\; [1 + \exp(V_2^{(t)} - V_1^{(t)} + \delta_{12})] \quad (46)$$

$$p(Y^{(t)}=2) = 1 \;/\; [1 + \exp(V_1^{(t)} - V_2^{(t)} + \delta_{21})] \quad (47)$$

where $V_i^{(t)}$ is the deterministic utility component of choice alternative i at time t, $i=1,2$;

δ_{12} and δ_{21} are threshold values.

This implies that an optimal model may be derived by defining the probability that the first alternative will be chosen, given that the individual is indifferent. The authors postulate two models for the indifference state. First, they postulate that if alternative 1 and alternative 2 are equally preferred, alternative 1 will be chosen with probability θ, and alternative 2 will be chosen with probability $(1-\theta)$. Thus, it follows that:

$$p(c_t=1) = p(Y^{(t)}=1) + \theta p(Y^{(t)}=3)$$

$$= \theta\left\{\exp(V_1^{(t)}-V_2^{(t)}+\delta_{21}) \;/\; [1+\exp(V_1^{(t)}-V_2^{(t)}+\delta_{21}]\right\} +$$

$$(1-\theta)\left\{1 \;/\; [1+\exp(V_2^{(t)}-V_1^{(t)}-\delta_{12})]\right\} \qquad (48)$$

The second model is based on the indifference postulate that the most recently chosen alternative will also be chosen at time t. Hence we have:

$$p(c_t=1|c_{t-1}=2) = p(Y^{(t)}=1)$$
$$= 1 \;/\; [1+\exp(V_2^{(t)}-V_1^{(t)}+\delta_{12})] \qquad (49)$$

and

$$p(c_t=1|c_{t-1}=1) = p(Y^{(t)}=1) + p(Y^{(t)}=3)$$
$$= \exp(V_1^{(t)}-V_2^{(t)}+\delta_{21}) \;/\; [1+\exp(V_1^{(t)}+\delta_{21})] \qquad (50)$$

Another interesting development is the beta logistic model (Heckmann and Willis, 1977). This model provides predictions for both the mean probability of choosing a particular alternative and the distribution of the choice probabilities around the mean. The original model is based on the assumptions that the exogenous variables are constant over time and the absence of state/time dependence. Heterogeneity is introduced by defining a subgroup (g) of a sample in which all individuals have exactly the same values on all exogenous variables included in the model. The distribution of the probabilities for such a group represents heterogeneity. Given these assumptions the beta logistic model assumes that the mean probability can be modelled in terms of a conventional logit model, and that the distribution of probabilities is a beta distribution. The dichotomous model may be written as:

$$E(p_g) = \frac{\exp\{\mathbf{X}_g' \, (\boldsymbol{\beta}_1 - \boldsymbol{\beta}_2)\}}{1 + \exp\{\mathbf{X}_g' \, (\boldsymbol{\beta}_1 - \boldsymbol{\beta}_2)\}} \qquad (51)$$

where $E(p_g)$ is the mean probability of subgroup g choosing an alternative;
$\mathbf{X}_g$ is a set of exogenous variables for subgroup g;
$\boldsymbol{\beta}_1$ and $\boldsymbol{\beta}_2$ are vectors of parameter values.

The form of the heterogeneity is given by:

$$f(p_g|\alpha_{1g},\alpha_{2g}) = \frac{\Gamma(\alpha_{1g} + \alpha_{2g})}{\Gamma(\alpha_{1g})\,\Gamma(\alpha_{2g})}\, p_g^{\alpha_{1g}-1} \, (1-p_g)^{\alpha_{2g}-1} \tag{52}$$

where $\alpha_{1g} = \exp\{\mathbf{X}_g' \boldsymbol{\beta}_1\}$ (53)

$\alpha_{2g} = \exp\{\mathbf{X}_g' \boldsymbol{\beta}_2\}$ (54)

The original specification of the dichotomous beta logistic model has been generalized by Davies (1984) and Davies and Pickles (1984) to incorporate feedback effects and time-varying exogenous variables. They applied the model successfully to the study of residential mobility. The dichotomous model may be extended to the polytomous case by replacing the binomial distribution of the logistic model by the multinomial distribution and the beta distribution describing heterogeneity by the Dirichlet distribution, its multivariate equivalent. Dunn and Wrigley (1985) provide an application of this model in a study of spatial shopping behaviour.

The above models of dynamic choice behaviour can all be considered as extensions of the econometric methodology associated with discrete choice models when using panel data. Leonardi (1983) approached the problem from a different angle and developed a theoretical model. In particular, he assumed that a utility or disutility is associated both with a transition from one alternative to another and with staying with a particular alternative. In addition he assumed a discount rate of utilities over time. It follows that the utility associated with a transition at time $t + \Delta$ is given by:

$$U_j^{(t+\Delta)} = (1 - \alpha\Delta)\ [v_{ij} + V_j^{(t+\Delta)} + \varepsilon_j] \tag{55}$$

where α is the discount rate, $(\alpha > 0)$;

v_{ij} is the utility (disutility) associated with a transition from i to j;

$V_j^{(t+\Delta)}$ is the total expected utility for a process started in i at time $t+\Delta$;

ε_j is an error term.

Likewise, the gain in utility of remaining in i starting from $t + \Delta$ equals:

$$U_i^{(t+\Delta)} = (1 - \alpha\Delta)\ [V_i^{(t+\Delta)} + \varepsilon_i] \tag{56}$$

and he will gain $h_i^{(t)}\Delta$, during his stay in i in $(t,t+\Delta)$, assuming decisions to move are made at the end of the time interval. Assuming utility-maximizing behaviour and independently and identically Gumbel distributed error terms, Leonardi shows that the probability of moving from i to j in $(t,t+\Delta)$ is equal to:

$$p_{ij}^{(t,t+\Delta)} = \frac{M_j \exp\{\beta\ [v_{ij} + V_j^{(t+\Delta)}]\}}{\sum_{j'} M_{j'} \exp\{\beta\ [v_{ij'} + V_{j'}^{(t+\Delta)} + \exp[\beta V_i^{(t+\Delta)}]\}} \tag{57}$$

where M_j is the number of sampled alternatives.

Leonardi also demonstrated that this choice process can also be formulated in terms of an optimal control problem.

Another interesting theoretical model has been put forward by De Palma. Basically he has been concerned with an expansion of classical models of individual behaviour by taking into account interindividual interaction and interdependence of individual decisions (Deneubourg, De Palma and Lefèvre 1979; De Palma and Lefèvre, 1981, 1983; De Palma, 1983). The deterministic utility component consists of two parts: a part representing the absolute benefit associated with a choice and a part which measures the relative benefit associated with a choice given the choice behaviour of another actor. By assuming that each actor anticipates rationally the behaviour of the other actor and linear utility functions, De Palma derives a theoretical model which is able to describe such phenomena as repulsion, attraction and competition between actors. The model may be viewed as dynamic in the sense that time could be used as a basis for the utility function specification. If not, his model should be considered as an extension of conventional static discrete choice models.

The field of dynamic discrete choice models is developing rapidly. Apart from the contributions discussed above, several other important publications have appeared recently. For example, Avery *et al.* (1983) presented a multiperiod probit model; Hensher (1984) developed a quasi-dynamic choice model for automobile demand; Dagsvik (1983) introduced a dynamic extension of Thurstonian and Lucian choice models; Meyer and Sathi (1985) discussed a dynamic model of consumer choice involving product learning; and Miller and O'Kelly (1983) used a dynamic logit model to estimate shopping destination choice. The reader is referred to these publications for further details.

5. CONCLUDING REMARKS

It is evident that the modelling of dynamic aspects of choice behaviour is getting increasingly more attention in geography and related disciplines. It is to be expected therefore that in the next few years most important advances will be made in this area. While much progress has already been made, it is also true that existing models are still based on very rigorous assumptions such as stationarity. In any case, the various effects discussed for conventional discrete choice models still have to be incorporated into their dynamic counterparts. In addition, comparative testing of the models is important to learn their characteristics and predictive performance.

Urban Planning Group
Faculty of Architecture, Building and Planning
Eindhoven University of Technology
The Netherlands

BIBLIOGRAPHY

Avery, R.B., L.P. Hansen, and V.J. Hotz: 1983, 'Multiperiod probit models and orthogonality condition estimation', *International Economic Review* **24**, 21-35.

Broom, D. and N. Wrigley: 1983, 'Incorporating explanatory variables into stochastic panel-data models of urban shopping behavior', *Urban Geography* **4**, 244-257.

Burnett, P.: 1974, 'A three state Markov model of choice behavior within spatial structures', *Geographical Analysis* **6**, 53-68.

Burnett, P.: 1975, 'A Bernoulli model of destination choice', *Transportation Research Record* **527**, 33-44.

Burnett, P.: 1977, 'Tests of a linear learning model of destination choice: applications to shopping travel by heterogeneous groups', *Geografiska Annaler* **B 59**, 95-108.

Burnett, P.: 1978, 'Markovian models of movement within urban spatial structures', *Geographical Analysis* **10**, 142-153.

Chatfield, C., A.S.C. Ehrenberg, and G.J. Goodhardt: 1966, 'Progress on a simplified model of stationary purchasing behaviour', *Journal of the Royal Statistical Society Series* **A 129**, 317-367.

Crouchley, R., R.B. Davies, and A. Pickles: 1982, 'A re-examination of Burnett's study of Markovian models of movement', *Geographical Analysis* **14**, 260-262.

Daganzo, C.F. and Y. Sheffi: 1982, 'Multinomial probit with time series data: unifying state dependence and serial correlation models', *Environment and Planning* **A 14**, 1377-1388.

Dagsvik, J.K.: 1983, 'Discrete dynamic choice: an extension of the choice models of Thurstone and Luce', *Journal of Mathematical Psychology* **27**, 1-43.

Davies, R.B.: 1984, 'A generalized beta-logistic model for longitudinal data', *Environment and Planning* **A 16**, 1375-1386.

Davies, R.B. and A.R. Pickles: 1984, 'Accounting for omitted variables in the analysis of panel and other longitudinal data', Papers in Planning Research 77, University of Wales, Institute of Science and Technology, Cardiff.

De Palma, A.: 1983, 'Incomplete information, expectation, and subsequent decisionmaking', *Environment and Planning* **A 15**, 123-130.

De Palma, A. and C.E. Lefèvre: 1981, 'A probabilistic search model', *Journal of Mathematical Sociology* **8**, 43-60.

De Palma, A. and C.E. Lefèvre: 1983, 'Individual decision making in dynamic collective systems', *Journal of Mathematical Sociology* **9**, 103-124.

Deneubourg, J.L., A. De Palma, and D. Kahn: 1979, 'Dynamic models of competition between transportation modes', *Environment and Planning* **11**, 665-673.

Dunn, R., S. Reader, and N. Wrigley: 1983, 'An investigation of the assumptions of the NBD model as applied to purchasing at individual stores', *Applied Statistics* **32**, 249-259.

Dunn, R. and N. Wrigley: 1985, 'Beta-logistic models of urban shopping center choice', *Geographical Analysis* **17**, 95-113.

Ehrenberg, A.S.C.: 1968, 'The practical meaning and usefulness of the NBD/LSD theory of repeat buying', *Applied Statistics* **17**, 17-32.

Ehrenberg, A.S.C.: 1972, *Repeat Buying: Theory and Application*, North-Holland, Amsterdam.

Farquhar, P.H. and V.R. Rao: 1976, 'A balance model of evaluating subsets of multiattribute items', *Management Science* **22**, 528-539.

Ginsberg, R.B.: 1971, 'Semi-Markov processes and mobility', *Journal of Mathematical Sociology* **1**, 233-263.

Ginsberg, R.B.: 1972, 'Incorporating causal structure and exogenous information within probabilistic models, with special reference to choice, gravity, migration and Markov chains', *Journal of Mathematical Sociology* **2**, 88-103.

Ginsberg, R.B.: 1973, 'Stochastic models of residential and geographic mobility for heterogenous populations', *Environment and Planning* **A 5**, 113-124.

Ginsberg, R.B.: 1978, 'The relationship between timing of moves and choices of destination in stochastic models of migration', *Environment and Planning* **A 10**, 667-679.

Ginsberg, R.B.: 1979, 'Tests of stochastic models of timing in mobility histories', *Environment and Planning* **A 11**, 1387-1404.

Givon, M.: 1984, 'Variety seeking through brand switching', *Marketing Science* **3**, 1-22.

Goodhardt, G.J., A.S.C. Ehrenberg, and C. Chatfield: 1984, 'The Dirichlet: a comprehensive model of buying behaviour', *Journal of the Royal Statistical Society Series* **A 147**, 621-655.

Halperin, W.C.: 1985, 'The analysis of panel data for discrete choices', in P. Nijkamp, H. Leitner and N. Wrigley (eds.), *Measuring the Unmeasurable: the Analysis of Qualitative Spatial Data*, Martinus Nijhoff, The Hague.

Halperin, W.C. and N. Gale: 1984, 'Towards behavioural models of spatial choice: some recent developments', in D.E. Pitfield (ed.), *Discrete Choice Models in Regional Science*, Pion, London.

Heckman, J.J.: 1981, 'Statistical models for discrete panel data', in C.F. Manski and D. McFadden (eds.), *Structural Analysis of Discrete Data with Econometric Applications*, MIT Press, Cambridge, Massachusetts.

Heckman, J.J. and R. Willis: 1977, 'A beta-logistic model for the analysis of sequential labor force participation by married women', *Journal of Political Economy* **85**, 27-58.

Hensher, D.A.: 1984, 'Model specification for a quasi-dynamic discrete-continuous choice automobile demand system in discrete time using panel data', Working Paper No 11, School of Economic and Financial Studies, Marquarie University, North Ryde.

Hensher, D.A. and N. Wrigley: 1984, 'Statistical modelling of discrete choices with panel data', Working Paper No 16, Dimensions of Automobile Demand Project, School of Economic and Financial Studies, Marquarie University, North Ryde.

Herniter, J.: 1971, 'A probabilistic market model of purchase timing and brand selection', *Management Science* **18**, 102-113.

Jeuland, A.: 1978, 'Brand preference over time: a partially deterministic operationalization of the notion of variety seeking', in S. Jain (ed.), *Research Frontiers in Marketing: Dialogues and Directions*, American Marketing Association, Chicago.

Jeuland, A.B., F.M. Bass, and G.P. Wright: 1980, 'A multibrand stochastic model compounding heterogenous Erlang timing and multinomial choice processes', *Operations Research* **28**, 255-277.

Johnson, L. and D.A. Hensher: 1982, 'Application of multinomial probit to a two-period panel data set', *Transportation Research* **A 16**, 457-464.

Jones, J.M. and F.S. Zufryden: 1980, 'Adding explanatory variables to a consumer purchase behavior model: an exploratory study', *Journal of Marketing Research* **17**, 323-334.

Jones, J.M. and F.S. Zufryden: 1982, 'An approach for assessing demographic and price influences on brand purchase behavior', *Journal of Marketing* **46**, 36-46.

Kau, K.A. and A.S.C. Ehrenberg: 1984, 'Patterns of store choice', *Journal of Marketing Research* **21**, 399-409.

Krishnan, K.S. and M.J. Beckmann: 1979, 'Dynamic disaggregate choice models with an application in transportation', *Decision Science* **10**, 218-231.

Lattin, J.H. and L. McAlister: 1985, 'Using a variety seeking model to identify substitute and complementary relationships among competing products', *Journal of Marketing Research* **9**, 141-151.

Leonardi, G.: 1983, 'An optimal control representation of a stochastic multistage multiactor choice process', in D.A. Griffith and A.C. Lea (eds.), *Evolving Geographical Structures: Mathematical Models and Theories for Space-Time Processes*, Martinus Nijhoff, The Hague.

Massy, W.F., D.B. Montgomery, and D.G. Morrison: 1970, *Stochastic models of buying behaviour*, MIT Press, Cambridge.

McAlister, L.: 1979, 'Choosing multiple items from a product class', *Journal of Consumer Research* **6**, 213-224.

McAlister, L.: 1982, 'A dynamic attribute satiation model of variety seeking behavior', *Journal of Consumer Research* **9**, 141-151.

McAlister, L.: 1984, 'A similarity based Markov model of variety seeking behavior, its implications and a linear programming technique to estimate its parameters', MIT Working Paper 1539-84, A.P. Sloan School of Management, Cambridge, Massachusetts.

McAlister, L. and E.A. Pessemier: 1982, 'Variety-seeking behavior: an interdisciplinary review', *Journal of Consumer Research* **9**, 311-322.

Meyer, R.J. and A. Sathi: 1985, 'A multiattribute model of consumer choice during product learning', *Marketing Science* **4**, 41-61.

Miller, E.J. and M.E. O'Kelly: 1983, 'Estimating shopping destination models from travel diary data', *The Professional Geographer* **35**, 440-449.

Paull, A.E.: 1978, 'A generalized compound Poisson model for consumer purchase panel data analysis', *Journal of the American Statistical Association* **73**, 706-713.

Pessemier, E.A.: 1985, 'Varied individual behavior: some theories, measurement and models', *Multivariate Behavioral Research* **20**, 69-94.

Tardiff, T.J.: 1979, 'Definition of alternatives and representation of dynamic behaviour in spatial choice models', *Transportation Research Record* **723**, 25-30.

Tversky, A.: 1977, 'Features of similarity', *Psychological Review* **84**, 327-352.

Wrigley, N.: 1980, 'An approach to the modelling of shop-choice patterns: an explanatory analysis of purchasing patterns in a British city', in D.T. Herbert and R.J. Johnston (eds.), *Geography and the Urban Environment: Progress in Research and Applications*, Volume 3, John Wiley, Chichester, Sussex.

Wrigley, N. and R. Dunn: 1984a, 'Stochastic panel-data models of urban shopping behaviour: 1.purchasing at individual stores in a city', *Environment and Planning* **A 16**, 629-650.

Wrigley, N. and R. Dunn: 1984b, 'Stochastic panel-data models of urban shopping behaviour: 2.multistore purchasing patterns and the Dirichlet model', *Environment and Planning* **A 16**, 759-778.

Wrigley, N. and R. Dunn: 1984c, 'Stochastic panel-data models of urban shopping behaviour: 3.the interaction of store choice and brand choice', *Environment and Planning* **A 16**, 1221-1236.

Wrigley, N. and R. Dunn: 1985, 'Stochastic panel-data models of urban shopping behaviour: 4.incorporating independent variables into the NBD and Dirichlet models', *Environment and Planning* **A 17**, 319-332.

Zufryden, F.S.: 1977, 'A composite heterogeneous model of brand choice and purchase timing behavior', *Management Science* **24**, 121-136.

Zufryden, F.S.: 1978, 'An empirical evaluation of a composite heterogeneous model of brand choice and purchase timing behavior', *Management Science* **24**, 761-773.

JEAN-CLAUDE THILL AND ISABELLE THOMAS

IS SPATIAL BEHAVIOUR SERVICE-SENSITIVE? AN EMPIRICAL TEST

1. INTRODUCTION

The analysis of *public* facilities has led to the development of an extensive body of mainly theoretical work about location problems (see for example Beaumont, 1981; Hodgart, 1978; Thisse and Zoller, 1983). However, the need for empirical investigation was underlined in the early eighties (see Leonardi, 1981a, 1981b) in order to verify the relevance of the often simplistic behavioural assumptions of these models and several empirical studies have since been carried out (see Beguin *et al.*, 1982). In contrast, *private* facility location modelling has often relied on an empirical basis, and this has tended to develop through time (e.g. Coelho and Wilson, 1976; Dunn and Wrigley, 1985; Miller and Lerman, 1979; Timmermans *et al.*, 1984). Moreover, in terms of behavioural modelling, there is a need for further theoretical as well as empirical work on *trip tour structures*, in order to increase the operationality of public and private services location models. Several theoretical contributions have already demonstrated how spatial choice is influenced by multipurpose shopping (see references in Thill and Thomas, 1987). Attempts have also recently been made to integrate explicitly the practice of multipurpose journeys in shopping models (e.g. O'Kelly, 1983a, 1983b).

Where consumer spatial behaviour is modelled on an empirical basis, large data sets are required. In order to reduce as far as possible the collection of expensive data, it seems useful to know to what extent trip chaining models might be applied to a large variety of decision contexts. The temporal variability of travel patterns has already been reported (e.g. Kostyniuk and Kitamura, 1985) and the effects of urban spatial structure on spatial behaviour are also well documented in the literature (see Bentley *et al.*, 1977; Borgers and Timmermans, 1985; Fotheringham, 1983; Horton and Reynolds, 1971). However, as far as we know, the sensitivity of the spatial behaviour (trip structure, choice of spatial alternatives) to the type of service visited has not yet been investigated. Is there one behavioural structure associated with all types of services or is behaviour service-specific? To this end, we test the following hypothesis:

Spatial behaviour associated with the visits to the service of type *J*	=	Spatial behaviour associated with the visits to the service of type *K*

J. Hauer et al. (eds.), Urban Dynamics and Spatial Choice Behaviour, 27–41.

Because data are required for the same urban area, only two types of services are compared in this chapter: pharmacies and post offices. The design and content of the data bases for these services are described in Section 2; Section 3 presents the statistical method on which the analysis relies, and the results of the sensitivity analyses are summarized in Section 4. Finally, conclusions are drawn in Section 5.

2. THE DATA

To test the sensitivity of spatial behaviour to the type of service visited, it is necessary to have reliable empirical estimates of individual behaviour as well as of the locational and socioeconomic characteristics of the users. Data on the spatial behaviour associated with the two services under review (pharmacies and post offices) were collected in a fairly typical medium-sized Belgian municipality (Namur). The sampling area is the city centre, defined as homogeneous in terms of sociodemographic characteristics as well as overall spatial behaviour variables (Thill, 1986; Thomas, 1983, 1986). It includes 64,680 inhabitants, covers 29.3 square kilometres and is divided into five administrative districts (the pre-1977 Belgian communes). The retailing system is hierarchically structured, combining nucleations and ribbons; the city is known to have a regional function (Sporck and Goossens, 1985).

Both of the chosen services offer commodities for which they do (stamps and other postal services; drugs) or do not (financial and auxiliary services; pharmaceutical products) have a monopoly. For all pharmaceutical items and most postal services, users are free to choose the facility they wish to patronise. The price of postal services does not vary with the location of post offices, but this inelasticity is not fully verified for the goods sold in pharmacies: one-third of the pharmacists offer price discounts of up to 10%, and there is no fixed price for over-the-counter products (however these only represent less than 10% of the turnover). The Belgian post office is a public service; on the other hand, the pharmacy is privately owned but its organisation is strongly constrained by the requirements of the Public Health Department. Given the numerous similarities on the supply side, a comparison of the patronage of both services and of the related spatial behaviour is relevant.

Two surveys were conducted in Namur, one for each type of service. Both are based on a multistage temporal stratification. Interviews were carried out simultaneously with users of the ten existing post offices (Thomas, 1986) and with customers of 23 pharmacies (selected from a total of 52) (Thill, 1986). The two questionnaires were specially designed to provide similar information on consumers' characteristics, demand structure, trip patterns (ordering, location, purpose of stops, transportation mode), and perception of the services. 5,388 valid postal questionnaires and 2,156 responses from the survey of the pharmacies are available.

Because of the conditions in which the two surveys were conducted, the data sets can reasonably be considered as independent. These two large samples are highly representative of the populations from which they are drawn. Tests of comparison between parameters or distributions computed on very large data sets will often lead to significant results; a very slight difference is always statistically significant because of the number of individuals included in each sample. Yet in order to give meaning to the statistical tests, in this chapter the analysis is based on reduced sample sizes. Two uniform random subsamples are taken from the total initial samples: their sizes are similar to, and correspond to, 10% of the pharmacy sample (i.e. 217 observations). Both reduced data sets are still representative of the populations they are drawn from; the random reduction does not affect the reliability of the data. The resulting samples will be compared using a set of behavioural variables common to both sets of actual users.

The comparison is applied to a variety of personal characteristics of the consumers, to the structure of the individual demand, and finally to the travel patterns of the respondents. The individual demand is particularly interesting with respect to the analysis of spatial behaviour.

We now examine the *variables* selected for analysis. The sociodemographic profile of the individual is compromised of their age, sex and socioeconomic status. Age is expressed as a continuous and a categorical variable; the three categories identify the life cycle of the individuals (under 40, 41 to 64 and over 65 years old). Professional status is depicted by two (working, not working) or four (blue-collar, white-collar, retired, other) categories. Individual demand is commonly defined by the quantity of service consumed per unit of time; it is here measured by the number of visits to a facility a consumer makes per month. It is alternatively used as a continuous or as a classification variable (once a month or less, more than once a month).

Consumers' spatial behaviour is analysed here using the *sequencing* of activities observed during journeys that include a visit to one of the services (pharmacies, post offices). A journey is defined as a series of movements made between successive destination choices over some period of time. It may be depicted as a decision process which elaborates through time a series of stages, each stage being identified by location and land-use attributes. This concept comprises complex relations and thus poses modelling problems (Thill and Thomas, 1987). It is represented here by a set of measurable components. Timing of the journey is expressed by the departure time from the origin (residence or workplace), the interview time at the service and the arrival time at the destination point (residence or workplace). These three time variables are categorised into four classes: morning, midday, afternoon and evening. The extent of the journey is measured by its duration (number of minutes actually spent between the beginning and the end of the complete journey) and its length. This latter value is expressed by two variables: the number of supplementary stops that occur during the journey to one of the two facilities and the origin-service-destination distance. This latter measure is a rough estimate of the distance actually travelled.

The successive stages of the tour process (origin, stop at the place of interview, other stops, destination) are classified in two different ways. The first way classifies the stages into five or six locational categories which refer to the administrative division of the urban area (5 communes and, if the case arises, the 'elsewhere' category). The second way provides an alternative three-level classification which corresponds to a commercial hierarchy based on the counts of the retail and service outlets in the urban area of Namur. In addition, stops are clustered into three classes according to the purpose for which trips are made: shopping, service, or other (including leisure and education). Two complementary attributes of journeys are studied because of their known impact on the travel pattern: the transportation mode which is represented by three categories (car, foot, other); the nature of the extremities of the tour (residence or work); which leads to three types of journeys (residence-residence, work-work, other).

Finally, in a broader temporal framework, two additional variables are considered. The first is a binary variable expressing the loyalty to a service unit (1: full loyalty, 0: otherwise). The second variable corresponds to the general reasons underlying the choice of the specific facility visited (proximity of residence or workplace, proximity of other land-uses, other reasons).

3. THE METHOD

Customer behaviour is represented by a set of empirically measured variables. The similarity of the values found in both data bases is measured using adequate statistical tests: mean values are compared by Student's t test, and variable distributions are compared by Chi-square tests. Correlation coefficients computed for each sample, between behavioural variables and candidate explanatory variables, are also statistically tested. Because of the statistical reliability of both data bases, levels of significance up to 0.10 are accepted; when the levels are greater than 10%, the tested hypothesis is rejected. Unfortunately, there is no statistical test to compare levels of association (e.g. contingency coefficients) between two categorical variables. The similarity of the associations is indicated by the proximity of the pair-wise values of the Chi-square statistic and its associated significance level.

4. THE EMPIRICAL RESULTS

The respondents are known to represent the populations of customers of each type of service in the city of Namur. We will first test the following hypothesis: post offices and pharmacies are visited by identical segments of the urban population. Therefore, the *socioeconomic profile* of the users is compared: the age, sex and professional status variables have identical means (Table I) and/or distributions (Table II). In all cases but one, equalities are not statistically rejected; only the age means are slightly different (5.7% significance level) when comparing the total subsamples (A and B in Table I). However, both mean values belong to the 40-44 years age group and the difference is no more significant when the professional use of the postal service is excluded from the study (subsample C in

Table I). As a result, users' characteristics do not vary with the nature of the service considered; there is one common consumer profile.

A related question is to test whether both services are consumed at the same monthly frequency. Both the mean values and the statistical distributions of the demand variable are identical when restricted to private use; the consumers go as often to pharmacies as to post offices (Tables I and II) - two or three times per month. The inclusion of professional use of the postal service significantly increases the mean number of monthly visits.

Supply of both services is known to exhibit similar features, and demand has been shown to be similar in its characteristics. Moreover, the same level of loyalty to a specific service unit is reached in both patronages: 75.1% of the interviewed customers always patronise the same pharmacy while 76.7% always visits the same post office. Consequently, one might expect that the same homogeneity characterises trip structure. In a first step, this hypothesis is tested by comparisons of means as well as of statistical distributions of variables describing the individual travel patterns of both data bases.

Customers of pharmacies have a higher propensity to undertake *multistop journeys*: 67.7% of the individuals associate other purposes with their visit to a pharmacy, against 55.3% of the individuals visiting a post office. Moreover, the two data bases disclose

TABLE I
Comparison of mean values

Variables	A mean	B mean	t	α	C mean	t	α
Monthly frequency of visit	2.2	2.9	2.259	0.024	2.4	0.687	0.493
Age	44.0	40.8	1.902	0.058	41.5	1.467	0.143
Number of supplement. stops	1.3	0.8	3.809	0.000	0.9	3.388	0.001
Trip tour duration	81.2	64.5	2.102	0.036	65.6	1.906	0.057
Trip tour length (m)	2635	2172	1.910	0.057	2234	1.591	0.113

A: data set related to the pharmacies;
B: data set related to the postal service;
C: data set restricted to the private use of the postal service;
t: Student's t statistic;
α: significance level.

different individual mean numbers of supplementary stops: 1.3 and 0.8 stops (Table I). Surprisingly, although the number of stops per journey is different, the trip tour length - whether expressed in minutes or in metres - is only slightly different (Table I). Location as well as the purpose of stops also show totally different patterns - even though the location of the visited facilities does not (Table II). The patronage of the pharmacy tends to be combined with shopping activities while post offices are more frequently visited at the same time as service or leisure facilities. The majority of stops made by pharmacy customers is concentrated in the city centre, i.e. where shopping opportunities are more numerous. On the contrary, stops by postal users are scattered over space, in particular outside the urban area of Namur, even though postal users show a higher tendency to live and to begin their journey in the area under investigation (Table IV).

The three combinations of the extremity types (residence or work) are not equally represented in the full samples, but they are when the postal data set is restricted to the private use of the postal service. More than 75% of the journeys are loops with the residence as base, but the workplace is the unique base of only one journey out of ten. The spatial distribution of the origins of journeys including both a pharmacy and a post office do not coincide, in contrast to the place of the residence of both types of consumers (Tables II and IV). This latter variable, however, is more a personal attribute of the users than a trip characteristic. Proximity of residence as well as of workplace is the main reason given by postal users for the choice of a spatial alternative (70.8%). The same trend is observed in the actual choices: 68.8% of the postal users go to the closest available opportunity. However, only 39.4% of the pharmacy customers (see Table II for the statistical test) go to the closest facility. This difference can be explained by the smaller total number of post offices (10) compared with pharmacies (52) in the study area. Customers of the pharmacies chiefly base their choice on other criteria than the mere proximity of the origin or destination place (Table II and IV). The higher proportion of pharmacy customers giving the proximity of other land-uses as a reason (20.7%) is consistent with their higher propensity to make multipurpose journeys. The nature of the service itself explains the predominance of non-spatial reasons, such as basic health advice, discounts, or range of articles.

There is no difference in the transportation modes used which are not specifically related to the use of a particular service - their choice refers to a broader decision context (Table II). Almost half of the journeys are made on foot (Table IV). This questions the validity of many studies based on transportation surveys that collect data only on car movements. The usefulness of work in accounting for pedestrian movements is thus confirmed (e.g. Borgers and Timmermans, 1986). The timing of the journeys (departure, interview and arrival times) also varies between the two data bases; this is partly caused by the restricted opening times of the public facilities - as shown in the postal location model (Thomas, 1984), post offices would gain by remaining open later into the afternoon. Users tend to go to the post office earlier in the day because of administrative constraints (opening times).

TABLE II
Comparison of variables' distribution

Variables	Cat.	Pharmacy/Post Offices Users		Pharmacy/Private Postal Users	
		χ^2	α	χ^2	α
Age	3	1.464	0.481	1.034	0.596
Sex	2	1.079	0.299	0.944	0.331
Profession	2	2.214	0.137	0.036	0.849
	4	3.599	0.308	2.275	0.517
Frequency of visit	2	2.839	0.092	1.114	0.291
Location of service	3[1]	1.430	0.489	2.900	0.235
	5[2]	6.770	0.149	6.417	0.170
Type of journey	3	9.480	0.009	1.704	0.427
Transport	3	0.673	0.714	0.788	0.474
Number of sup. stops	2	10.920	0.001	7.180	0.007
Departure time	4	12.076	0.007	8.335	0.040
Interview time	4	20.002	0.000	21.304	0.000
Arrival time	4	9.343	0.025	10.290	0.016
Location of residence	6	5.286	0.382	6.677	0.246
Location of origin	6	16.083	0.013	12.873	0.045
Location of stops	6	43.059	0.000	45.199	0.000
Loyalty	2	0.048	0.827	0.152	0.697
Spatial choice reason	3	109.980	0.000	97.407	0.000
Using closest fac.	2	41.818	0.000	36.900	0.000
Purpose of stops	6	21.792	0.001	21.154	0.001

Cat.: Number of categories;
α: Significance level;
[1] Levels in the commercial hierarchy;
[2] Communes.

The computed levels of association between candidate explanatory variables (socioeconomic as well as behavioural) and the trip tour length (whether expressed in minutes, metres or number of supplementary stops) do not systematically differ in the two data sets (Tables III and V). This is especially true when socioeconomic variables are introduced into the analysis. Expressing the number of supplementary stops by a two- or a three-category variable does not change the interpretation of the results. In contrast to the univariate analysis, no clear-cut explanation can be formulated. In particular, some noteworthy results are that:

1. The extent of the journey is never significantly related (at the 5% level) to the frequency of visits to pharmacies or to post offices.
2. The age of the individual generally has no effect on the decision to make supplementary stops, nor on the extent of the journeys (customers of pharmacies, however, travel farther when they are younger). This independence contradicts previous observations made by Bentley *et al.*, (1977) and Hemmens (1970). In contrast to the customers of pharmacies, users of post offices decide on the extent of their journeys on the basis of their sex and economic status. Male and working people tend to make single-stop trips while female and non-working people readily undertake longer journeys (see e.g. Bentley *et al.*, 1977; Horowitz, 1982).

TABLE III
Comparison of simple correlation coefficients

	Frequency of visits	Age	Trip tour length (m)	Trip tour duration
Age	NS[1] NS[2] - NS[3]	-	-	-
Trip tour length (m)	NS NS - NS	-0.25 NS - NS	-	-
Trip tour duration	-0.13[5] NS - NS	NS NS - NS	0.30 0.30 - 0.27	-
Number of supplement. stops	NS NS - NS	NS 0.11[5] - NS	0.21 0.19 - 0.17[4]	0.43 0.55 - 0.54

[1] Association computed for the users of the pharmacies.
[2] Association computed for postal users.
[3] Association computed for private use of postal services.
[4] Significance level between 0.01 and 0.05.
[5] Significance level between 0.05 and 0.10.

TABLE IV
Description of the data bases (in percent)

TRANSPORTATION MODE	Ph.	P.O.
Foot	49.3	46.5
Car	42.4	42.9
Other transportation modes	8.3	10.6

TIME OF INTERVIEW	Ph.	P.O.
Morning	40.9	43.4
Midday	12.9	12.8
Afternoon	30.4	40.2
Evening	16.6	3.7

COMMUNAL STOP LOC.	Ph.	P.O.
Namur City Centre	66.9	43.5
Belgrade	0.7	4.2
Bouge	5.2	7.3
Jambes	16.7	25.7
Saint-Servais	6.3	5.2
Elsewhere	2.6	14.1

COMMUNAL RESIDENCE LOCATION		
Namur City Centre	27.7	31.1
Belgrade	3.2	6.9
Bouge	6.4	8.2
Jambes	19.4	19.6
Saint-Servais	14.8	13.7
Elsewhere	28.6	20.6

LOCATION SERV. UNITS	Ph.	P.O.
Namur City Centre	51.2	50.2
Belgrade	2.3	6.9
Bouge	11.1	9.1
Jambes	21.2	22.8
Saint-Servais	14.3	11.0

COMMUNAL ORIGIN LOCATION		
Namur City Centre	31.8	38.4
Belgrade	2.3	5.5
Bouge	6.0	8.2
Jambes	20.3	19.6
Saint-Servais	15.2	13.7
Elsewhere	22.6	14.6

TYPE OF JOURNEY	Ph.	P.O.
Residence-Work or Work-Residence	15.6	14.2
Work-Work	6.6	10.1
Residence-Residence	77.7	75.8

MOTIVES OF SPATIAL CHOICE		
Proximity of work or residence	37.8	70.8
Other proximity	20.7	14.6
Other reasons	60.8	23.3

STOP PURPOSES	Ph.	P.O.
Shopping	50.6	42.4
Services	28.6	34.0
Leisure	20.8	23.6

FREQUENCY OF VISIT	Ph.	P.O.
1/month or less	53.5	48.4
More than 1/month	46.5	51.6

3. The transportation mode is not significant in explaining trip tour length of postal users. This counter-intuitive result contrasts with observations in the pharmacy data base (in which journeys by car have more stages than expected), and many other studies (e.g. Bentley *et al*., 1977; Horowitz, 1976; Horowitz, 1982).
4. Journeys are more likely to be single-purpose when the workplace is one of the two extremities. Surprisingly, Hanson (1980) has pointed to an opposite result in the Uppsala survey.
5. Journeys to central pharmacies and post offices (city centre, upper levels of the commercial hierarchy) have a higher probability of being multipurpose. The location of the homeplace has no effect on the extent of the journey. On the other hand, the location of the origin of the trip has a significant influence on the number of stops made by postal customers - individuals coming from the suburban area have a higher

TABLE V
Comparison of Chi-square values observed in both data bases between the number of supplementary stops and several explanatory variables

		PHARMACY sample		POSTAL sample (private use)	
NUMBER OF STOPS (2 cat.) with	number of categories	χ^2	α	χ^2	α
Profession	3	0.002	0.963	10.434	0.001
Sex	2	1.378	0.240	9.729	0.002
Age	3	0.004	0.998	3.716	0.156
Type of journey	3	4.895	0.087	28.499	0.000
Transport mode	3	13.166	0.001	1.422	0.491
Location of service unit	3[1]	9.577	0.008	6.986	0.030
	5[2]	11.303	0.020	13.575	0.009
Location of residence	6	6.624	0.250	9.387	0.101
Location of origin	6	9.518	0.146	12.894	0.024

[1] Levels of the commercial hierarchy.
[2] Communes.

propensity to undertake multipurpose journeys, whereas the behaviour of pharmacy customers is not influenced by the location of their trip origin. This leads to the conclusion that spatial behaviour shows some variability with the relative situation of customers and service units in the urban area (see the references mentioned in the introduction).

Finally, we will consider a more integrated view of spatial behaviour, which is achieved by representing the journey structure as the result of a decision process. The associated pooled *transition matrices* are computed for linkages between land-uses (Table VI) as well as stop locations (Table VII). The two land-use (or purpose) transition matrices are compared by a statistical Chi-square test based on the raw values. The test leads to a significant difference at the 5% level. Once more, the spatial behaviour is shown to be ruled by separate logics according to the service patronised. However, the size and structure of the urban shopping areas of Namur mean that when location transition matrices are derived they are not comparable statistically, but they do show homogeneity in the spatial behaviour of both types of customers. Individuals tend to stay in the same shopping area during their multistage journey.

Chi-square statistics are also computed separately for each land-use transition matrix (Collins, 1975; Kitamura, 1983), enabling us to test the Markov property. In journeys including a visit to a post office or a pharmacy, the Chi-square tests are significant at the 5% level, representing the interdependence of successive land-use

TABLE VI
Purpose transition matrices constructed for both data sets

Pharmacy sample journeys	SHOPPING	SERVICES	LEISURE	Total
SHOPPING	0.11	0.19	0.02	0.32
SERVICES	0.23	0.22	0.09	0.55
LEISURE	0.02	0.07	0.04	0.13
Total	0.36	0.48	0.15	1.00

Postal sample journeys	SHOPPING	SERVICES	LEISURE	Total
SHOPPING	0.07	0.12	0.02	0.21
SERVICES	0.28	0.29	0.13	0.71
LEISURE	0.00	0.05	0.02	0.08
Total	0.35	0.47	0.18	1.00

states. This would be an element in favour of Markov models, quite apart from their other numerous drawbacks (Thill and Thomas, 1987). It is also interesting to know whether the interdependence previously advanced can be connected with some ordering of stop-purpose choices. Therefore, the symmetry of the two-stop location transition matrices is tested by means of pair-wise tests of proportion (Kitamura, 1983). None of the three tests computed for the pharmacy sample is significant: there is no clear-cut sequencing of stop purposes. On the other hand, for the sample of post office patronage, the three pairs of successive purposes tested are statistically symmetric at the 5% significance level. The typical sequence of purposes is shopping-service-leisure. No statistical test can be applied to linkages between stop locations because the associated pooled transition matrices are too sparse. Symmetry is, however, observed in both data sets: this means that there is no particular sequencing of stop locations.

TABLE VII
Location transition matrices constructed for both data sets

Pharmacy sample journeys	JAMBES	SERVAIS	BELGRADE	BOUGE	NAMUR	ELSEWHERE	Total
JAMBES	0.13	0.00	0.00	0.00	0.05	0.00	0.18
ST-SERVAIS	0.00	0.05	0.00	0.00	0.01	0.00	0.07
BELGRADE	0.00	0.00	0.00	0.00	0.00	0.00	0.01
BOUGE	0.00	0.00	0.00	0.05	0.00	0.00	0.06
NAMUR	0.02	0.02	0.00	0.01	0.62	0.01	0.67
ELSEWHERE	0.00	0.00	0.00	0.00	0.00	0.00	0.01
Total	0.15	0.07	0.01	0.07	0.68	0.02	1.00

Postal sample journeys	JAMBES	SERVAIS	BELGRADE	BOUGE	NAMUR	ELSEWHERE	Total
JAMBES	0.23	0.00	0.00	0.00	0.03	0.02	0.28
ST-SERVAIS	0.00	0.05	0.00	0.00	0.00	0.00	0.06
BELGRADE	0.01	0.00	0.03	0.00	0.01	0.00	0.06
BOUGE	0.00	0.00	0.00	0.06	0.01	0.00	0.08
NAMUR	0.01	0.00	0.00	0.00	0.45	0.03	0.50
ELSEWHERE	0.00	0.00	0.00	0.00	0.01	0.00	0.01
Total	0.22	0.07	0.04	0.07	0.52	0.05	1.00

5. CONCLUSION

A question regarding urban consumer behaviour is addressed in this paper: is spatial behaviour (and especially trip tour patterns) specific to the type of service visited? The hypothesis of non-sensitivity is tested using statistical comparisons of the characteristics of spatial behaviour in the context of the patronage of two specific urban services, i.e. pharmacies and post offices. The analysis is undertaken for a given urban area; it shows that both types of service are similar in terms of their supply and demand features as well as the profiles of their customers. But, surprisingly, the descriptions and the univariate explanations of the respective spatial behaviours disclose quite a few dissimilarities - they affect the travel patterns and spatial choice criteria of the consumer. However, features of spatial behaviour, when placed in a broader decision context, are generally identical (e.g. modal and residential choices). It follows that location modellers cannot assume the ubiquity of spatial choice patterns, even when supply and demand features of the service under review are quite similar. In particular, trip-chaining models cannot be indiscriminately applied to a variety of planning circumstances.

The present analysis would benefit from being extended in order to investigate, in more detail, the logic underlying the specificity of the individual's behaviour in space and decisions pertaining to the spatial, temporal and activity pattern of journeys. Furthermore, the suspected specificities of multipurpose travel behaviour need to be confirmed by investigations of other urban services in both similar and dissimilar urban areas.

ACKNOWLEDGEMENTS

The authors are very grateful to H. Beguin, D. Peeters, J. Charlier and P. Densham for their comments on the first draft of the paper.

Department of Geography
Florida Atlantic University
Boca Raton
Florida, USA

Institute of Geography
Catholic University of Louvain
Louvain-la-Neuve
Belgium

BIBLIOGRAPHY

Beaumont, J.: 1981, 'Location-allocation problems in a plane: a review of some models', *Socio-Economic Planning Science* **15**, 217-229.

Beguin, H., P. Hansen, and J-F. Thisse (eds.): 1982, 'Où construire les équipements collectifs?' *Recherches Economiques de Louvain* **48**, 211-373.
Bentley, G., A. Bruce, and D. Jones: 1977, 'Intra-urban journeys and activity linkages', *Socio-Economic Planning Science* **11**, 213-220.
Borgers, A. and H. Timmermans: 1985, 'Effects of spatial arrangement and similarity on spatial choice behaviour', paper presented at the Fourth European Colloquium on Theoretical and Quantitative Geography, Veldhoven.
Borgers, A. and H. Timmermans: 1986, 'A model of pedestrian route choice and demand for retail facilities within inner-city shopping areas', *Geographical Analysis* **18**, 115-128.
Coelho, J. and A. Wilson: 1976, 'The optimum location and size of shopping centres', *Regional Studies* **10**, 413-442.
Collins, L.: 1975, *An Introduction to Markov Chain Analysis*, Norwich, Geo Abstracts, Catmog 10.
Dunn, R. and N. Wrigley: 1985, 'Beta-logistic models of urban shopping center choice', *Geographical Analysis* **17**, 95-113.
Fotheringham, A.: 1983, 'A new set of spatial-interaction models: the theory of competing destinations', *Environment and Planning* **A 15**, 15-36.
Hanson, S.: 1980, 'The importance of the multipurpose journey to work in urban travel behavior', *Transportation* **9**, 229-248.
Hemmens, G.: 1970, 'Analysis and simulation of urban activity patterns', *Socio-Economic Planning Science* **4**, 53-66.
Hodgart, R.: 1978, 'Optimizing access to public services', *Geographical Analysis* **2**, 17-48.
Horowitz, A.: 1982, 'A comparison of socioeconomic and structural determinants of trip tour length', *Papers of the Regional Science Association* **50**, 185-195.
Horowitz, J.: 1976, 'Effects of travel time and costs on the frequency and structure of automobile travel', *Transportation Research Record* **592**, 1-5.
Horton, F. and D. Reynolds: 1971, 'Effects of urban spatial structure on individual behavior', *Economic Geography* **47**, 36-48.
Kitamura, R.: 1983, 'A sequential, history-dependent approach to trip-chaining behavior', *Transportation Research Record* **944**, 13-22.
Kostyniuk, L. and R. Kitamura: 1985, 'Trip chains and activity sequences: test of temporal stability', *Transportation Research Record* **987**, 29-39.
Leonardi, G.: 1981a, 'A unifying framework for public facility location problems. Part I: a critical overview and some unsolved problems', *Environment and Planning* **A 13**, 1001-1028.
Leonardi, G.: 1981b, 'A unifying framework for public facility location problems. Part II: some new models and extensions', *Environment and Planning* **A 13**, 1085-1108.
Miller, E. and S. Lerman: 1979, 'A model of retail location, scale and intensity', *Environment and Planning* **A 11**, 177-193.
O'Kelly, M.: 1983a, 'Multipurpose shopping trips and the size of retail facilities', *Annals of the Association of American Geographers* **73**, 231-239.
O'Kelly, M.: 1983b, 'Impacts of multistop, multipurpose trips on retail distributions', *Urban Geography* **4**, 173-190.

Sporck, J. and M. Goossens: 1985, 'Le réseau urbain. Les zones d'influence des villes et la hiérarchie urbaine', *Bulletin Trimestriel du Crédit Communal de Belgique* **154**, 191-204.

Thill, J-C.: 1986, 'La composante géographique de l'offre et de la demande des pharmacies en milieu urbain', in *Mélanges de Géographie Urbaine. Hommage au Professeur Sporck*, Liège

Thill, J-C. and I. Thomas: 1987, 'Towards conceptualizing trip chaining behaviour: a review', *Geographical Analysis* **19**, 1-18.

Thisse, J-F. and H. Zoller (eds.): 1983, *Locational Analysis for Public Facilities*, Amsterdam, North-Holland, Studies in Mathematical and Managerial Economics, volume 31.

Thomas, I.: 1983, 'Amélioration du découpage géographique des agglomérations urbaines', *Espace Géographique* **12**, 207-214.

Thomas, I.: 1984, 'Towards the simplification of the location models for public facilities: the example of the postal service', *Papers of the Regional Science Association* **55**, 47-58.

Thomas, I.: 1986, *La Localisation Optimale des Services Publics. Une Méthode Opérationnelle et son Application au Service Postal.* Louvain-la-Neuve, Cabay.

Timmermans, H., R. van der Heijden, and H. Westerveld: 1984, 'Decisionmaking between multiattribute choice alternatives: a model of spatial shopping behaviour using conjoint measurements', *Environment and Planning* **A 16**, 377-387.

Mark D. Uncles

USING CONSUMER PANELS TO MODEL TRAVEL CHOICES

1. INTRODUCTION

This paper takes another look at some of the models which have been developed to describe and predict repeated choices. These models are used to predict decisions, such as which shopping centre to use, which shop to visit, and what brand to buy from the assortment on offer. All these decisions are routinely made by people going about their daily business, and collectively they give rise to definite patterns of behaviour.

What we do here is apply these models to travel decisions by shoppers. This represents something of a shift from studies of low-involvement decisions, like the selection of brands from a supermarket shelf, to situations which might be seen as more purposeful. The models are described in Section 2, and in Section 3 both two-alternative and multi-alternative cases are illustrated. Since the latter are of recent origin it is particularly important to assess their scope by reporting on new applications.

Using a consumer panel, decisions about travel choice are monitored over a period of several weeks. This enables us not only to calculate the probability of a choice but also to study the spread of probabilities among shoppers. By extending the basic model we can do a number of tests, such as checking for omitted variables, assessing how well the predictions fit observed values, and seeing if there is temporal dependence. Some results are presented in Section 4.

To anticipate what is reported later, our results show that there is less of a difference than might be expected between very low-involvement decisions and travel choices. There are some differences, however, such as the lower rate of switching between alternatives, this we ascribe to varying circumstances among shoppers, rather than to temporal effects like feedback or learning. The significance of this is that repeated choices can be studied using fairly simple dynamic models.

2. MODELS OF REPEATED CHOICE

2.1. *Choice between Two Alternatives*

Consider the situation where data have been collected about how shoppers reach their destinations, and say we know how often each person travels by bus. For the ith shopper, if the probability of travelling by bus, p_i, is constant and independent of what has gone before the choice sequence can be described as a Bernoulli process. Thus, a person's probability of going by bus t out of n weeks ($t = 0,1,\ldots,n$) is

J. Hauer et al. (eds.), Urban Dynamics and Spatial Choice Behaviour, 43–57.

$$f(t|n,p_i) = \binom{n}{t} p_i^t (1-p_i)^{n-t} \tag{1}$$

It follows that for a *homogeneous group* of shoppers, where $p_i = \bar{p}$ for all $i = 1,2,\ldots,N$ and the error variance is zero, their choice probability will be described by the Binomial distribution. This means that a cross-section estimate of $\bar{p}$, the choice probability of a "representative" shopper, can be used to predict average choice sequences for the whole group.

For a *heterogeneous group*, by contrast, the p_i will be distributed in some manner, and the problem arises as to how this should be characterised. A pragmatic solution is to assume a functional form for this distribution and estimate the associated parameters from panel data. The Beta distribution, for instance, gives

$$f(p_i|a_{1i},a_{2i}) = \frac{\Gamma(a_{1i} + a_{2i})}{\Gamma(a_{1i})\,\Gamma(a_{2i})} p_i^{a_{1i}-1} (1-p_i)^{a_{2i}-1} \tag{2}$$

where a_{1i} and a_{2i} are parameters of the mixing distribution for shopper i, both of which exceed zero, and $\Gamma(\cdot)$ is the Gamma function.

The mean probability of choosing to travel by bus is

$$\text{mean}(p_i) = \frac{a_{1i}}{a_{1i} + a_{2i}} \tag{3}$$

and other moments are

$$\text{mode}(p_i) = \frac{a_{1i} - 1}{a_{1i} + a_{2i} - 2} \tag{4}$$

if a_{1i} and $a_{2i} > 1$, and

$$\text{variance }(p_i) = \frac{a_{1i}\, a_{2i}}{(a_{1i} + a_{2i} + 1)\,(a_{1i} + a_{2i})^2} \tag{5}$$

In effect, a_{1i} and a_{2i} are shape parameters which describe the spread of probabilities across individuals. For instance, if a_{1i} exceeds one, and a_{2i} is less than one, the distribution has a *J* shape with a single mode at 1. By looking at these shape parameters we are able to study heterogeneity in considerable detail, and learn how well the model represents the behaviour of shoppers. Where the shape is bimodal, for instance, the amount of variation is high, and this suggests that an influential variable has been left out - an important determinant of whether a shopper travels by bus, say.

Further diagnostic information is available. Consider the case where the probability of travelling by bus in $t = 0,1,\ldots,n$ weeks and not in $k = n-t$ weeks is

$$p_i(t,n) = B(a_{1i}+t, a_{2i}+k) \;/\; B(a_{1i}, a_{2i}) \tag{6}$$

where $B(\cdot)$ is the Beta function.

Now the goodness of each fit can be gauged by comparing predictions from (6) against observed values. Furthermore, by summing over i we get the mean predicted probability

$$\text{mean}[p(t,n)] = \sum_{i=1}^{N} p_i(t,n) \;/\; N \tag{7}$$

Then this can be compared with the average of observed choices. Conditional probabilities can be derived as well (Kahn *et al.*, 1986). The usefulness of these as diagnostic tests is illustrated in Section 4.

The next step is to postulate that the parameters a_{1i} and a_{2i} are functions of exogenous variables, $\mathbf{x}_i'$, such as disposable income and attitudes to shopping. Thus

$$\left.\begin{aligned} a_{1i} &= e^{\mathbf{x}_i'\boldsymbol{\alpha}_1} \\ a_{2i} &= e^{\mathbf{x}_i'\boldsymbol{\alpha}_2} \end{aligned}\right\} \tag{8}$$

and $\boldsymbol{\alpha}_1$ and $\boldsymbol{\alpha}_2$ are vectors of parameter estimates.

We now have a Beta-logistic formulation, where the mean probability of choice is based on the logit function

$$\text{mean}(p_i) = \frac{e^{\mathbf{x}_i'(\boldsymbol{\alpha}_1 - \boldsymbol{\alpha}_2)}}{1 + e^{\mathbf{x}_i'(\boldsymbol{\alpha}_1 - \boldsymbol{\alpha}_2)}} \tag{9}$$

To estimate the model, maximum likelihood methods must be used (see Heckman and Willis, 1977). Apart from this there are no major new principles in making the extension to find higher moments and diagnostics. Thus, when the goodness of fit is tested, the probability of choosing bus travel t out of n weeks still has the same form as Equation (6), but now a_{1i} and a_{2i} are based on the exponential functions as in (8).

2.2. *Choice among Several Alternatives*

Typically shoppers are able to choose between several alternatives, so not only do they travel by bus they might also use a metro system, or drive, or walk. Recently Dunn and Wrigley (1985) have shown how to extend Equations (1) and (2) to handle the situation where there are $j = 1,2,\ldots,J$ alternatives. They suggests that the probability of making choice j in t out of n weeks is

$$f(\mathbf{t}|n,\mathbf{p}_i) = \begin{bmatrix} n \\ t_1, t_2, \ldots, t_J \end{bmatrix} p_{1i}^{t_1} \, p_{2i}^{t_2} \cdots p_{Ji}^{t_J} \tag{10}$$

From the general form of (10) it is straightforward, in principle at least, to obtain mean probabilities, higher moments and diagnostics. For instance, the mean probability of choosing alternative j is derived from individual parameter values, a_{ji}, and from their share of all a_{ji} terms, the sum of which we shall call S_i

$$\text{mean } (p_{ji}) = a_{ji} \,/\, S_i \tag{11}$$

What this gives rise to is a multinomial version of the Beta-binomial model, i.e. the Dirichlet model (Goodhardt *et al*., 1984; see also Wrigley and Dunn, 1985). As in Equation (8), the parameters of this model are estimated from exogenous variables

$$\begin{aligned} a_{1i} &= e^{\mathbf{x}_i'\boldsymbol{\alpha}_1} \\ a_{2i} &= e^{\mathbf{x}_i'\boldsymbol{\alpha}_2} \\ &\vdots \\ a_{Ji} &= e^{\mathbf{x}_i'\boldsymbol{\alpha}_J} \end{aligned} \tag{12}$$

This is the basis for calibrating the so-called Dirichlet-logistic model. Now the mean probability of choosing alternative j is found from

$$\text{mean}(p_{ji}) = \frac{a_{ji}}{S_i} = \frac{e^{\mathbf{x}_i'\boldsymbol{\alpha}_j}}{\sum_{s=1}^{J} e^{\mathbf{x}_i'\boldsymbol{\alpha}_s}} \tag{13}$$

In the examples that follow, both the Beta-logistic and Dirichlet-logistic models are used to study travel choices, also a number of diagnostic tests are performed.

3. TRAVEL CHOICE ON GROCERY SHOPPING TRIPS

3.1. *Choice of whether to Travel by Car or Not*

Shops can be reached using several forms of travel, notably car travel and walking. The relative amount of usage varies, depending on (i) cultural factors (such as heavier car usage in the US), (ii) the supply of transport services (for example, the efficiency and coverage of local rapid transit systems), and (iii) average distances travelled. These factors mean that for shoppers in a British city something like 56% of all grocery trips are on foot, 32% are car-borne and a further 12% are by bus. On major trips the car-borne component rises to roughly 44% and bus usage falls off a lot.

Within the overall setting of travel usage, an individual's choice depends on both socio-demographic factors (disposable income, the value of time, the amount of discretion a person has, and so forth) and specific features of each trip (such as the need to buy in bulk and the weight of goods carried). Here we study the choices made by roughly 300 grocery shoppers when faced with two alternatives: either to travel by car, or to walk or use a local bus. These data relate to a sub-sample of continuous reporters, observed for 5 weeks, on the 1982 Cardiff Consumer Panel (Wrigley *et al.*, 1985; Uncles, 1985, 1987).

Bearing in mind all these possible influences on an individual's choice, we start by calibrating the Beta-logistic model with just four exogenous variables: income and household size, and two binary variables for work status and deep freezer ownership. The full model is shown in Table I. Beside each of the parameter estimates are asymptotic t values and at conventional significance levels (i.e. $t > 1.9$) most estimates are acceptable.

The likelihood ratio between the full model and partial models (obtained by removing each variable in turn) gives a test statistic distributed as χ^2 with 2 degrees of freedom (one degree for each parameter removed). Most variables are significant at the 90% level, except working status which is only significant at 85%.

The findings are consistent with what we know about travel decisions already. Thus, the probability of travelling by car goes up

TABLE I
Estimates for a two-alternative model of travel choice

Variable	Parameter estimates		Asymptotic t values		Likelihood ratio test	Signif. levels
	car	other	car	other		(%)
Constants	-1.92	0.33	3.7	0.6		
Income	0.35	-0.16	3.4	1.5	55	99
Size	-0.20	-0.11	2.1	1.1	5	90
Deep freezer	0.16	-0.26	0.6	1.0	5	90
Working	0.58	0.37	2.0	1.3	4	85

as income rises whereas walking and local bus usage become less likely. A larger household will buy more goods, which means the amount of shopping activity rises; moreover, many shoppers in this situation are full-time housewives and are able to devote more time to shopping tasks. Those employed full-time, by contrast, have a high valuation of time and more disposable income, consequently they rely on car travel, especially when buying in bulk at edge-of-town superstores and hypermarkets. Deep freezer ownership has a less significant effect, but even here the signs are plausible, showing that owners choose car travel because this mode alone offers an easy way to carry bulky goods which are then stored in the freezer.

To see the implications of these results it is worth looking in detail at some household types. Four are presented in Table II. Among low income couples (Type 2) the mean probability of choosing to travel by car is only 0.13, whereas this rises above 0.70 for high income families whose members work (Types 3 and 4).

Within sub-groups there is further variation, for example Types 1 and 3 have bimodal Beta distributions and large variances of 0.12 and 0.10 (see the final two columns of Table II). Members of these households nearly always, or hardly ever, choose a car on major shopping journeys and only a few lie between these extremes. Small rich families (Type 4) almost always drive to shops. By contrast, households whose income is low (Type 2) tend to travel by foot or bus.

The heterogeneity we see in these examples may be due to omitted variables, rather than purely random variations in taste. The U-shaped mixing distribution for Types 1 and 3, for instance, might arise from factors which favour shopping by foot and because retail outlets are situated nearby, in which case a measure of proximity to shops might improve the model.

TABLE II
Mixing distribution for a two-alternative model of travel choice

Household type	$\hat{a}_{1i}$	$\hat{a}_{2i}$	Mean($\hat{p}_i$)	Mode($\hat{p}_i$)	Var($\hat{p}_i$)
1	0.56	0.53	0.51	bimodal	0.12
2	0.14	0.95	0.13	bimodal	0.05
3	0.76	0.31	0.71	bimodal	0.10
4	1.75	0.69	0.72	1	0.06

Definition of household types

Type	Income	Size	Deep freezer	Working
1	average	average	yes	yes
2	low	low	no	no
3	high	high	yes	yes
4	high	low	no	yes

3.2. *Choice between Car, Walk and Bus*

So far we have looked at the choice between two alternatives on major grocery trips, now we consider a wider range of alternatives on all grocery trips - it might be the case that differences between walking and public transport are greater than differences between these two and car. Moreover, when incidental trips (such as local topping up visits and convenience shopping) are included, the share of trips by car falls to 32% from a figure of 44% on major trips.

A Dirichlet-logistic model is calibrated with five exogenous variables, including household income and size (both ordinal) and three binary variables to describe the shopper's work status (employed/not employed), deep freezer ownership (present/absent), and whether the consumer is elderly (yes/no). Shown in columns 1 to 3 of Table III are the parameter estimates for each of j = 1,2,3 choice alternatives.

Significance tests show that income is a good predictor of car choice, whereas work status is most significant as a predictor of bus travel (many working shoppers buy their goods on the journey home from work and often this involves bus travel). The likelihood ratio statistic, distributed as χ^2 with 3 degrees of freedom, confirms the importance of household income, followed by ownership of a deep freezer and work status.

With rising household income the chance of car travel increases and the propensity to travel by bus falls. This is wholly consistent with the view that car travel is a superior good - it is seen as more flexible and more comfortable than the alternatives. The influence of income over pedestrian movement is more equivocal; this is because local shops are reached on foot, almost regardless of income. However, bulky goods are rarely bought on these trips, so ownership of a deep freezer reduces the chance of walking and raises the probability of travel by car.

Just as the spread of choice probabilities among households can be studied for the two-alternative model, so too can it be studied for the multi-alternative model. Thus, a low income elderly person living alone (Type 1 in Table IV) is quite likely to choose bus travel (the

TABLE III
Estimates for a multi-alternative model of travel choice

Variable	Parameter estimates			Asymptotic *t* values			Likelihood ratio test	Signif. levels
	car	walk	bus	car	walk	bus		(%)
Constants	-2.37	0.19	-1.15	5.8	0.5	2.9		
Income	0.52	-0.03	-0.12	7.0	0.4	1.6	101	99
Size	-0.17	0.04	-0.01	2.2	0.5	0.1	11	98
Deep freezer	0.17	-0.32	-0.03	0.9	1.6	0.1	9	96
Working	0.31	0.23	0.43	1.5	1.1	1.9	4	80
Elderly	0.24	0.04	0.40	0.6	0.1	1.1	2	-

mean probability is 0.22), although on local trips walks are even more important (so the mean probability is 0.69). What this highlights is the double deprivation facing many elderly people: they rely on public transport and walking, which confines them to local shops where prices tend to be higher and where the product range is narrower.

For each form of travel, choice probabilities map out distinctive shapes: the probability of an elderly person walking to the shops decribes a J-shaped curve with a single mode at 1, whereas car choice is reverse J-shaped and the mode is close to 0. This too shows how elderly people rely on local pedestrian travel, either by choice or because of immobility.

By way of contrast, consider an affluent household - income is high, the shopper works and this gives rise to time constraints, but goods are bought in bulk and stored in a freezer (Type 2 in Table IV). There is less variation here: car travel is by far the most likely choice. Parameters for income, freezer ownership and paid employment all contribute positively to the choice of car, to give a mean probability of 0.74 and a mode approaching 1. Whereas the probability of bus travel describes a reverse J-shaped curve, and this serves to underline how the use of public transport is unlikely among affluent households.

TABLE IV
Mixing distributions for a multi-alternative model of travel choice

Household type	$\hat{a}_{ji}$	$\hat{S}_i-\hat{a}_{ji}$	Mean($\hat{p}_{ji}$)	Mode($\hat{p}_{ji}$)	Var($\hat{p}_{ji}$)
$i = 1$					
j = Car	0.17	1.68	0.09	bimodal	0.03
j = Walk	1.27	0.58	0.69	1	0.08
j = Bus	0.41	1.44	0.22	bimodal	0.06
$i = 2$					
j = Car	3.46	1.21	0.74	0.92	0.03
j = Walk	1.01	3.66	0.22	0.04	0.03
j = Bus	0.20	4.47	0.04	0	0.01

Definition of household types

Type	Income	Size	Deep freezer	Working	Elderly
1	low	single	no	no	yes
2	high	nuclear	yes	yes	no

4. TESTING THE MODELS OF REPEATED CHOICE

In this section we have another look at the first example. Travel choices for the main weekly shopping trip are observed over 5 weeks. This means the number of trips is fixed - something which contrasts with the previous example where the number of trips varies amongst households from week to week. In substantive terms, the incidence of car travel is somewhat greater on main trips than is generally the case. From a practical modelling viewpoint, the ability to fix the number of trips enables us to test the model for goodness of fit and to assess the independence of choices through time.

4.1. *Goodness of Fit*

The simplest test is to compare observed and predicted probabilities. Each week the choice is binary (car/no car), so the comparison of probabilities over five weeks can be read off from a *binomial tree.* Such a tree is shown in Figure 1. Actual proportions are along the top of each branch, and mean predicted probabilities below; for example, in the first week 44% of consumers travel by car (these follow the "1" branch) compared with 56% who go by some other means (the "0" branch). Equations (6) and (7) are used to find the predicted probabilities, following the method of Heckman and Willis (1977).

The fit between predicted and actual values is close; for instance, we predict a 46% chance of car travel in the first week, a 35% chance of making this choice twice in the first two weeks, and even after 5 weeks a 24% chance that car travel is always chosen. Deviations from actual values amount to only a few percentage points, such as the 2% over-prediction of car choice in week 1, and an average difference of merely 0.8% among those making the same choice over all weeks. These differences of only a few percent represent no more than half-a-dozen shoppers from a sample of about 300.

These points are reiterated if we look at the results in a slightly different way (Table V). The actual number of shoppers who travel by car on every main trip is 25% (24% predicted), the number who travel by car for four out of five main trips is 10% (12% predicted), and so on. Continuous choices are dominant - shoppers always travel by car or never do so - while the least likely pattern is to choose car on just two or three occasions. Furthermore, it is among the switching patterns that the difference between actual and predicted values is greatest.

In general, what all these results confirm is the goodness of fit. They also show how travel choices are more rigid than other choices, like the decision of where to shop, what product to buy and which brand to select. In these other cases loyalty is low and shoppers often switch allegiance between competing alternatives. Nevertheless, from our study of heterogeneity it is likely that the greater degree of loyalty for travel decisions arises from variation among people rather than from temporal dependence or feedback, in which case it differs from many high-involvement decisions, like job and home choice. This is what we unravel next.

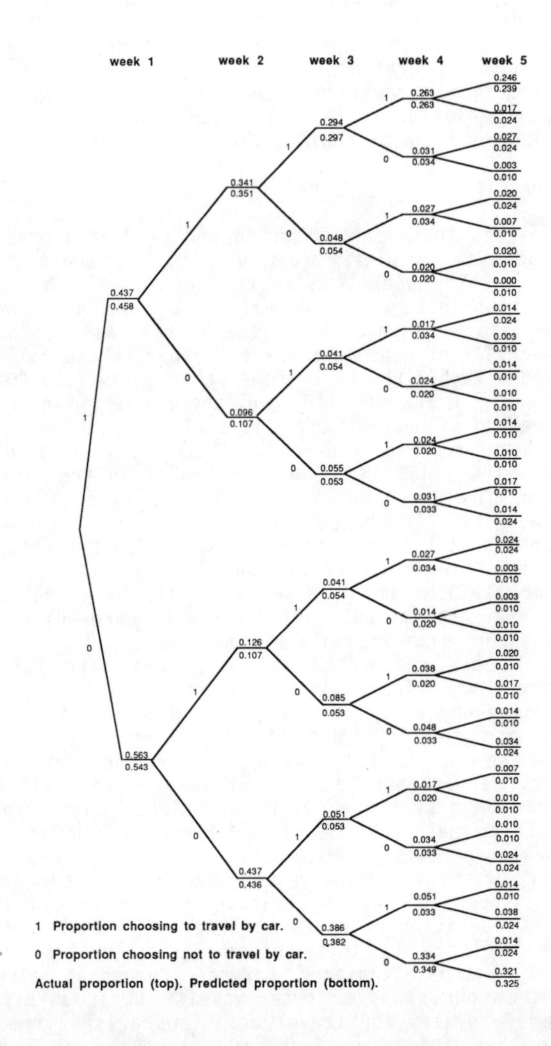

Fig. 1: Actual and predicted car choice probabilities, paths over 5 weeks.

4.2. *Intertemporal Dependence*

The Beta-logistic and Dirichlet-logistic models are based on the zero-order assumption, whereby current choices are independent of previous history and selected probabilities are constant. Now we see how far this assumption is valid, first by comparing different transition probabilities, and then by looking at conditional probabilities. In both exercises we are attempting to find out the relative effects of heterogeneity and intertemporal dependence.

4.2.1. *Transitions*

If the observed transitions from one choice to another are roughly constant and their absolute values are close to the Beta-logistic transitions we ought to accept the zero-order assumption. On the other hand, if the fit is poor, or equivocal, transitions from alternative stochastic processes ought to be tested too.

Generally, the probability that a shopper who chooses g in week t and then switches to m in week $t+n$ is denoted by $p_{gm}^{(n)}$. In our example there are two possible choices each week: car travel, "1", and all other options, "0", giving a total of four transition probabilities

$$\mathbf{p}^{(n)} = \begin{bmatrix} p_{00}^{(n)} & p_{01}^{(n)} \\ p_{10}^{(n)} & p_{11}^{(n)} \end{bmatrix} \tag{14}$$

This is the n-step transition matrix. Since row probabilities sum to one we need only consider $p_{11}^{(n)}$ and $p_{01}^{(n)}$ (i.e. always choosing car and switching into car travel).

Actual n-step choice probabilities and those predicted by the Beta-logistic model are shown in Table VI (rows 1 and 2). The base period, t, is week 1 and $t+n$ refers to transitions for each subsequent

TABLE V
Actual and predicted distribution of weeks with car travel

Number of weeks with car travel	Percent of sample Actual	Predicted	Discrepancy (percentage points)
5	25	24	0.7
4	10	12	1.8
3	9	10	0.6
2	11	10	1.2
1	12	12	0.4
0	32	33	0.4
Total	100	100	

week. Constancy is assumed in the zero-order model, so all the transitions in line 2 are identical: the probability of continually choosing car travel is 0.77, while the probability of switching into car travel is 0.20. As $p_{11}^{(n)}$ greatly exceeds $p_{01}^{(n)}$, switching between alternatives is less likely than remaining in the same state; however, we cannot yet say how this comes about.

The observed transitions in line 1 oscillate, though they always lie reasonably close to the constant Beta-logistic probabilities and are always about the same absolute size. We can see just how close this fit is when a contrast is drawn with outcomes from another process. For instance, in line 3 of Table VI the results for a first-order Markov chain are given, using the matrix

$$\mathbf{M} = \mathbf{p}^{(1)} = \begin{bmatrix} 0.776 & 0.224 \\ 0.220 & 0.780 \end{bmatrix} \tag{15}$$

for t = week 1 and t+1 = week 2. The Markov effect soon operates, with $p_{11}^{(n)}$ falling from 0.78 during weeks 1 & 2 to 0.55 in weeks 4 & 5. But, this is totally at odds with what we observe in line 1, to the extent that by week 5 the discrepancy for $p_{11}^{(n)}$ is about 30%. On this evidence the Markov model is not a credible alternative. What we see instead is that the Beta-logistic model fits much better; moreover, it is easier to calibrate because no explicit account of temporal dependence is needed.

TABLE VI
Transition probabilities for car travel

Week = $t + n$	Week 2 (%)	Week 3 (%)	Week 4 (%)	Week 5 (%)
$p_{11}^{(n)}$				
1. Actual	78	77	76	85
2. Beta-logistic	77	77	77	77
3. Markov	78	66	59	55
$p_{01}^{(n)}$				
1. Actual	22	16	24	19
2. Beta-logistic	20	20	20	20
3. Markov	22	35	42	46

t = week 1; t+n = weeks 2 to 5

4.2.2. *Conditional Probabilities*

Another way to decide whether temporal dependence ought to be specified in our models is to study conditional probabilities. Consider two groups of shoppers: those in group *A* travel by car in week *t* and *t*-2, but not in *t*-1 (i.e. $p_{1|10}$); while group *B* comprises those who travel by car in week *t* and *t*-1, but not in *t*-2 (i.e. $p_{1|01}$). In a pure heterogeneity model, of the Beta-logistic type, the proportion of consumers in each group should remain fairly similar through time. If, however, there really is state dependence then choice probabilities in week *t* will fall among members of group *A*, possibly to a level below those for *B* (Heckman and Willis, 1977, pp. 52-53).

Observed probabilities from the binomial tree are used to calculate the conditional probabilities. On average, 48% of shoppers in group *A* choose car travel, compared to 40% among members of group *B* (Table VII). By choosing car in *t*-1 the chance of doing so again in the current period is marginally *lower* than if a car had been chosen in *t*-2; this is the converse of what we would expect if there was true dependence. This relationship holds fairly steadily through time, so once again the simplest model - without having explicit feedback terms - will suffice.

5. DISCUSSION

This paper shows how to predict travel choices using a zero-order model, for both two-alternative situations (i.e. whether or not to go shopping by car) and multi-alternative cases (i.e. choice between car, bus and walking). There are several conclusions.

First, the spread of probabilities seems to be governed by factors such as household income and whether a person works. The existence of such within group conformity is to be expected, not only as a reflection of what people from similar backgrounds can afford, but also because shoppers often simplify the decision process by referring to their peer group. There is further variation, however, which embodies the random effect of tastes and possibly the effect of omitted variables (nearness to local shops for instance).

TABLE VII
Conditional probabilities for car travel

Actual car choice probabilities in week *t* for shoppers who have travelled by car in one of the preceding weeks

Group		Conditional probability in week *t* Week 3 (%)	Week 4 (%)	Week 5 (%)	Average
A	$p_{1\|10}$	43	49	52	48
B	$p_{1\|01}$	33	37	49	40

Second, travel choices are repeated over successive weeks, especially on major trips where car usage is of greater importance. This level of apparent loyalty arises because of the conformity within groups - for example wealthier families tend to travel by car - rather than because of feedback or learning effects. In practical terms this means decisions can be modelled as if the process is zero-order (it may not even matter if the underlying process is like this, especially if prediction is the main aim).

Third, conceptually travel choice lies mid-way on a scale that runs from low-involvement decisions, like which brand of butter to buy, to high-involvement decisions, like the deliberations lying behind a career choice. In many ways travel choice shares many of the same features of store choice - more involved than buying butter and yet still a frequently made decision which has low monetary significance (Uncles and Ehrenberg, 1987).

Overall, this work brings to the fore the issue of how dynamic models should be developed. From our findings it is clear that there should be a presumption in favour of less elaborate models, at least for the type of decision underlying people's regular behaviour and their movement around a city. There is some intuitive support for this view: often there is little prior thought before a decision is made, the outcome has low monetary significance anyway, and the risk of a wrong decision is slight (Olshavsky and Granbois, 1979).

A concern for simplicity does not absolve us from continuing to test the models, nor from re-specifying them. By finding new applications, where the models hold again, we learn about their robustness and their shortcomings. Alternative specifications must continue to be tested too, either by making new distributional assumptions or by working with non-parametric methods (Dalal *et al.*, 1984; Davies, 1984; Dunn *et al.*, 1987; Reader and Uncles, 1987). It may turn out that the parameter estimates and substantive conclusions hardly alter - or at least they are less of a problem than data error - in which case the simplest models ought to be used.

A thorough knowledge of the patterns that arise from these individual choices is needed if we are to comment on the provision of transport facilities, the performance of stores, and the success of shopping centres. It is these patterns that the models and empirical work in this paper begin to describe.

ACKNOWLEDGEMENTS

For advice and encouragement I would like to thank Neil Wrigley and Richard Dunn. Financial Support came from the Economic and Social Research Council, and from the Centre for Marketing and Communication at the London Business School.

London Business School
London
United Kingdom

BIBLIOGRAPHY

Dalal, S.R., J.C. Lee, and D.J. Sabavala: 1984, 'Prediction of individual buying behavior: a Poisson-Bernoulli model with arbitrary heterogeneity', *Marketing Science* **3**, 352-367.

Davies, R.B.: 1984, 'A generalised Beta-logistic model for longitudinal data with an application to residential mobility', *Environment and Planning* **A 16**, 1375-1386.

Dunn, R., S. Reader, and N. Wrigley,: 1987, 'A non-parametric approach to the incorporation of heterogeneity into repeated polytomous choice models of urban shopping behaviour', *Transportation Research* **A 21**, 327-345.

Dunn, R. and N. Wrigley: 1985, 'Beta-logistic models of urban shopping centre choice', *Geographical Analysis* **17**, 95-113.

Goodhardt, G.J., A.S.C. Ehrenberg, and C. Chatfield: 1984, 'The Dirichlet: a comprehensive model of buying behaviour', *Journal of the Royal Statistical Society* **A 147**, 621-655.

Heckman, J.J. and R.J. Willis: 1977, 'A Beta-logistic model for the analysis of sequential labour force participation by married women', *Journal of Political Economy* **85**, 27-58.

Kahn, B.E., M.U. Kalwani, and D.G. Morrison: 1986, 'Measuring variety-seeking and reinforcement behaviors using panel data', *Journal of Marketing Research* **23**, 89-100.

Olshavsky, R.W. and D.H. Granbois: 1979, 'Consumer decision making - fact or fiction?' *Journal of Consumer Research* **6**, 93-100.

Reader, S. and M.D. Uncles: 1987, 'The collection and analysis of consumer data', in M.D. Uncles (ed.), *Longitudinal Data Analysis: Methods and Applications*, Pion, London.

Uncles, M.D.: 1985, 'Models of consumer shopping behaviour in urban areas: analysis of the Cardiff Consumer Panel', Unpublished PhD Thesis, Department of Geography, University of Bristol, Bristol.

Uncles, M.D.: 1987, 'A Beta-logistic model of mode choice: goodness of fit and intertemporal dependence', *Transportation Research* B **21**, 195-205.

Uncles, M.D. and A.S.C. Ehrenberg: 1987, 'Patterns of store choice: new evidence from the USA', in N. Wrigley (ed.), *Store Choice, Store Location and Market Analysis*, Routledge & Kegan Paul, London.

Wrigley, N. and R. Dunn: 1985, 'Stochastic panel-data models of urban shopping behaviour: 4. Incorporating independent variables into the NBD and Dirichlet models', *Environment and Planning* **A 17**, 319-331.

Wrigley, N., C.M. Guy, R. Dunn, and L.G. O'Brien: 1985, 'The Cardiff Consumer Panel: methodological aspects of the conduct of a long-term panel survey', *Transactions of the Institute of British Geographers* New Series **10**, 63-76.

RICHARD B. DAVIES

ROBUSTNESS IN MODELLING DYNAMICS OF CHOICE

1. INTRODUCTION

During the fifteen to twenty years that statistical analysis has been extensively taught and used in Geography, there have been several changes in the emphasis placed upon different statistical approaches. These changes have reflected the evolving consensus on theoretical issues in statistical inference both within social science and elsewhere. Hypothesis testing was the dominant theme initially, although it has some obvious weaknesses as a data analysis technique and, arguably, has been a major impediment to the general acceptance of the unique rationality of statistical inference. Not only do the results of hypothesis tests depend upon arbitrary decisions on significance levels, but they also tend to be of meagre interest within a research programme. If the null hypothesis is rejected the researcher merely has some (probabilistic) confirmation of what he anticipated; if the null hypothesis is not rejected it may just be that the test has low power for the sample size analysed. The exaggerated claims for hypothesis testing did not help and, in particular, attempts to 'dress up' hypothesis testing as synonymous with scientific method, typical of social science statistics textbooks of that era, were unconvincing even to those with the most limited grasp of the philosophy of science.

A change in emphasis from hypothesis to significance testing afforded some improvement; by reporting the significance level (p-value) achieved by a test statistic, the researcher could direct attention at the strength of the evidence for rejecting the null hypothesis. In this approach, there was no question of overwhelming evidence of rejection (e.g. a p-value of 0.00001) having the same status as a marginal rejection (e.g. a p-value of 0.049). But the basic contradiction remained. The researcher is interested in evidence about the research hypothesis; the significance test merely assesses the plausibility of the null hypothesis given an outcome at least as extreme (in some sense) as that actually obtained.

More recently, we have witnessed a major change of emphasis with statistical modelling now very much in vogue. In statistical modelling, the analysis is directed at distinguishing systematic variation attributable to explanatory variables from 'random' variation due to other effects. Significance testing conventionally plays a part in statistical modelling. In particular, it provides criteria for assessing parsimony. But the logical rationale of statistical modelling is, in general, far more closely related to the data analysis requirements of the social science researcher. This rationale is based upon likelihood and, in effect, provides the researcher with a new theoretical framework for inference. Inference in likelihood methods is not based upon the weak logical links between evidence about the plausibility of something you do not believe to be true (the null hypothesis) and drawing conclusions about research

J. Hauer et al. (eds.), Urban Dynamics and Spatial Choice Behaviour, 59–80.

hypotheses. Nor is it based upon the analysis of outcomes that did not occur (for example, the dubious supplementation of the actual outcome with those more extreme). Rather, likelihood methods are explicitly concerned with assessing the evidence provided by the data. See, for example, Vincent and Haworth (1984) and Pickles (1985).

Statistical modelling may be formalised as a structured process and techniques for each stage (specification, calibration, criticism) are now well documented and well supported by software, notably GLIM (Baker and Nelder, 1979) and GENSTAT (Genstat 5, 1986). However, in practice, statistical modelling is rarely as straightforward as portrayed in textbooks; worked examples tend to have a misleading simplicity. Often the examples are chosen to illustrate a specific point but, even when examples have not been consciously selected to avoid potentially confusing complexity, constraints on the data that can be readily reproduced in a textbook tend to result in simple models performing well on parsimony and diagnostic criteria. Typically, the data are grouped, there are few explanatory variables, and the sample size is small. In contrast, the serious empirical researcher tends to confront large datasets with many potential explanatory variables and a unique data vector on each individual.

The main problem is that, with a large array of data, it is generally possible to obtain support for more and more complex statistical models. This is exacerbated by two factors. First, with ungrouped data the researcher sometimes has no way of reliably assessing absolute goodness of fit and can only assess parsimony by comparison with more complex alternatives. Second, to investigate the empirical support for alternative theories, models should be 'nested' to obtain independent tests of each possible source of variation (Selvin and Stuart, 1966). Attempts to follow the systematic rules of statistical modelling may therefore lead the conscientious researcher into complex probability models and take his empirical work beyond the scope of package software. This is clearly undesirable; such complications would militate strongly against the widespread adoption of statistical modelling in social science research whatever the theoretical merits of the approach. The question then arises as to whether it is appropriate to follow this route to seemingly ever more complicated models. As evidenced elsewhere in this volume (Pickles and Davies, 1989), comparative complexity is sometimes not readily avoided without serious risk of inferential error, although careful attention to sampling design may help. But in other circumstances, inferences may be confidently made on the basis of simple models even though more complex models may result in considerable goodness-of-fit improvements. A pragmatic approach (for example, Wrigley and Dunn, 1984) is to treat simple statistical models as 'bench marks' against which empirical regularities and irregularities may be compared. The more formal approach is to recognise that the calibrated model is over-simplified and, in effect, incorrect and to fully allow for this misspecification in the precise inferential procedures adopted. This paper is concerned with the latter approach for event and recurrent choice processes.

The statistical theory which informs conventional statistical modelling methods is not applicable to misspecified models; it assumes that the probability model is correctly specified. Methods of inference for misspecified models have to be based upon alternative, or at least more general, theoretical foundations. An appropriate formulation, termed 'pseudo-likelihood', has been established in recent years (White, 1982; Gourieroux *et al*, 1984) although the theoretical developments data back at least to Cox (1961). The main statistical results appropriate to applied work are summarised in Section 2. This section also reports some related results on 'quasi-likelihood', a robust statistical modelling approach originally proposed by Wedderburn (1974). Quasi-likelihood is particularly important in the context of this paper for the insight it provides into misspecified Poisson and Binomial models. This insight has already been exploited in the geography literature by Baxter (1985). The remainder of the paper presents examples of how misspecified models sometimes provide a suitable basis for inference even if the precise nature of the misspecification is unknown. The Poisson regression model is illustrated in Section 3; the Binomial Logistic model in Section 4. Section 5 concludes the paper.

2. GENERAL RESULTS

Consider the statistical modelling of responses from a sample of size N. Let the response vector for case i be $\mathbf{y}_i$ (which may, of course, be a scaler) and the value of m explanatory variables be given by the vector $\mathbf{x}_i$. Additionally, let the log-likelihood for case i under a specific model be $\ell_i(\boldsymbol{\beta}|\mathbf{y}_i,\mathbf{x}_i)$ where $\boldsymbol{\beta}$ is a vector of, say, K structural parameters. In simple formulations, K would equal m+1 with one parameter for each explanatory variable and one parameter as the constant or intercept term in the model. More complex models, such as competing risk models, would require more parameters. Maximum likelihood parameter estimates $\hat{\boldsymbol{\beta}}$ may be obtained by maximising the total log-likelihood or, equivalently, by solving the first order conditions (score equations):

$$\partial\ell/\partial\beta_k = \sum_{i=1}^{N} \partial\ell_i(\boldsymbol{\beta}|\mathbf{y}_i,\mathbf{x}_i)/\partial\beta_k = 0, \qquad k=0,1,\ldots,K-1$$

If the sample size N were to be increased, the $\hat{\boldsymbol{\beta}}$ would approach, in probability, a limit vector $\boldsymbol{\beta}^*$. It is a well-known result that, assuming the model to be correct, $\boldsymbol{\beta}^*$ is the vector of true parameter values. That is, $\hat{\boldsymbol{\beta}}$ is a consistent estimator of the structural parameters. Consistency is generally regarded as a minimum requirement of an acceptable estimation procedure and is an important characteristic of maximum likelihood methods.

If the log-likelihood is based upon a misspecified model, Cox (1961) shows that the limit vector $\boldsymbol{\beta}^*$ is given by the solution of the equations

$$E[N^{-1}\partial\Omega/\partial\beta_k] = 0, \qquad k=0,1,\ldots,K-1 \tag{1}$$

where the expectation E is under the correct model. Some assumption about the correct model is required to exploit this very general result. One of the simplest and least restrictive assumptions is that

$$E(y_i) = f(\boldsymbol{\alpha},\mathbf{x}_i)$$

where $\boldsymbol{\alpha}$ is an unknown vector of structural parameters and $f(\cdot)$ is an algebraic function chosen on the basis of statistical logic and, hopefully, prior theory. It is emphasised that this very partial, first moment, specification of the correct model requires no assumptions about error distributions and focuses solely upon the relationship between the explanatory variables and the expected (mean) response. It is clearly insufficient for conventional inference but with judicious selection of the misspecified model it is generally possible to ensure that the solution of Equations (1) is given by

$$E(y_i) = f(\boldsymbol{\beta}^*,\mathbf{x}_i) \tag{2}$$

In this case, the structural parameter estimates from the misspecified model would have a limit vector given by $\boldsymbol{\beta}^* = \boldsymbol{\alpha}$. It directly follows that $\hat{\boldsymbol{\beta}}$ are consistent estimators of the 'true' structural parameters and that the misspecified model may be used to make inferences about the effect of explanatory variables. We will refer to such a model as a pseudo-likelihood model, noting that this is a more restrictive definition than sometimes adopted.

Inference is rarely based just upon point estimates of parameters; we are usually interested in whether each estimate suggests a systematic relationship between the appropriate explanatory and response variables or whether it could reasonably be attributed to sampling variation. Significance testing and confidence interval estimation for structural parameters are often based upon estimated standard errors. For a correctly specified model, it is a standard result that $(\hat{\boldsymbol{\beta}}-\boldsymbol{\beta}^*)$ are asymptotically multivariate Normal with covariance matrix $\mathbf{V} = \mathrm{H}^{-1}$ where $\mathbf{H}$ is the Fisher information matrix whose elements are given by

$$h_{uv} = E[-\partial^2\Omega/\partial\beta_u\partial\beta_v]$$

and evaluated at the point $\boldsymbol{\beta} = \boldsymbol{\beta}^*$. For estimating standard errors, $\mathbf{H}$ is estimated consistently by some matrix $\hat{\mathrm{H}}$, usually the working Fisher information matrix or the observed information matrix given by

$$\hat{h}_{uv} = E[-\partial^2\Omega/\partial\beta_u\partial\beta_v] \tag{3}$$

and

$$\hat{h}_{uv} = [-\partial^2 \ell / \partial\beta_u \partial\beta_v] \tag{4}$$

respectively, both evaluated at the point $\boldsymbol{\beta} = \hat{\boldsymbol{\beta}}$. The square roots of the trace of $\hat{\mathbf{V}}$ (= $\hat{\mathbf{H}}^{-1}$) then provide estimates of the standard errors of the structural parameter estimates.

For the misspecified model satisfying Equation (2), it appears that $(\hat{\boldsymbol{\beta}} - \boldsymbol{\beta}^*)$ is still asymptotically multivariate Normal but the covariance matrix is given by

$$\mathbf{V} = E(\mathbf{H}^{-1})E(\mathbf{H}^*)E(\mathbf{H}^{-1}) \tag{5}$$

where expectation is under the correct model, $\mathbf{H}$ is again the Fisher information matrix, and $\mathbf{H}^*$ is the inner cross-product matrix with

$$h^*_{uv} = [(\partial\ell/\partial\beta_u)\cdot(\partial\ell/\partial\beta_v)]$$

and evaluated at $\boldsymbol{\beta} = \boldsymbol{\beta}^*$. See Cox (1961) and Huber (1967). Estimation of standard errors is more complicated but quite feasible. The $E(\mathbf{H}^{-1})$ in Equation (5) may be consistently estimated by the observed information matrix (4). Following Zeger *et al* (1985), the $E(\mathbf{H}^*)$ may be consistently estimated by $\hat{\mathbf{H}}^*$ where

$$\hat{h}^*_{uv} = [\sum_{i=1}^{N} (\partial\ell_i/\partial\beta_u)\cdot(\partial\ell_i/\partial\beta_v)]$$

evaluated at $\boldsymbol{\beta} = \hat{\boldsymbol{\beta}}$. Thus the covariance matrix $\mathbf{V}$ and hence standard errors may be estimated without any further assumptions about the correct model.

Quasi-likelihood is relevant to studies of robustness because it has properties similar to log-likelihood and may, in specific circumstances, be used to estimate structural parameters without specifying the probability model in full (Wedderburn, 1974; McCullagh, 1983). It is, however, less general than the above analysis because it requires a parametric specification, not only of the expected value of the response, but also of its variance. Nevertheless, this is still a modest requirement in comparison to the specification of a parametric form for the error distribution in a conventional statistical modelling exercise. The quasi-likelihood function $Q = \sum_{i=1}^{N} Q_i$ is defined by

$$\partial Q_i/\partial E(y_i) = [y_i - E(y_i)]/\mathrm{Var}(y_i)$$

where $E(y_i)$ and $\mathrm{Var}(y_i)$ are the mean and variance, respectively, of the response variable. Score-type functions are readily obtained from this definition,

$$\partial Q_i/\partial\beta_k = [\partial Q_i/\partial E(y_i)]\cdot[\partial E(y_i)/\partial\beta_k]$$

$$= \frac{[y_i - E(y_i)]}{\mathrm{Var}(y_i)} \cdot \frac{\partial E(y_i)}{\partial\beta_k}, \qquad k=0,1,\ldots K-1 \tag{6}$$

and may be used to obtain maximum quasi-likelihood estimates of the structural parameters. Second derivatives may be used to obtain estimated standard errors, as in Equations (3) and (4).

3. THE POISSON MODEL

3.1. *Theoretical Development*

The Poisson model is one of the most basic statistical models and is readily calibrated using GLIM and other software packages. The response variable is a count of outcomes assumed to occur independently and the probability of case i having n_i outcomes is given by

$$P(n_i|\lambda_i) = \lambda_i^{n_i} \exp(-\lambda_i)/(n_i)!$$

where λ_i is the rate of outcome for i. Explanatory variables are usually incorporated into the model as a log-linear function of λ_i; this ensures that the outcome rate is always positive:

$$\lambda_i = \exp(\mathbf{x}_i'\boldsymbol{\beta})$$

The log-likelihood and score functions are readily derived and are, respectively,

$$\mathfrak{L} = \sum_{i=1}^{N} \mathfrak{L}_i = \sum_{i=1}^{N} [n_i\mathbf{x}_i'\boldsymbol{\beta} - \exp(\mathbf{x}_i'\boldsymbol{\beta})]$$

and

$$\partial\mathfrak{L}/\partial\beta_k = \sum_{i=1}^{N} x_{ik}\,[n_i - \exp(\mathbf{x}_i'\boldsymbol{\beta})] \tag{7}$$

Consider a counting process which is not Poisson but for which $E(n_i) = \exp(\mathbf{x}_i'\boldsymbol{\alpha})$. To emphasise the generality of this specification we list some of the models it could represent in Appendix 1. If a Poisson model is fitted to data from any such process, then Equations (1) and (7) indicate that the maximum likelihood parameter estimates would have a limiting vector $\boldsymbol{\beta}^*$ given by the solution to the simultaneous equations

$$E\{N^{-1} \sum_{i=1}^{N} x_{ik}[n_i - \exp(\mathbf{x}_i'\boldsymbol{\beta}^*)]\} = 0, \qquad k=0,1,\ldots m$$

Substituting $E(n_i) = \exp(\mathbf{x}_i'\boldsymbol{\alpha})$ in these equations gives the solution $\boldsymbol{\beta}^* = \boldsymbol{\alpha}$. That is, the $\hat{\boldsymbol{\beta}}$ consistently estimate $\boldsymbol{\alpha}$ and the conventional Poisson model gives pseudo-maximum likelihood parameter estimates for any process with $E(n_i) = \exp(\mathbf{x}_i'\boldsymbol{\alpha})$. Standard errors for the parameter estimates may be calculated from Equation (5). For a Poisson process, the observed information matrix (which, incidentally, is identical to the working Fisher information matrix) is given by $\hat{\mathbf{H}}$ with elements

$$\hat{h}_{uv} = [-\partial^2 \ell/\partial\beta_u \partial\beta_v]_{\boldsymbol{\beta}=\hat{\boldsymbol{\beta}}} = \sum_{i=1}^{N} x_{iu}x_{iv}\exp(\mathbf{x}_i'\hat{\boldsymbol{\beta}})$$

and the expected inner cross product matrix may be estimated by $\hat{\mathbf{H}}^*$ with elements

$$\hat{h}^*_{uv} = [\sum_{i=1}^{N} (\partial\ell_i/\partial\beta_u)\ (\partial\ell_i/\partial\beta_v)]_{\boldsymbol{\beta}=\hat{\boldsymbol{\beta}}}$$

$$= \sum_{i=1}^{N} x_{iu}x_{iv}[n_i - \exp(\mathbf{x}_i'\hat{\boldsymbol{\beta}})]^2$$

If the Poisson model were correct, the standard errors would be estimated by the square root of the diagonal elements of $\hat{\mathbf{V}} = \hat{\mathbf{H}}^{-1}$. In effect, the leading diagonal of $\hat{\mathbf{V}}^* = \hat{\mathbf{H}}^{-1}\ \hat{\mathbf{H}}^*\ \hat{\mathbf{H}}^{-1}$ gives corrected estimates of standard errors, allowing for the misspecification. It is emphasised that the precise nature of the misspecification need not be known.

Quasi-likelihood methods give an alternative and, for those who rely upon software packages for statistical modelling, a more useful insight into the Poisson model. For the Poisson model, the variance is equal to the mean. That is, $\mathrm{var}(n_i) = E(n_i)$. From Equation (6), the score function for the quasi-likelihood of the Poisson model is thus given by

$$\partial Q/\partial\beta_k = \sum_{i=1}^{N} x_{ik}[n_i - \exp(\mathbf{x}_i'\boldsymbol{\beta})] \qquad (8)$$

which is identical to the score function (Equation 7) of the conventional Poisson likelihood. Consequently, likelihood and quasi-likelihood provide the same parameter estimates. Consider now the computational implications of relaxing the stringent variance/mean relationship of the Poisson model by putting $\mathrm{var}(n_i) = \eta^2 E(n_i)$ where η^2 is an unknown scaling factor. The quasi-likelihood score function is now given by

$$\partial Q/\partial \beta_k = \sum_{i=1}^{N} x_{ik}[n_i - \exp(\mathbf{x}_i'\boldsymbol{\beta})]/\eta^2 \tag{9}$$

The only difference between Equations (8) and (9) is the divisor η^2. As parameter estimates are obtained by equating the score function to zero, this divisor makes no computational difference and the parameter estimates from the relaxed model must be identical to those that would be obtained by assuming a Poisson process. It follows that, although misspecified, the Poisson model provides consistent parameter estimates if the data are generated by a process with mean given by $E(n_i) = \exp(\mathbf{x}_i'\boldsymbol{\beta})$ and variance given by $\mathrm{var}(n_i) = \eta^2\exp(\mathbf{x}_i'\boldsymbol{\beta})$. This result is not surprising as we have already seen that estimation is consistent with only the former assumption. However, quasi-likelihood provides new results on standard errors. From Equation (9), the second derivatives of the quasi-likelihood function are identical to those of a Poisson model apart from the divisor η^2. Each element in the inverse of the working information matrix will therefore differ from that in a conventional Poisson analysis by a multiplying factor of η^2 and the standard errors will differ from the Poisson estimates by a multiplying factor of η. To correct for misspecification, it is therefore only necessary to multiply the conventionally obtained standard errors for a Poisson model by an estimate of η. Different methods have been proposed for estimating η (McCullagh, 1983). The usual estimate is given by

$$\hat{\eta}^2 = \frac{1}{N-m} \sum_{i=1}^{N} \frac{[n_i - \text{Estimate of } E(n_i)]^2}{\text{Estimate of } \mathrm{var}(n_i)} \tag{10}$$

where m is the number of parameters estimated. It effectively assumes that $E(n_i)$ is approximately linear in $\boldsymbol{\beta}$ and that the estimate of $\mathrm{var}(n_i)$ differs negligibly from its true value (Wedderburn, 1974). Equation (10) is the generalised Pearson statistic calculated by GLIM divided by N-m. However, for the Poisson model it simplifies to $\hat{\eta}^2 = \chi^2/(N-m)$ where χ^2 is the Pearson chi-square statistic calculated in most statistical packages.

In effect, the quasi-likelihood approach leads to a scaling factor method for correcting for misspecification. It is interesting to note that the use of scaling factors to correct Poisson standard errors for the effects of over-dispersion (and under-dispersion) has a long history in statistics although, in the past, the justification was largely pragmatic.

3.2. *Empirical Example: Store Visit Frequency*

Most of the modelling methods used by market researchers to analyse purchase incidence are based upon the Poisson model. The Negative Binomial Distribution, derived as a Gamma mixture of Poisson processes (see Appendix 1), has proved to be particularly useful (Ehrenberg, 1971) although the assumption that individual purchase incidence timings are Poisson is known to be strictly untenable and has encouraged more complex formulations (e.g. Chatfield and Goodhardt, 1973; Zufryden, 1977). Indeed, market research provides a good illustration of the statistical modelling problems discussed in the Introduction. Sample sizes tend to be large and significance testing criteria do not provide an effective limit to the model complexity which can find empirical support. These problems are exacerbated by the paucity of substantive theory; model development must rely heavily upon empirical analysis. In this section we illustrate the alternative to full statistical modelling. Specifically, we apply a Poisson regression model to the geographical analogue of purchase incidence, namely store visits, and examine the effect of correcting for the misspecification inevitable in such a simple model.

The data cover four weeks of shopping behaviour for 275 households in the Guy/Wrigley Cardiff Consumer Panel (Guy *et al*, 1983). The response variable is the number of visits to small stores for grocery purchases and the explanatory variables are a dummy variable for freezer ownership, a dummy variable for the panellist in full-time employment, the logarithm of household size, the logarithm of household income, and the logarithm of distance to the nearest supermarket (or larger food store). More details of these data are given in Davies and Pickles (1987). The first two columns of Table I show the maximum likelihood parameter estimates and the conventionally obtained standard errors, respectively, for a Poisson model. The results are as would be obtained through using GLIM or other appropriate software package and indicate that all but the income variable are significant at the 5% level. However, the pseudo-likelihood estimates of standard errors incorporating, as we have seen, a general correction for misspecification, provide a very different picture; only the supermarket accessibility variable is significant at the 5% level. The conventional Poisson model is clearly misleading.

The quasi-likelihood estimated standard errors are listed in the fourth column of Table I. The calculated Pearson chi-square was 1276.3 with 269 degrees of freedom for the Poisson model, giving a scaling factor of 2.18. Although based upon rather more restrictive assumptions, the estimates are encouragingly similar to the computationally complex pseudo-likelihood estimates. The final two columns in Table I list the results obtained in calibrating a Negative Binomial regression model on the same data. This may be seen as a direct statistical modelling development of the Negative Binomial distribution applications in marketing research. It was suggested for

TABLE I
Analysis of local store visit frequency

	POISSON				NEGATIVE BINOMIAL	
	Parameter	Estimated Standard Errors			Parameter	Estimated Standard
	Estimate	Poisson	Pseudo-likelihood	Quasi-likelihood	Estimate	Error
Freezer dummy	-0.126	0.051	0.102	0.112	-0.125	0.121
Panellist in Full-time Work Dummy	-0.126	0.062	0.130	0.136	-0.135	0.137
Logarithm of Household Size	0.140	0.047	0.103	0.103	0.131	0.104
Logarithm of Income	-0.040	0.044	0.096	0.097	-0.027	0.096
Logarithm of Distance to Supermarket	0.278	0.043	0.087	0.093	0.273	0.096
Constant Term	1.269	0.342	0.712	0.745	0.356	0.759
Negative Binomial Parameter					2.289	0.255
Log-Likelihood	-1139.223				-862.055	

store visits by Broom and Wrigley (1983) and subsequently applied to data from the Cardiff Consumer Panel by Wrigley and Dunn (1985). There is a substantial improvement in the log-likelihood over the Poisson model (corresponding to a likelihood ratio chi-square of 554.32 with one degree of freedom). However, the parameter estimates and their standard errors are very similar to those obtained for the pseudo-likelihood and quasi-likelihood approaches. Some caution must be exercised in drawing conclusions from these comparisons as the Negative Binomial model may itself be misspecified. But the standard error similarities do suggest that there are no significant efficiency gains from calibrating the much better-fitting but more complex model. This has been formally proved for modest over-dispersion by Cox (1983).

4. THE LOGISTIC MODEL

4.1. *Theoretical Development*

There are two simple statistical models in widespread use for modelling binary response variables: the logistic and probit models. The logistic and probit distributions are very similar in shape and, in practice, it makes little difference which is chosen. In this paper we concentrate on the former because it is analytically more tractable. Let the response vector $\mathbf{y}_i$ represent the sequence of T_i binary outcomes for case i. It is convenient and involves no loss of generality to code the two possible outcomes as zero and one. Thus, assuming that the same vector of explanatory variables is appropriate throughout the sequence, a logistic specification would give

$$P(y_{it} = 1|\mathbf{x}_i) = p_i = \exp(\mathbf{x}_i'\boldsymbol{\beta})/[1 + \exp(\mathbf{x}_i'\boldsymbol{\beta})] \tag{11}$$

If the outcomes are independent over time (that is, the process is Binomial) and, of course, independent between cases, the log-likelihood is easily shown to be

$$\ell = \sum_{i=1}^{N} \sum_{t=1}^{T_i} \{y_{it}\mathbf{x}_i'\boldsymbol{\beta} - \log[1 + \exp(\mathbf{x}_i'\boldsymbol{\beta})]\} \tag{12}$$

This is identical to the log-likelihood that would be obtained if there were N groups with T_i cases in the ith group. Consequently, as pointed out by Allison (1982), the longitudinal model may be calibrated by treating the data as if they were obtained from a large cross-section of $\sum_{i=1}^{N} T_i$ cases and using standard software packages such as GLIM and SPSSX (SPSSX, 1983).

However, this Binomial model is unrealistic on two general counts. First, it does not allow for any variation between individuals other than that due to included explanatory variables; second, it does not allow for any temporal dependence in outcomes arising, for example, from learning. Consider a longitudinal binary process more complex than the Binomial model but for which

$$E(y_{it}) = \exp(\mathbf{x}_i'\boldsymbol{\alpha})/[1 + \exp(\mathbf{x}_i'\boldsymbol{\alpha})] \quad (13)$$

Some examples of the processes that conform to this relatively unrestrictive first moment specification are given in Appendix 2. Following the logic explained in Section 2, this specification has been chosen so that the maximum likelihood parameter estimates obtained by calibrating the Binomial model are consistent estimators of the parameters $\boldsymbol{\alpha}$; the Binomial model provides pseudo-likelihood estimation for the model (13). This is readily demonstrated. The score function for the Binomial model is obtained by differentiating the log-likelihood function (12), giving

$$\partial\ell/\partial\beta_k = \sum_{i=1}^{N} \sum_{t=1}^{T_i} x_{ik}(y_{it} - p_i)$$

Substituting in Equation (1) gives

$$E\left[N^{-1} \sum_{i=1}^{N} \sum_{t=1}^{T_i} x_{ik}(y_{it} - p_i^*)\right] = 0, \qquad k=0,1,\ldots m$$

where $p_i^* = \exp(\mathbf{x}_i'\boldsymbol{\beta}^*)/[1 + \exp(\mathbf{x}_i'\boldsymbol{\beta}^*)]$ and $\boldsymbol{\beta}^*$ is the limiting vector of the maximum likelihood parameter estimates $\hat{\boldsymbol{\beta}}$. From Equation (13), it is clear that these equations have the solution $\boldsymbol{\beta}^* = \boldsymbol{\alpha}$; $\hat{\boldsymbol{\beta}}$ are consistent estimators of $\boldsymbol{\alpha}$.

Standard error calculations for this pseudo-likelihood approach require estimates of both the observed information matrix (again, incidentally, identical to the working Fisher information matrix) and the inner cross-product matrix. The former is given by $\hat{\mathbf{H}}$ with elements

$$\hat{h}_{uv} = [-\partial^2\ell/\partial\beta_u\partial\beta_v]_{\boldsymbol{\beta}=\hat{\boldsymbol{\beta}}}$$

$$= \sum_{i=1}^{N} T_i x_{iu} x_{iv} \exp(\mathbf{x}_i'\hat{\boldsymbol{\beta}})/[1 + \exp(\mathbf{x}_i'\hat{\boldsymbol{\beta}})]^2$$

and the latter is given by $\hat{\mathbf{H}}^*$ with elements

$$\hat{h}_{uv}^* = \left[\sum_{i=1}^{N} (\partial\ell_i/\partial\beta_u)(\partial\ell_i/\partial\beta_v)\right]_{\boldsymbol{\beta}=\hat{\boldsymbol{\beta}}}$$

$$= \sum_{i=1}^{N} x_{iu} x_{iv} (n_i - T_i\hat{p}_i)^2$$

where $n_i = \sum_{t=1}^{T_i} y_{it}$ and $\hat{p}_i = \exp(\mathbf{x}_i'\hat{\boldsymbol{\beta}})/[1 + \exp(\mathbf{x}_i'\hat{\boldsymbol{\beta}})]$. From Equation (5), standard errors correcting for misspecification are given by the square roots of the diagonal elements of $\hat{\mathbf{V}}^* = \hat{\mathbf{H}}^{-1}\, \hat{\mathbf{H}}^*\, \hat{\mathbf{H}}^{-1}$.

As for the Poisson model, the estimation of standard errors for the pseudo-likelihood approach is not possible using standard package software. However, a quasi-likelihood analysis again provides a simpler, although less general, method for correcting conventionally obtained standard errors for the misspecified model. The results are particularly straightforward, closely paralleling the Poisson results, if the sequence of outcomes is the same length for each case. This is often the situation in social science research and we will proceed with the assumption that $T_1 = T_2 = \ldots = T_N = T$ but note that even if this assumption is relaxed, a quasi-likelihood calibration may be effected by published GLIM Macros (Williams, 1982). It is convenient to reformulate the misspecified model as Binomial with T trials and outcome probability p_i given by Equation (11). In this context, the response variable is $n_i = \sum_{t=1}^{T} y_{it}$, the number of 'successes'. The log-likelihood and score function are given, respectively, by

$$\ell = \sum_{i=1}^{N} [n_i \log(p_i) + (T-n_i)\log(1-p_i)]$$

and
$$\partial\ell/\partial\beta_k = \sum_{i=1}^{N} x_{ik}\,(n_i - Tp_i) \tag{14}$$

Simple algebra will confirm that these are identical to the log-likelihood and score functions obtained by taking the vector $\mathbf{y}_i$ as the response variable.

Applying standard results (see, for example, Johnson and Kotz, 1969, p.51), the mean and variance of the response n_i for the Binomial model are given by $E(n_i) = Tp_i$ and $\mathrm{var}(n_i) = Tp_i(1-p_i)$. From Equation (6), it follows that the quasi-likelihood score function is given by

$$\partial Q/\partial\beta_k = \sum_{i=1}^{N} \left\{ \frac{[n_i - E(n_i)]}{\mathrm{var}(n_i)} \frac{\partial E(n_i)}{\partial\beta_k} \right\}$$

$$= \sum_{i=1}^{N} \left\{ \frac{[n_i - Tp_i)}{Tp_i(1-p_i)} \cdot \frac{T\partial p_i}{\partial\beta_k} \right\}$$

$$= \sum_{i=1}^{N} x_{ik}(n_i - Tp_i) \tag{15}$$

Comparison of Equations (14) and (15) shows that this quasi-likelihood formulation is computationally indistinguishable from the conventional likelihood approach. As for the Poisson model, the variance/mean relationship of the Binomial model may be relaxed by introducing an unknown scaling factor η^2 and writing

$$E(n_i) = Tp_i \text{ and } \operatorname{var}(n_i) = \eta^2 Tp_i(1-p_i) \tag{16}$$

The quasi-likelihood score function for this more general formulation is given by

$$\partial Q/\partial \beta_k = \sum_{i=1}^{N} x_{ik}(n_i - Tp_i)/\eta^2$$

and, again, this differs from the conventional likelihood score function by just the computationally irrelevant divisor η^2; maximum likelihood parameter estimates for the Binomial model will be identical to, and may therefore be interpreted as, quasi-maximum likelihood estimates of the more general model defined by Equations (16). However, exactly as for the Poisson model, the η would have to be estimated from the data and used to correct the conventionally obtained standard errors for misspecification. From Equation (10),

$$\hat{\eta}^2 = \frac{1}{N-m} \sum_{i=1}^{N} \frac{(n_i - T\hat{p}_i)^2}{T\hat{p}_i(1 - \hat{p}_i)}$$

$$= \chi^2/(N-m)$$

where χ^2 is the generalised Pearson statistic calculated by GLIM.

4.2. *Empirical Example: Voting Behaviour*

Conventional statistical modelling of longitudinal data on voting behaviour provides a vivid example of the complexities of social processes and the difficulties that arise in attempting to obtain a parsimonious representation when the sample size is large. In particular, there is clear evidence of serial autocorrelation, unexplained heterogeneity even when a considerable number of explanatory variables were included in the model, and nonstationarity arising from swings of popular support. A realistic model would therefore be a composite of several of the models listed in Appendix 2. This leads to major problems over analytical tractability and practical applications have been severely constrained by computational considerations (Davies and Crouchley, 1985). However, the robust modelling of voting propensity using the inferential procedures of the previous section is relatively straightforward.

The Binomial analysis reported in Table II is based upon data from the British Election Study (ESRC Data Archive No. 1533) for the General Elections of February and October 1974. The data utilised in the analysis are for English constituencies in which a Liberal candidate stood in both elections and the sample size is 833 individuals. The response variable is the binary Liberal/other vote choice for the two elections and the explanatory variables are age, two social grade dummy variables ('Managerial' consisting of social grades A and B; 'manual' consisting of social grades C2 and D), a sex dummy variable, and a dummy variable for trade union membership. This subset of the British Election Study data is discussed in more detail in Crouchley *et al* (1984). The first two columns of Table II show the maximum likelihood parameter estimates and their standard errors for a Binomial model, as would be routinely obtained using GLIM or other appropriate software package. Age, sex, and trade union membership are all clearly significant at the 5% level and the managerial dummy variable is marginally significant at this level. The pseudo-likelihood standard error estimates, listed in the third column, are not as dramatically different as in the Poisson example of Section 3. Nevertheless, this general correction for the consequences of misspecification is sufficient to challenge some of the inferences from the simple Binomial model; only age remains unequivocally significant at the 5% level.

The quasi-likelihood parameter estimates are listed in the final column of Table II. The generalised Pearson chi-square was 1325.8 with 827 degrees of freedom. This gave a scaling factor of 1.27. The estimated standard errors based upon this scaling factor are again markedly similar to the pseudo-likelihood estimates.

TABLE II
Binomial analysis of liberal voting in the general elections of 1974

VARIABLE	PARAMETER ESTIMATE	ESTIMATED STANDARD ERRORS		
		Binomial	Pseudo-Likelihood	Quasi-Likelihood
Age × 0.1	-0.145	0.038	0.047	0.048
Social Grade Dummies:				
Managerial	0.335	0.167	0.214	0.211
Manual	-0.156	0.141	0.176	0.178
Sex:				
Female Dummy	0.328	0.127	0.165	0.160
Trade Union Membership Dummy	-0.374	0.168	0.216	0.213
Constant Term	-0.704	0.219	0.285	0.277
Log Likelihood	-847.076			

5. CONCLUDING COMMENTS

In trying to develop realistic models of the dynamics of social processes, the analyst is often constrained by both the practical computational problems and the difficult statistical theory issues raised by complex models. This paper has demonstrated that, in defined circumstances, valid inference about the propensity of a choice or outcome is possible using the simplest statistical models. Such models will inevitably be misspecified but it is possible to correct for the effects of the misspecification. The general, pseudo-likelihood, method of correction is not currently possible using available computer software packages but a theoretically more restrictive approach based upon quasi-likelihood appears to provide very similar results and may be routinely operationalised using GLIM. It is emphasised that these methods do not obviate the need for research into more comprehensive but operational models of recurrent behaviour. They are relevant when interest focuses upon propensity; inference on other characteristics such as loyalty, learning, and heterogeneity, require more complex approaches. Finally, it is noted that analysis in this paper has been confined to data with time-constant explanatory variables. The methods can be extended for application to data with time-varying explanatory variables but this raises several more difficult problems which are currently being addressed.

ACKNOWLEDGEMENTS

The author is grateful to the ESRC Survey Archive for supplying the voting behaviour data. Neither the Survey Archive nor the original depositors bear any responsibility for the analysis and the interpretation of the results presented. The author is also grateful to his colleague, Dr. C.M. Guy, for permitting access to the Cardiff Consumer Panel data.

Centre of Applied Statistics
University of Lancaster
Lancaster
England

and

Department of Town Planning
University of Wales Institute of Science and Technology
Cardiff
Wales

BIBLIOGRAPHY

Allison, P.D.: 1982, 'Discrete-time methods for the analysis of event histories', in S. Leinhardt (ed.), *Sociological Methodology*, Jossey-Bass, San Francisco.

Baker, R.J. and J.A. Nelder: 1979, *The GLIM System: Release 3*, Numerical Algorithms Group, 256 Banbury Road, Oxford, England.

Baxter, M.J.: 1985, 'Quasi-likelihood estimation and diagnostic statitics for spatial interaction models', *Environment and Planning A* **17**, 1627-1635.

Blumen, I., M. Kogan, and P.J. McCarthy: 1955, 'The industrial mobility of labor as a probability process', *Cornell Studies of Industrial and Labour Relations* **6**, Cornell University.

Broom, D. and N. Wrigley: 1983, 'Incorporating explanatory variables into stochastic panel-data models of urban shopping behaviour', *Urban Geography* **4**, 244-257.

Chatfield, C. and G.J. Goodhardt: 1973, 'A consumer purchasing model with Erlang inter-purchase times', *Journal of the American Statistical Association* **68**, 828-835.

Cliff, A.D. and J.K. Ord: 1981, *Spatial Processes: Models and Applications*, Pion, London.

Cox, D.R.: 1961, 'Tests of separate families of hypotheses', in the *Proceedings of the Fourth Berkeley Symposium on Mathematical Statistics and Probability*, University of California Press, Berkeley CA, 105-123.

Crouchley, R., R.B. Davies, and A.R. Pickles: 1984, 'Specification errors in the analysis of voting', mimeo; presented to a meeting of the Social Research Section of the Royal Statistical Society, London, October 1984.

Davies, R.B.: 1984, 'A generalised Beta-Logistic model for longitudinal data', *Environment and Planning* **A 16**, 1375-1386.

Davies, R.B.: 1986a, 'Mass point methods for dealing with nuisance parameters in longitudinal studies', in R. Crouchley (ed.), *Longitudinal Data Analysis*, Sage, Newbury Park.

Davies, R.B.: 1986b, 'Poisson and Binomial models of behaviour; the practical consequences of extra-variation', mimeo.

Davies, R.B. and R.Crouchley: 1985, 'The determinants of party loyalty: a disaggregate analysis of panel data from the 1974 and 1979 general elections in England', *Political Geography Quarterly* **4**, 307-320.

Davies, R.B. and A.R. Pickles: 1987, 'A joint trip-timing, store choice model including feedback and inventory effects and nonparametric control for omitted variables', *Transportation Research A* **21**, 345-362.

Ehrenberg, A.S.C.: 1972, *Repeat Buying: Theory and Applications*, North Holland Publishing Co., Amsterdam.

GENSTAT: 1986, *Genstat, version 5*, Numerical Algorithms Group, 256 Banbury Road, Oxford, England.

Gourieroux, C., A. Monfort, and A. Trognon: 1984, 'Pseudo maximum likelihood methods: theory', *Econometrica* **52**, 681-700.

Guy, C.M., N. Wrigley, L.G. O'Brien, and G. Hiscocks: 1983, *The Cardiff Consumer Panel: A Report on the Methodology*, Papers in Planning Research No. 68, Department of Town Planning, UWIST.

Heckman, J.J. and R.J. Willis: 1977, 'A Beta-Logistic model for the analysis of sequential labor force participation by married women', *Journal of Political Economy* **85**, 27-58.

Huber, P.J.: 1967, 'The behaviour of maximum likelihood estimates under nonstandard conditions', in the *Proceedings of the Fifth Berkeley Symposium on Mathematical Statistics and Probability*, University of California Press, Berkeley CA, 221-233.

Johnson, N.L. and S. Kotz: 1969, *Distributions in Statistics: Discrete Distributions*, Wiley, New York.
McCullagh, P.: 1983, 'Quasi-likelihood functions', *The Annals of Statistics* **11**, 59-67.
Pickles, A.R.: 1985, *An Introduction to Likelihood Analysis*, Geo Books, Norwich.
Pickles, A.R. and R.B. Davies: 1989, 'Inference from cross-sectional and longitudinal data for dynamic behavioural processes', in J. Hauer, H. Timmermans and N. Wrigley (eds.), *Urban Dynamics and Spatial Choice Behaviour*, D. Reidel, Dordrecht.
Selvin, H.C. and A. Stuart: 1966, 'Data dredging procedures in survey analysis', *American Statistician* **20**, 20-23.
SPSSX: 1983, *User's Guide*, McGraw Hill, New York.
Vincent, P. and J. Haworth: 1984, 'Statistical inference: the use of the likelihood function', *Area* **16**, 131-146.
Wedderburn, R.W.M.: 1974, 'Quasi-likelihood functions, generalised linear models, and the Gauss-Newton method', *Biometrika* **61**, 439-447.
White, H.: 1982, 'Maximum likelihood estimation of misspecified models', *Econometrica* **50**, 1-25.
Williams, D.A.: 1975, 'The analysis of binary responses from toxicological experiments involving reproduction and teratogenicity', *Biometrics* **31**, 949-952.
Williams, D.A.: 1982, 'Extra-binomial variation in logistic linear models', *Applied Statistics* **31**, 144-148.
Wrigley, N. and R. Dunn: 1984, 'Stochastic panel-data models of urban shopping behaviour: 1. Purchasing at individual stores in a single city', *Environment and Planning* **A 16**, 629-650.
Wrigley, N. and R. Dunn: 1985, 'Stochastic panel-data models of urban shopping behaviour: 4. Incorporating independent variables into the NBD and Dirichlet models', *Environment and Planning* **A 17**, 319-331.
Zeger, S.L., K-Y Liang, and S.G. Self: 1985, 'The analysis of binary longitudinal data with time-independent covariates', *Biometrika* **72**, 31-38.
Zufryden, F.S.: 1977, 'A composite heterogeneous model of brand choice and purchase timing behaviour', *Management Science* **24**, 121-136.

APPENDIX 1. SOME MODELS WITH POISSON PSEUDO-LIKELIHOODS

In this Appendix we list some counting models for which $E(n_i) = \exp(\mathbf{x}_i'\boldsymbol{\alpha})$ and for which the Poisson regression model of Section 3 therefore provides an appropriate pseudo-likelihood formulation.

i. A General Mixture Model

The general Poisson mixture model may be written as

$$P(n_i) = \int_{-\infty}^{+\infty} \lambda_i^{n_i} \exp(-\lambda_i) dF(\varepsilon_i)/(n_i)! \tag{17}$$

where $\lambda_i = \exp(\mathbf{x}_i'\alpha+\varepsilon_i)$ and $F(\varepsilon_i)$ is the distribution function of the error term ε_i through the population. This is derived from the Poisson model by including an individual specific error term in the linear predictor to represent the effects of omitted variables and then removing the error term from the model by integrating ('mixing') over its distribution function. This assumes, of course, that the error term is independent of the explanatory variables included in the model. Recent work on the nonparametric characterisation of mixing distributions permits the operationalisation of this general model without any parametric assumptions about the distribution $F(\varepsilon_i)$. See, for example, Davies (1986a). The mean of the mixture model is given by

$$E(n_i) = \int_{-\infty}^{+\infty} \exp(\mathbf{x}_i'\boldsymbol{\alpha}+\varepsilon_i) dF(\varepsilon_i)$$

$$= \int_{-\infty}^{+\infty} w_i \exp(\mathbf{x}_i'\boldsymbol{\alpha}) dG(w_i)$$

where $w_i = \exp(\varepsilon_i)$ and $G(w_i)$ is the distribution function of this transformed error term. As the linear predictor $\mathbf{x}_i'\boldsymbol{\alpha}$ includes a constant term, there is no loss of generality in putting $E(w) = 1$, giving

$$E(n_i) = \exp(\mathbf{x}_i'\boldsymbol{\alpha})$$

ii. The Negative Binomial Regression Model

The Negative Binomial regression model may be written as

$$P(n_i) = \frac{\Gamma(c+n_i)}{\Gamma(c)\ (n_i)!} \left[\frac{\exp(\mathbf{x}_i'\boldsymbol{\alpha})}{c + \exp(\mathbf{x}_i'\boldsymbol{\alpha})} \right]^{n_i} \left[\frac{c}{c + \exp(\mathbf{x}_i'\boldsymbol{\alpha})} \right]^{c} \tag{18}$$

giving, as a standard result, the required first moment $E(n_i) = \exp(\mathbf{x}_i'\boldsymbol{\alpha})$. The model may be interpreted as relaxing the assumptions of the Poisson model in two alternative and quite distinct ways. First, it may be interpreted as relaxing the assumption that variation in the outcome rate between cases is fully explained by the explanatory variables; it allows for unexplained heterogeneity. In this context, the Negative Binomial model is a special case of the general mixture model (*i*) above. Specifically, Equation (18) may be derived from Equation (17) by assuming a Gamma parametric form given by

$$dG(w_i)/dw_i = w_i^{c-1}\, c^c \exp(-cw_i)/\Gamma(c) \tag{19}$$

for the distribution of the transformed $w_i = \exp(\varepsilon_i)$ error term. Second, the Negative Binomial model may be seen as relaxing the Poisson independence assumption. One formal derivation postulates outcomes occuring in clusters with the number of clusters Poisson-distributed and the number of outcomes in each cluster conforming to the logarithmic series distribution. This 'true contagion' derivation is familiar to geographers from quadrat count methods (e.g. Cliff and Ord, 1981, Chapter 4), although without explanatory variables.

iii. Poisson with Added Zeros

This is a dichotomous finite mixture model defined by

$$P(n_i=0) = \phi + (1-\phi)\exp(-\lambda_i)$$

$$P(n_i=y) = (1-\phi)\lambda^y{}_i\exp(-\lambda_i)/(y)!, \qquad y\geq 1$$

It may be conceptualised as a 'mover-stayer' model with an unknown proportion ϕ of the population having a zero probability of recording any outcome (stayers or non-participants) and a proportion $(1-\phi)$ having Poisson outcomes at a rate $\lambda_i = \exp(\mathbf{x}_i'\boldsymbol{\alpha})$. The use of this type of model in the social sciences dates back to Blumen *et al* (1955). The mean of the distribution is given by

$$E(n_i) = (1-\phi)\exp(\mathbf{x}_i'\boldsymbol{\alpha})$$

$$= \exp[\mathbf{x}_i'\boldsymbol{\alpha} + \log(1-\phi)]$$

This is of the required form except that the constant term α_o is aliased with the term $\log(1-\phi)$. This is of no consequence if interest focuses upon the effect of the explanatory variables.

iv. Negative Binomial with Added Zeros

This is defined by

$$P(n_i=0) = \phi + (1-\phi)\psi_i(0)$$

$$P(n_i=y) = (1-\phi)\psi_i(y), \qquad y\geq 1$$

where $\psi_i(\cdot)$ is the Negative Binomial probability given in Equation (18). It may be derived from the mixture model (17) by assuming that the error term $w_i = \exp(\varepsilon_i)$ has a 'spiked Gamma' distribution. That is, a proportion ϕ of the population are stayers or non-participants with $w_i = 0$ and for the remainder the error term has the Gamma distribution given by Equation (19). The mean of the distribution is exactly as for the Poisson with added zeros.

APPENDIX 2. SOME MODELS WITH BINOMIAL LOGISTIC PSEUDO-LIKELIHOODS

For all the models listed in this Appendix the ith response is given by a sequence of T_i outcomes $y_{i1}, y_{i2}, \ldots$ with each outcome either zero or one. Moreover, the probability of any $y_{it} = 1$ is given by $E(y_{it}) = \exp(\mathbf{x}_i'\boldsymbol{\alpha})/[1 + \exp(\mathbf{x}_i'\boldsymbol{\alpha})]$, a sufficient condition for the pseudo-likelihood formulation of Section 4 to provide an appropriate method for estimating the structural parameters. We will write $n_i = \sum_{t=1}^{T_i} y_{it}$, as in Section 4.

i. Type I Mixture Model: the Beta-Logistic

As proposed by Williams (1975) and Heckman and Willis (1977), this model is given by

$$P(n_i) = \frac{\Gamma(a+b)\ \Gamma(a+n_i)\ \Gamma(b+T_i-n_i)}{\Gamma(a)\ \Gamma(b)\ \Gamma(a+b+T_i)}$$

where $a = \exp(\mathbf{x}_i'\boldsymbol{\gamma}_1)$ and $b = \exp(\mathbf{x}_i'\boldsymbol{\gamma}_2)$. The model may be derived from the Binomial model by assuming that $P(y_{it}=1)$ has the Beta distribution

$$f(p_i) = \Gamma(a+b) p_i^{\,a-1} (1-p_i)^{b-1} / \Gamma(a)\Gamma(b)$$

The first moment of interest is given by

$$\begin{aligned} E(y_{it}) &= a/(a+b) \\ &= \exp[\mathbf{x}_i'(\boldsymbol{\gamma}_1-\boldsymbol{\gamma}_2)]/\{1 + \exp[\mathbf{x}_i'(\boldsymbol{\gamma}_1-\boldsymbol{\gamma}_2)]\} \end{aligned}$$

which is of the required form. Clearly, a pseudo-likelihood approach could only estimate the differences between the appropriate γ-parameters. However, this is sufficient to make inference about the significance of explanatory variables and, in any case, the parameter extravagance of the Beta Logistic model is unnecessary (Davies, 1984) and difficult to justify when there is no specific interest in the higher moments of the distribution.

ii. Type II Mixture Model: the Logistic Normal

This is given by

$$p(n_i) = \int_{-\infty}^{+\infty} p_i^{\,n_i} \, (1-p_i)^{T_i-n_i} \, d\Phi(\varepsilon_i)$$

where $p_i = \exp(\mathbf{x}_i'\boldsymbol{\alpha}+\varepsilon_i)/[1 + \exp(\mathbf{x}_i'\boldsymbol{\alpha}+\varepsilon_i)]$ and $\Phi(\cdot)$ is the distribution function of a Normal distribution with mean zero and variance σ^2. The first moment of interest for this mixture model is given approximately by

$$E(y_{it}) = \exp(r\mathbf{x}_i'\boldsymbol{\alpha})/[1 + \exp(r\mathbf{x}_i'\boldsymbol{\alpha})]$$

where r is a scaling factor depending upon the unknown error variance σ^2. This result depends upon the similarity between the logistic and probit distributions. See Davies (1986b). The scaling factor is of no practical consequence because the structural parameters of any logistic model can only be estimated 'up to a scalar constant'. It does not, for example, affect significance testing as the standard errors are subject to the same scaling.

iii. First-Order Markov Chain

One formulation of a first-order Markov chain, namely

$$E(y_{it}) = \exp(\mathbf{x}_i'\boldsymbol{\alpha})/[1 + \exp(\mathbf{x}_i'\boldsymbol{\alpha})], \quad \text{corr}\,(y_{it}, y_{i,t-1}) = \rho$$

gives the required first moment immediately. See, for example, Zeger *et al* (1985).

iv. Nonstationary Binomial

A simple nonstationary formulation is given by

$$E(y_{it}) = \exp(\mathbf{x}_i'\boldsymbol{\alpha}+\tau_t)/[1 + \exp(\mathbf{x}_i'\boldsymbol{\alpha}+\tau_t)]$$

where the nonstationary scaling factor τ_t is Normally distributed with an expected value of zero and is independent between time periods. This is similar to the Type II Mixture model above except that the 'nuisance' parameter, τ_t, varies between time periods rather than individuals. The same computational logic gives

$$E(y_{it}) = \exp(r\mathbf{x}_i'\boldsymbol{\alpha})/[1 + \exp(r\mathbf{x}_i'\boldsymbol{\alpha})]$$

as an approximation to the first moment. Again, the multiplicative factor r is of no practical consequence.

ANDREW R. PICKLES AND RICHARD B. DAVIES

INFERENCE FROM CROSS-SECTIONAL AND LONGITUDINAL DATA FOR DYNAMIC BEHAVIOURAL PROCESSES

1. INTRODUCTION

Geographers have long been aware of some of the problems encountered in attempting to determine causality from cross-sectional data. The distinction between pattern and process soon became apparent within point pattern analysis, where several alternative, but theoretically plausible, models were shown to often lead to the same cross-sectional spatial pattern. Such equifinality, as it is known, is common throughout all areas of study. Though results from such analyses may be justifiably presented as descriptions of the data, this is rarely viewed as sufficient. It is generally expected that both model form and parameter estimates should have explanatory significance. The temptation to interpret parameters is strong. In many behavioural studies of cross-sectional data, in which the models used invariably include so-called "explanatory" variables, such a temptation has frequently proved irresistible. Is such interpretation justifiable? Does a parameter estimate actually measure the impact of an explanatory variable? Formally, all statistical inference about a parameter is made conditionally upon the model being correct. In view of the potential equifinality problem, interpretation may be thought of as, at best, one of a possible set of "correct inferences", each made from a different underlying model. But what is the underlying model in a typical cross-sectional behavioural analysis? Is it as plausible as some alternative model which could give rise to a radically different interpretation? This widely recognised problem appears to receive little serious attention in practice. A blind faith in the robustness of the results to the underlying model assumed appears to have been accepted under the guise of pragmatism.

Doubts as to the validity of this practice are most common where the process under study is suspected of being non-equilibrium, or in some sense dynamic. For example, in the analysis of housing consumption, it has been recognised that moving costs are substantial and that this may lead to households persisting to occupy sub-optimal housing. Such disequilibrium or sub-optimality has tempted some, for example Murie (1974), to analyse the housing consumption of recent movers, since such households may be more easily considered to exhibit optimality, though little consideration has been given as to the possible broader impact of the sample selection implied.

This paper argues that such doubts are not only well founded but that some form of dynamic complication or endogeneity is commonplace within many apparently simple behavioural processes. To make the discussion more concrete, the argument is illustrated within the context of housing mobility. Section 2 briefly introduces maximum likelihood estimation and inference, tools which prove essential for a consideration of all but the simplest of dynamic models. Section 3 gives a derivation of a model of household mobility, discussing and contrasting the assumptions that might be made in estimating the

J. Hauer et al. (eds.), Urban Dynamics and Spatial Choice Behaviour, 81–104.

parameters of the model for cross-sectional and longitudinal data. Section 4 summarises the radically different results that can be obtained from analyses making these different assumptions when applied to simulated data for the same known process. The clear potential of a correctly specified longitudinal analysis to uncover the underlying process is powerfully illustrated. Some results obtained from "real world" household mobility data for Leeds, England, are also discussed. Section 5 relates the findings of the previous section to the concept of endogeneity, a concept originating in time series analysis. The remaining sections detail three areas of difficulty, identification, initial conditions and sample design, that can still obstruct us in an attempt to represent and measure the population process.

2. MAXIMUM LIKELIHOOD ESTIMATION AND INFERENCE

From the use of the ordinary regression model researchers are familiar with the notion of "least squares", the minimisation of the sum of squared residuals, as a "best fit" criterion. Though there is sometimes scope for discussion as to what defines "best fit", it is generally agreed that the least squares criterion is not appropriate for many more complex models.

The principle of Maximum Likelihood is not only an intuitively reasonable criterion for parameter estimation but can be applied over an enormous range of models. Likelihood estimation requires that we must first construct an explicit probability model, something which whilst essential within the more familiar statistical approaches often remains implicit. A probability model is a conceptual mechanism of the process we are examining which specifies the general form of the relationship expected between variables but leaves certain "parameters" to be estimated from the data, just as the constant and slope coefficients are estimated within ordinary regression. Given values of the parameters, say a vector $\boldsymbol{\beta}$, the probability model is able to assign a measure of probability to the occurence of any particular response, y, for any value of the explanatory variables, say a row vector $\mathbf{x}$. More formally we could write

$$\Pr[y] = f(\mathbf{x};\boldsymbol{\beta}) \tag{1}$$

which shows that a probability model is fundamentally a probability function $f(\cdot;\cdot)$. Maximum Likelihood (ML) estimation examines alternative values for the parameter vector $\boldsymbol{\beta}$, and selects those values which allow the model to assign the highest probability of occurence to the observed values of y. That is, for a sample of n independent observations $\{y_i,\mathbf{x}_i\}$, $i=1,\ldots,n$, the likelihood function L is maximised where

$$L = \prod_{i=1}^{n} \Pr[y_i] = \prod_{i=1}^{n} f(\mathbf{x}_i;\boldsymbol{\beta}) \tag{2}$$

Loosely speaking, the parameter values selected are those which minimise the discrepancy between observed and model predicted values.

Indeed, for a number of simple situations ML estimates are identical to least squares estimates.

Inference is usually straightforward since under relatively mild conditions, ML estimators can be shown to be approximately normally distributed in large examples. This allows the use, with care, of the familiar *t*-test type statistic, amongs others, for inference concerning values within the parameter vector β. Pickles (1985) provides a more extensive introduction to likelihood together with several geographical examples of its use.

3. A MODEL OF HOUSEHOLD MOBILITY

A simple model of household residential mobility may be derived using random utility theory. A household i may be considered as choosing between staying in its current dwelling a, or moving to the best identified and available alternative dwelling b. The utility derived from the current dwelling, u_{ai}, may be thought of as a function of a deterministic component v_{ai}, whose value depends upon observed characteristics of the household and dwelling, and also a random or error component, e_{ai}, accounting for unobserved characteristics and tastes. For the alternative dwelling, we usually have no information about its characteristics but can assume it has been drawn from a population of potential dwellings, possessing a distribution of housing characteristics. Selection of the alternative dwelling also means that various moving and transaction costs will be incurred. The utility of the alternative dwelling, u_{bi}, may then be summarised by a function of a deterministic component, v_{bi}, that represents housing market conditions and moving costs for household i, and a random component associated with the particular dwelling selected, e_{bi}. Thus,

$$\left.\begin{aligned} u_{ai} &= f_a(v_{ai}, e_{ai}) \\ u_{bi} &= f_b(v_{bi}, e_{bi}) \end{aligned}\right\} \quad (3)$$

The household will choose to move if $u_{bi} > u_{ai}$. If the utility functions are additive in the errors and appropriately monotonic, then the probability of moving $\Pr[u_{bi} > u_{ai}]$ is given by

$$\Pr[u_{bi} > u_{ai}] = \Pr[e_{bi} - e_{ai} > v_{ai} - v_{bi}] \quad (4)$$

If the random components can be assumed independently and identically distributed according to an extreme value density, then $e_{bi} - e_{ai}$ has a logistic density. If the deterministic components are linear functions of the differences between observed characteristics, housing market conditions and moving costs described within the row vector $\mathbf{x}_i$, this gives rise to the familiar logit model in which

$$\Pr[u_{bi} > u_{ai}] = \exp(\alpha + \mathbf{x}_i\boldsymbol{\beta})/[1 + \exp(\alpha + \mathbf{x}_i\boldsymbol{\beta})] \tag{5}$$

This is a standard result (see, for example, Dhrymes, 1978, 340-344). Alternatively, the random components may be assumed normally distributed, $e_{bi} - e_{ai}$ then has a normal density and a probit model is obtained where

$$\Pr[u_{bi} > u_{ai}] = \Phi(\alpha + \mathbf{x}_i\boldsymbol{\beta}) \tag{6}$$

and Φ is the cumulative density function of a standard normal distribution. Such logit and probit models have been applied to cross-sectional mobility data using standard computer programs. Although mobility itself involves movement and is therefore at a trivial level, dynamic, the model appears quite static. The mobility process develops along a path which is determined solely externally to the model. The probability is simply a function of the value of the exogenous covariates existing at that time. The inclusion of some lagged exogenous variables does not alter this essentially static nature. A more dynamic process exists when the development of the process depends upon its own past behaviour. The response variable, in this case mobility, then depends upon both exogenous variables and variables describing past behaviour, which are referred to as endogenous. It has often been suggested that social and sentimental ties increase with time in the same dwelling, resulting in "cumulative inertia" (Huff and Clark, 1978), a declining probability of movement with increasing duration of stay. Moss (1979), amongst others, has proposed the inclusion among the covariates of a duration variable to proxy for these unobserved but duration dependent social and psychological costs. It has been suggested this model could be used to analyse mobility within a single year, from cross-sectional data which included information on the time of the last move preceding the interview data.

These same models could also be applied to longitudinal or panel data, analysing the probability of movement within each of a series of moves for each household. The only extension required would seem to be the addition of time subscripts and the writing of an observed mobility history as the product of the probabilities of the observed moves and no-moves within each time period. For example, were household i to have moved in Years 1 and 3, then

$$\Pr[u_{bi1} > u_{ai1},\ u_{bi2} \le u_{ai2},\ u_{bi3} > u_{ai3}]$$

$$= \Pr[u_{bi1} > u_{ai1}]\ \Pr[u_{bi2} \le u_{ai2}]\ \Pr[u_{bi3} > u_{ai3}] \tag{7}$$

and using the probit model of Equation (6)

$$= \prod_{t=1}^{3} \left\{ [\Phi(\alpha + \mathbf{x}_{it}\boldsymbol{\beta})]^{y_{it}} [1 - \Phi(\alpha + \mathbf{x}_{it}\boldsymbol{\beta})]^{1-y_{it}} \right\} \tag{8}$$

where $y_{it} = \begin{cases} 1 \text{ if } u_{bit} > u_{ait} \\ 0 \text{ otherwise} \end{cases}$

The sample likelihood is then given by

$$L = \prod_{i=1}^{n} L_i = \prod_{i=1}^{n} \prod_{t=1}^{3} \left\{ [\Phi(\alpha + \mathbf{x}_{it}\boldsymbol{\beta})]^{y_{it}} [1 - \Phi(\alpha + \mathbf{x}_{it}\boldsymbol{\beta})]^{1-y_{it}} \right\} \tag{9}$$

This is referred to as the simple pooled cross-sectional model, since the same results are obtained if information from each year are pooled and thought of as one large cross-section.

However, a brief consideration of the theoretical development of the model would suggest that such an immediate extension is not easily justified. The decomposition imposed in Equation (7) assumes that movement probabilities are conditionally independent from year to year; that given values of the exogenous and endogenous covariates, mobility in one year is independent of mobility in another. The decomposition implies that the error components of the utilities, e_{ait} and e_{bit}, are independent. As described these errors include many unobserved characteristics of the household and their perceptions of the desirability and cost of movement. These can be expected to possess considerable persistence over time. The errors are, therefore, likely to exhibit strong serial correlations, not independence, suggesting the above models to be misspecified.

A more plausible approach is to decompose each error into a persistent term, constant over time, and an independent time-varying term

$$e_{ait} = \mu_{ai} + \nu_{ait} \tag{10}$$

$$e_{bit} = \mu_{bi} + \nu_{bit} \tag{11}$$

If the ν's are assumed normally distributed then the movement probability is now given by a probit model of the following form

$$\Pr[u_{bit} > u_{ait}] = \Phi(\alpha_i + \mathbf{x}_{it}\boldsymbol{\beta}) \tag{12}$$

$$\text{where } \alpha_i = \alpha + \mu_{bi} - \mu_{ai} \tag{13}$$

This is the same as Equation (6) but the shared constant α has been replaced by individual specific constants $\{\alpha_i\}$. Such a model cannot be applied directly because there are too many parameters to be estimated. Even as the size of the sample of households increases so too does the number of α_i parameters, leading to both computational and statistical difficulties (Neyman and Scott, 1948). There are a large number of ways of dealing with the problem of eliminating these incidental or nuisance parameters, as the $\{\alpha_i\}$ are called, from our problem (Basu, 1977). In practice only two of these methods, marginal and conditional likelihood, are applicable, and of these conditional likelihood is highly restrictive in empirical scope (see Chamberlain, 1979; Pickles and Davies, 1983).

Marginal likelihood assumes that the value of the individual specific constants is distributed according to some distribution $G(\alpha)$ and, like the error in an ordinary regression model, is assumed independent of the included covariates. The probability of the example series of moves and no-moves is then given by

$$\Pr[u_{bi1} > u_{ai1}, u_{bi2} \leq u_{ai2}, u_{bi3} > u_{ai3}]$$

$$= \int \Pr[u_{bi1} > u_{ai1}|\alpha] \Pr[u_{bi2} \leq u_{ai2}|\alpha] \cdot \Pr[u_{bi3} > u_{ai3}|\alpha] dG(\alpha) \tag{14}$$

$$= \int \prod_{t=1}^{3} \left\{ [\Phi(\alpha + \mathbf{x}_{it}\beta)]^{y_{it}} [1 - \Phi(\alpha + \mathbf{x}_{it}\beta)]^{1-y_{it}} \right\} dG(\alpha) \tag{15}$$

$$\text{where } y_{it} = \begin{cases} 1 \text{ if } u_{bit} > u_{ait} \\ 0 \text{ otherwise} \end{cases}$$

This looks rather more complicated, especially in view of the fact that we have little empirical or theoretical evidence to suggest the form of the distributors $G(\alpha)$. However, the essential idea can be grasped by considering, for a moment, a hypothetical population in which a proportion, r, of the population has a value of α equal to k_1, and the remaining proportion $r_2 = 1-r_1$ has a value of α equal to k_2. On drawing an individual household at random from the population we have no certain knowledge as to which sub-population it belongs. We do know that it has a probability of r_1 of being from the first sub-population and r_2 from the second. The probability of any observed move/no-move history should therefore be calculated as the correspondingly weighted average of the probability of the history for a household from the first sub-population and that from the second. By assuming that the density $dG(\alpha)$ is concentrated into these "mass points" (i.e. discrete probabilities) at locations k_1 and k_2, the integration of Equation (15) is seen to be just simple weighted averaging;

$$\Pr[u_{bi1} > u_{ai1}, u_{bi2} \leq u_{ai2}, u_{bi3} > u_{ai3}]$$

$$= \sum_{j=1}^{2} r_j \left\{ \prod_{t=1}^{3} [\Phi(k_j + \mathbf{x}_{it}\boldsymbol{\beta})]^{y_{it}} [1 - \Phi(k_j + \mathbf{x}_{it}\boldsymbol{\beta})]^{1-y_{it}} \right\} \tag{17}$$

This apparent simplified hypothetical example takes on new significance in the light of results from mathematical statistics by Simar (1976), Laird (1978) and Lindsay (1983a, b). In this work the mass points are not interpreted as a finite mixture, that is they are not thought of as representing a real partition of the population into homogenous sub-populations. Instead, they are seen as forming a non-parametric representation of a continuous distribution. More surprising, the theory shows that for a wide range of data a complete non-parametric representation requires only a finite number of mass points. However, still more surprising is the recent empirical experience showing just how few mass points are required to achieve this complete non-parametric form (Davies and Crouchley, 1984; Pickles and Davies, 1985; Heckman and Singer, 1984a).

Whatever the particular technique selected for accounting for the nuisance parameters $\{\alpha_i\}$, the more basic point is that we can expect some inferential problems within pooled cross-sectional studies which assume them all equal. With a single cross-section of data α_i cannot be distinguished from $(e_{bit} - e_{ait})$, and so the cross-sectional and longitudinal models defined by Equations (6) and (12) might seem to be equivalent. We have a problem of equifinality. These issues are examined empirically in the following section.

4. AN EMPIRICAL ASSESSMENT OF CROSS-SECTIONAL, POOLED CROSS-SECTIONAL AND LONGITUDINAL MODELS

Comparing the results of various models is much more informative if we know what the correct results should be, which is very rarely the situation with real empirical data. Generating data by simulation provides such an opportunity, though this is often at the expense of "realism". We believe that the procedure adopted here achieves both objectives. A sample of 1073 households from the Michigan Panel Study of Income Dynamics was selected and the values of certain covariates obtained for each of the 12 years 1968-1979. The covariates selected were those found to be significant in a previous analysis of the data (Davies and Pickles, 1985a) and consisted of the age of the head of household, a measure of space standards called roomstress, and dwelling tenure. In addition, values for the unknown household-specific constants $\{\alpha_i\}$ were drawn from an appropriately defined normal distribution. From these data and suitable values for the parameter vector $\boldsymbol{\beta}$, hypothetical mobility histories were simulated according to the following equation

$$y_{it} = \begin{cases} 1 \text{ if } [\Phi(\alpha_i + \mathbf{x}_{it}\boldsymbol{\beta})] > c_{it} \\ 0 \text{ otherwise} \end{cases} \tag{18}$$

where c_{it} was a value drawn from a distribution uniform on (0,1). More details concerning the simulation process can be found elsewhere (Davies and Pickles, 1985b).

The 1073 twelve-year histories so generated were then analysed using the longitudinal model of Equation (12), the number of mass points being increased until no further improvement in the likelihood could be obtained. Five mass points were required and the results for the covariate parameters are presented in Table I. The true parameter values are given in the first column and reflect previous empirical findings that mobility increases as the result of both a shortage and an excess of roomspace, declines non-linearly with increasing age, and is reduced if the current dwelling is owner-occupied rather than rented. The second column gives the parameter values estimated by the model, which must be interpreted by reference to the standard errors given in parentheses below each one. The simulated data did not include any duration-of-stay effects, which is equivalent to a true value of zero for the duration-of-stay parameter. The success of the longitudinal model could hardly be more convincing. Column 3 shows t-statistics for a test of the parameter estimates against their true values. There is not even a hint of a significant difference.

The pooled cross-sectional model of Equation (8) could be applied to exactly the same data. However, to assess the relative performance of the simple cross-sectional model 12 x 1073 independent observations are required. The simulated cross-sectional data were made comparable to those used above by simulating 12 different histories for each household. In each case the values of the known covariates followed the same path given in the panel data, but the unknown household specific constant was different in each history. Then, from the first history the first year data were selected for analysis, from the second history the second year data, and so on. This allowed each household to contribute one observation for each year, as before, but ensured the independence of these observations.

A moment's hesitation is necessary before discussing the results from these models. Derivation of discrete choice models sometimes makes obscure reference to the fact that parameters are estimated up to a scalar multiple (e.g. Anas, 1982, 64) or more usually simply state that the distribution function for the error is assumed a standard normal or to have unit variance. What this means is that the parameter estimates depend not only upon the importance and variability of the regressor variable in question but also upon the residual variance of the data. The data and longitudinal model include separate terms for an omitted variable and time-varying error, with variances of 0.52 and 1 respectively, which must be summed to give a residual variance of 1.52 for the cross-sectional models. The usual estimates of parameters and standard errors must be scaled by a factor $\sqrt{1.52}$ to give quantities comparable to the parameter values used in generating the data. Since both parameters and standard errors are scaled by the same factor, such scaling does not alter any of the conclusions which would be reached using standard inferential procedures concerning the significance of particular variables in the model, for these are concerned with the size of the departure of parameter estimates from zero relative to the size of their standard errors. Related issues are discussed by Tardiff (1979).

TABLE I

Results of the residential mobility analyses including duration of stay

	Longitudinal			Pooled Cross-Sectional		Cross-Sectional	
Variable	True Parameter	Estimates	*t* about True	Scaled Estimates	*t* about True	Scaled Estimates	*t* about True
	Col. 1	Col. 2	Col. 3	Col. 4	Col. 5	Col. 6	Col. 7
Roomstress	-0.25	-0.248	0.02	0.039	3.91	0.107	4.85
		(0.090)		(0.074)		(0.074)	
$(\text{Roomstress})^2$	0.10	0.090	0.31	-0.014	4.15	-0.042	5.17
		(0.032)		(0.027)		(0.027)	
Age	-3.00	-3.066	0.22	-2.447	2.86	-2.289	3.67
		(0.302)		(0.193)		(0.193)	
Age^2	5.00	4.579	0.46	4.289	1.09	4.063	1.45
		(0.909)		(0.654)		(0.646)	
Owner	-1.00	-0.988	0.26	-0.669	8.28	-0.780	5.23
		(0.047)		(0.041)		(0.042)	
Duration	0.00	-0.013	0.15	-1.415	21.90	-1.385	21.52
		(0.086)		(0.064)		(0.064)	

The scaled estimates are presented in columns 4 and 6 of Table I, with the scaled standard errors in parentheses. In columns 5 and 7 t-test statistics are presented for tests of the estimates against their true values. The results are very poor. The t-test statistics indicate that for both models 5 of the 6 parameter estimates are very significantly different from their true values. Only the parameter estimate for age squared is acceptable. Furthermore, the models wrongly suggest roomstress to have no importance and duration effects to have substantial importance in the determination of residential mobility.

Why are the results so poor? Consider an unobserved but persistent household characteristic which depresses the rate of mobility. Such households, tending to move less frequently, will tend to have a longer duration since the previous move. The duration-of-stay variable is therefore causally linked to this unobserved variable, through the intermediate variable describing previous mobility decisions or, in other words, the response or dependent variable of our probit model. Since the unobserved characteristic is a component of the error term (e_{bit} - e_{ait}) we have a model in which the error term is correlated with an included explanatory variable. This is a serious specification problem both in ordinary regression and in such models as the probit model of the example. A longitudinal approach allows the time persistent omitted variables, through which the correlation to the error term arises, to be explicitly accounted for. The presence of an endogenous variable, such as duration of stay, together with the inevitable time persistant omitted variables, leads to problems in simple and pooled cross-section behavioural studies.

TABLE II
Results of the residential mobility analyses
excluding duration of stay

		Pooled Cross-section		Cross-section	
Variable	True Parameter Col. 1	Scaled Estimates Col. 2	t about True Col. 3	Scaled Estimates Col. 4	t about True Col. 5
Roomstress	-0.25	0.206 (0.071)	6.37	0.265 (0.071)	7.16
(Roomstress)^2	0.10	-0.089 (0.026)	7.19	0.113 (0.027)	7.99
Age	-3.00	-3.166 (0.189)	0.88	-2.950 (0.189)	0.27
(Age)^2	5.00	4.706 (0.643)	0.46	4.389 (0.683)	0.96
Owner	-1.00	-0.689 (0.039)	7.87	-0.798 (0.042)	4.85

An obvious response is that the endogenous variable should be removed from the cross-sectional models being estimated. Since there was no duration effect in the simulated data it could be suggested that we are somehow tricking the models into this specification error. Table II illustrates the results from the fitting of the models without the duration variable. This proves to offer no salvation. Although both age parameters are acceptably estimated, the others remain poor. At first glance the roomstress estimates might look much improved, and inference about the zero point would correctly suggest their importance within residential mobility; but unfortunately they are the wrong sign! The estimates obtained are wholly inconsistent with theory, for they give no minimum mobility at some optimum space standard, but a maximum.

The reason why these results are so poor is that the seemingly inocuous dummy variable for owner-occupation is in fact an endogenous variable. The choice of renting decreases strongly with age of the head of household (Pickles and Davies, 1985) as does mobility. In addition, those with large values of the omitted variable, and thus a high propensity to move, prefer the rented sector because of its lower moving costs. The end result is that the young and more mobile move often and choose to rent, so that in empirical data duration of stay on moving and tenure choice are correlated. This correlation was introduced into the simulated data. Endogeneity can also be expected for other reasons. A change of tenure normally can only occur when a move occurs. A household's current tenure does not reflect the current age of the head of household, but the age at the time of the previous move. A household with low movement propensity is more likely to be a renter household than one would have expected given the current age of the head of household. Thus the tenure dummy variable in part reflects past mobility behaviour.

Intuition would suggest that severity of the specification problems will depend upon the adequacy with which the distribution of the effects of omitted variables is accounted for. Table III presents some empirical results for an analysis of the migration life histories from age 21 of 633 persons resident in Leeds, England (Davies, 1984). The first column presents parameter estimates from a pooled cross-sectional logit model. The second column presents those from a model in which the effect of omitted variables is allowed to give rise to a beta-distribution in movement probabilities. This is a generalisation of the beta-logistic distribution proposed by Heckman and Willis (1977) and is fully explained in Davies (1984). The third column gives new results using the mass point approach already described.

Previously, in the simulation example, scaled parameter estimates were given for the pooled cross-section model. In practice, as here, we know neither the variance nor the distribution of the omitted variable and so cannot calculate the appropriate factor. However, it can quickly be seen that a factor that would appropriately scale the age and non-stationary estimates to those from the mass points model would not be appropriate for the duration estimates, and vice versa. The effects of endogeneity can be seen in this real data as in the simulation.

The beta-logistic model imposes some parametric restrictions upon the distribution of omitted variables. The parameter estimates are intermediate between those of the pooled cross-section and mass points models. This accords well with our expectations, suggesting that a beta distribution, whilst flexible, is not completely successful in accounting for the effects of omitted variables. It is of interest to note that the results from the pooled cross-section model suggest duration effects to be insignificant, those from the beta-logistic are equivocal ($p \sim 0.08$), whilst those from the mass points model suggest positive duration effects of "cumulative stress" (Huff and Clark, 1978) to be present ($p \sim 0.01$).

5. ENDOGENEITY

Behavioural researchers are, therefore, commonly dealing with explanatory variables which may be identified as endogenous with varying degrees of difficulty. There are those which are directly associated with previous behaviour, including not only any duration effect, but Markov type effects such as "party loyalty" within models of voting behaviour, or "state grew up in" within an interstate migration model. There are those variables which, in some sense, proxy for the previous behaviour. The proportion of a house purchase loan remaining to repay often simply reflects the duration of ownership. Variables describing attitudes to choice alternatives may reflect processes such as "cognitive dissonance" (Festinger, 1957) whereby post-choice rationalisation takes place to reinforce the belief that the correct choice was made or "alienation and apathy" as the duration of an undesirable experience, for example, unemployment, increases. Kalbfleisch and Prentice (1980, 124) discuss such endogenous proxy variables in the context of medical survival studies under the name of

Table III
Levels of control for the omitted variable:
a comparison of results using the Leeds data

Variable	Pooled Cross-section	Generalised Beta-logistic	Non-parametric Mass Point
Duration	1.06	2.80	3.92
$(\text{Duration})^2$	-0.62	-2.80	-4.25
Age	-7.87	-8.07	-9.55
$(\text{Age})^2$	5.19	5.27	7.17
Year	1.42	1.53	1.70
$(\text{Year})^2$	-1.22	-1.26	-1.41
$(\text{Year})^3$	0.37	0.37	0.41
$(\text{Year})^4$	-0.03	-0.03	-0.04
Log-likelihood	-2283.94	-2277.99	-2274.75

"internal" covariates. Finally, there are those variables such as current dwelling tenure within the examples, in which the endogenous characteristics are implicit occuring through some intermediate, but related, process. These will also occur elsewhere, such as studies of the influence of travel costs on the length of the journey to work. Choice of work location can generally only occur at the time of job changes, and reflect the travel cost surface which prevailed at that time. If there have been systematic changes in the travel cost surface over time, travel cost will behave as an endogenous variable.

More formal tests of endogeneity within longitudinal data have been discussed by several authors (Andersen, 1973; Chamberlain, 1979). Chamberlain suggests that such tests can be based on the tests used within time series to determine "strict exogeneity". The basic test is whether

$$E(y_{it}|x_{i1},\ldots,x_{it},\ldots,x_{iT}) = E(y_{it}|x_{i1},\ldots,x_{it}) \qquad (19)$$

or whether the estimation of the current value of y is not improved by knowledge of future values of x. A failure of the test may be the result of an element of $\mathbf{x}_i$ being truly endogenous or the result of both $\mathbf{x}_i$ and y being dependent upon an omitted variable $\{\alpha_i\}$. The particular test of interest is therefore

$$E(y_{it}|x_{i1},\ldots,x_{it},\ldots,x_{iT},\alpha_i) = E(y_{it}|x_{i1},\ldots,x_{it},\alpha_i) \qquad (20)$$

In the mobility example tenure choice will be dependent upon moving propensity α_i. Thus the tenure variable might be expected to fail the test implied by Equation (19) but pass that implied by Equation (20). However, as already discussed, tenure choice is restricted to occur at time of movement only. A future value of a tenure variable different from the current tenure gives information as to the occurence of a move in the intervening period. Equation (20) cannot, therefore, be expected to hold and tenure remain endogenous conditional upon the omitted variable.

6. THE LIMITS OF INFERENCE AND OMITTED VARIABLES

It is clear that the presence of omitted variables places severe limitations on the scope for inference within cross-sectional behaviour analysis. We have given no impression, as yet, of the problems that they pose for longitudinal studies. This is currently an area of vigorous research and, as we have seen with the use of mass points in the empirical example, has required the exploitation of new and difficult theory from within mathematical statistics. The results of the example analysis were most impressive and care was taken that the example should be quite typical of those found in practice. However, such optimistic results will not always be obtained in situations where either the form of the data or of the model are different. Unfortunately, the conditions necessary to achieve satisfactory inferences are often highly abstract and have not been agreed (see, for example, Lancaster and Nickell, 1980; Elbers and Ridder, 1982; Heckman and Singer, 1984b). The following discussion introduces some of the concerns of this work.

7. THE LIMITS OF INFERENCE - IDENTIFICATION

It is commonly argued that various kinds of behaviour are related, or, more generally, that events are interdependent. For example, in Britain it is argued that the financial burden of initial home purchase results in households delaying childbearing (Ineichen, 1981). Likewise, early childbearing may retard subsequent home purchase (Barbolet, 1969). The timing of a house purchase and childbearing should, therefore, be interdependent. A plausible way of examining this would be to study a sample of households and determine the time of first child birth and the time of first home purchase. A natural test of independence would be to group these times and then to construct a contingency table, as shown in Table IV, for hypothetical data.

The null hypothesis within the contingency table is independence, or whether the joint distribution of times to purchase and times to first birth $F(x,y)$ is given by the product of the marginal distributions $F_x(x)$ and $F_y(y)$, or

$$F(x,y) = F_x(x)F_y(y) \tag{21}$$

Expressing this in terms of probabilities of events in discrete time, the probability that home purchase will occur in time period x and first birth in time period y is simple the product of the probability of home purchase in x and the probability of first birth in y. In the above example a χ^2 of 70 is obtained, strongly suggesting that the events are not independent ($\chi^2_{.05}(16) = 26.30$). In fact the contingency table has been constructed as the sum of two tables (Table V), each of which gives absolutely no evidence of interdependence (χ^2 of 0.63 and 0.08 respectively).

TABLE IV
Contingency table for hypothetical data

		Time to First Birth					
		1	2	3	4	5	
Time	1	25	44	40	25	19	153
to	2	64	100	86	53	40	343
Home	3	91	127	100	58	42	418
Purchase	4	96	121	85	45	30	377
	5	227	247	143	61	31	709
		503	639	454	242	161	2000

Ignoring omitted variables or potential population heterogeneity, here represented by two groups, can give rise to spurious interdependency. Let us imagine that there is an unobserved explanatory variable which means that the distributions of x and y now depend upon household specific constants α_i and β_i. The hypothesis of independence that we want to test is now one of independence conditional upon the values of α_i and β_i.

$$F(x,y|\alpha_i,\beta_i) = F_x(x|\alpha_i)F_y(y|\beta_i) \tag{22}$$

We do not observe α_i nor β_i, nor can we directly estimate any of the above conditional distribution functions. However, for a sample of households drawn at random from the population we can estimate the marginal distributions $F(x,y)$ from the cell frequencies as before, noting that

$$F(x,y) = \int F(x,y|\alpha,\beta)\mathrm{d}G(\alpha,\beta) \tag{23}$$

where $G(\alpha,\beta)$ is the distribution of α and β values in the population. Written in terms of this marginal distribution the null hypothesis of conditional independence becomes

$$F(x,y) = \int F_x(x|\alpha)F_y(y|\beta)\mathrm{d}G(\alpha,\beta) \tag{24}$$

Unfortunately, without knowledge of the distribution $G(\alpha,\beta)$ this hypothesis can never be rejected. Any distribution $F(x,y)$ can be obtained from a suitable selection of $F_x(x|\alpha)$, $F_y(y|\beta)$ and $G(\alpha,\beta)$.

This is made clear by considering a trivial example of fatalistic life, discussed by Winship (1984) in which x and y are pre-determined times given by a household's values of α and β. Thus

$$\Pr[x|\alpha] = \begin{cases} 1 \text{ if } x = \alpha \\ 0 \text{ otherwise} \end{cases} \tag{25}$$

Table V
Underlying tables

	1	2	3	4	5	1	2	3	4	5
1	7	7	3	1	0	18	37	37	24	19
2	27	27	13	4	1	37	73	73	49	39
3	54	54	27	9	3	37	73	73	49	39
4	72	72	36	12	4	24	49	49	33	26
5	208	208	104	35	11	19	39	39	26	20

$$\Pr[y|\beta] = \begin{cases} 1 \text{ if } y = \beta \\ 0 \text{ otherwise} \end{cases} \tag{26}$$

which gives, using Equation (24)

$$F(x,y) = \int dG(\alpha,\beta) = G(\alpha,\beta) \tag{27}$$

The number of events in any (x,y) cell is entirely determined by the number of persons with corresponding (α,β) values. Any distribution $F(x,y)$ can be obtained by the appropriate choice of $G(\alpha,\beta)$. This example shows that it is not possible, in general, to distinguish interdependence of events from their mutual dependence upon other variables in the presence of omitted variables. This pessimistic conclusion concerning our inability to identify the process should be qualified where

(a) the model can be appropriately re-specified. In the example it was argued that the interdependency between events arose because of their financial burden within a limited budget. A more promising approach would be to examine the impact of financial stringency on these events directly;
(b) data is avaible for multiple events, for example births beyond the first;
(c) it is possible to make parametric assumptions restricting the form of $F(x)$, $F(y)$ or $G(\alpha,\beta)$. The inclusion of observed explanatory variables within such parametrisations may also improve identification (Elbers and Ridder, 1982).

None of these provide a panacea, though each may allow progress in particular circumstances. The problem becomes still more difficult where age, cohort and non-stationarity effects may also be present.

8. THE LIMITS OF INFERENCE - INITIAL CONDITIONS

In the presence of endogeneous variables current behaviour depends upon past behaviour. For a simple renewal process this dependency is limited to the period back to the preceding renewal. For processes with a Markovian dependency current behaviour depends, in theory, upon the whole of the previous behaviour, or until some point at which that behaviour was wholly exogenously determined.

If our observations begin at points of initialisation, such as the beginning of a period of dwelling occupation or at some exogenously determined start, such as initial household formation, then the first observed behaviour may be modelled simply as a function of contemporary values of explanatory variables. When, as is more often the case, the beginning of the survey data interrupts individuals partway through the process, then the problem is a good deal more difficult. In such cases several different solutions have been attempted.

The usual procedures for analysing such already operating processes is to analyse only the behaviour of individuals after their first observed circumstances. For the Markov process illustrated in Figure 1 the transitions and non-transitions between states within the observation period, represented by $t(1)$ to $t(5)$, would be modelled, given that the individual began in state 1. The occupancy of state 1 at $t(0)$ is not, of itself, to be explained. In a similar fashion, for the renewal process of Figure 1 the observed residual duration from d_1^* to $(d_1^*+d_1)$ is modelled together with the completed spell of length d_2 and right censored duration d_3^*. The duration d_1^*, equivalent to a duration of residence at time of first interview, is not explained.

Such an approach has certain consequences. Unless the process that generated the first observation is independent of the process generating the subsequent behaviour, we may be failing to analyse potentially useful information. For slow moving processes observed over a short duration the initial observed state may contain much more information about the process, summarising as it does the individual's progress from the start of the process to the first observation, than the small amount of change data that may have arisen within the observation period. However, to exploit this information may require more assumptions than we are willing to make. The most efficient approach is to use the same model of the process to determine both the

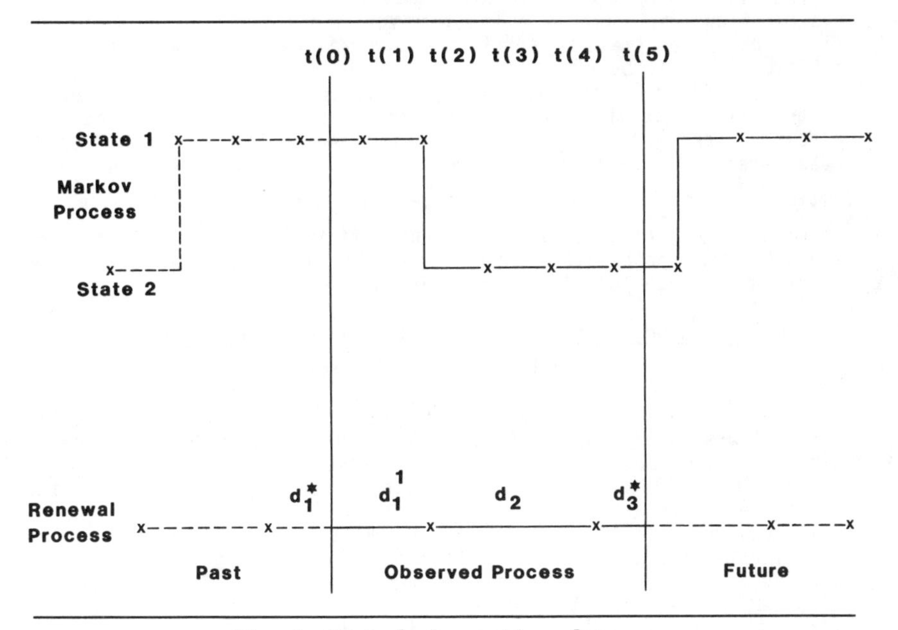

Fig. 1: Typical observation plans.

probability of the observed initial state and the changes in the observation period, for example by the calculation of steady state probabilities for the initial state of a Markov process or steady state densities for the recurrence times of a renewal process. Both of these require the typically unrealistic assumptions of process stationarity, implying no time variation in covariates, and that the process has been running for a long time. Little useful theory is available to guide any relaxation of these assumptions, but failure to do so can severely bias the results. In practice it seems that the best that can be achieved is to specify for the initial state a quite general sub-model possessing errors freely correlated with those in the main model and with no parameters shared between them. Though this accounts for the initial observed circumstances it requires the estimation of additional parameters with no clear gain in efficiency.

To ignore the information contained in the first observation can also result in biased estimates. For the procedures which analyse the data conditional upon the first observation it must be ensured that the model or likelihood which we use to calculate expected probabilities for subsequent behaviour take full account of what occupation of the first observed state implies for the value of explanatory variables. Where the process is a function solely of observed explanatory covariates, then knowledge of the first observed state can add nothing. But when unobserved/omitted explanatory variables are present then the first observed state carries information about their likely values which we cannot ignore. The distributions of omitted variables for individuals with different first observed states cannot be assumed to be independent and identically distributed.

Proposed solutions involve (i) estimating a set of different distributions for the omitted variables for different first observed circumstances. This was how the results were obtained for the Leeds data presented earlier, in which four such distributions were estimated for four groups of values for the initial duration d_1^* (Davies, 1984); (ii) estimating a single mixing distribution but using a conditional probability argument to derive a marginal likelihood function for each individual of the form

$$\frac{\int \text{likelihood for first observation and subsequent behaviour}}{\int \text{likelihood for first observation}}$$

For renewal processes the marginal likelihood can be quite straightforward (Lancaster, 1979) although Ridder (1984) identifies certain restrictions which must be imposed for the simplicity to occur. It should also be noted that it requires values for the covariates prior to the observation period, data which are not always collected or, if collected, whose reliability may be suspect.

The discussion of this section has been loosely set within the particular observational plan sketched in Figure 1. The consequences of conditioning upon the first observation, and the form that the first observation takes, will clearly depend upon the particular observational plan. This is, of course, intimately related to the sample design.

9. THE LIMITS OF INFERENCE - SAMPLE TO POPULATION GENERALISATION

A common failing of much empirical work is poor definition of the population to which the researcher's sample analysis is supposed to generalise. In longitudinal behavioural analysis this is an especially tricky problem since with the birth, death and ageing of the objects under study during the observation period, the concept of population becomes intrinsically dynamic. In general, it will always be necessary to include cohort, non-stationarity and ageing effects within such studies and to carefully consider the limitations of the sampling scheme. As mentioned in the introduction, there have been perceived advantages in the selective sampling of some participants, such as recent movers, for cross-sectional studies. In longitudinal studies, particularly of relatively rare events, sample selection might also seem attractive. In a study of interstate migration, for example, an analyst may have less interest in the very large proportion of people who stay put in the same state during a typical period of analysis. A selective sample from the population, obtained by sampling from the flows of inter-state migrants, might seem to have advantages. Alternatively, survey costs might be reduced through geographical clustering of the sample to those in a single state. Unfortunately, such schemes create their own problems. Marginal likelihood techniques require that the omitted variable should be independently and identically distributed and should not, therefore, be correlated with an included regressor variable. Chesher and Lancaster (1983) show that, in general, if this assumption is true for the population as a whole distributed across all states within a Markov type model, then it cannot be true for a sample drawn from amongst those within a particular state, nor for a sample drawn from any of the flows between states.

The impact this may have on an empirical conclusion can be illustrated through a consideration of an essentially similar problem long known within experimental design, called the regression effect (Campbell and Stanley, 1963, 10). An initial sample of y's are observed which are known to be a linear function of a covariate x and an error term e, uncorrelated with x. If a sub-sample is drawn whose membership depends upon the observed y value, then it will in general be the case that x and e within the sub-sample will be correlated. A new regression line estimated on the continued assumption of their independence can give rise to substantially different regression coefficients.

This has been illustrated in Figure 2 in which a y-value of zero has been taken as the threshold for sub-sample membership. Were y a latent variable observed only by means of an indicator variable representing state occupation, then such a sub-sample is entirely analogous to the sampling from among those in a particular state. Thus, if the distribution of the error or of omitted variables is assumed independent and identically distributed within the population, it cannot easily be assumed to be so for the sub-population within a state or flow between states. The depressed slope of the regression line estimated from the sub-sample, indicated by the dashed line, can become in the context of non-linear models of the sort more commonly applied to behavioural data, slopes with incorrect signs (Chesher and

Lancaster, 1984). Generalisation of the estimated slope parameter from the sub-sample to the sub-population occupying the particular state at the start of the observation plan remains possible, but generalisation to the whole population distributed across all states is not straightforward.

As yet our experience is insufficient to determine how robust an analysis might be to the scale of such effects to be expected in practice. At the end of the previous section it was noted that the procedure used in the analysis of the Leeds migration data overcame the initial conditions problem by estimating separate distributions for omitted variables within each of the four duration categories. This corresponds to a form of selective sampling of individuals such as that discussed above. If the distribution of the omitted variable

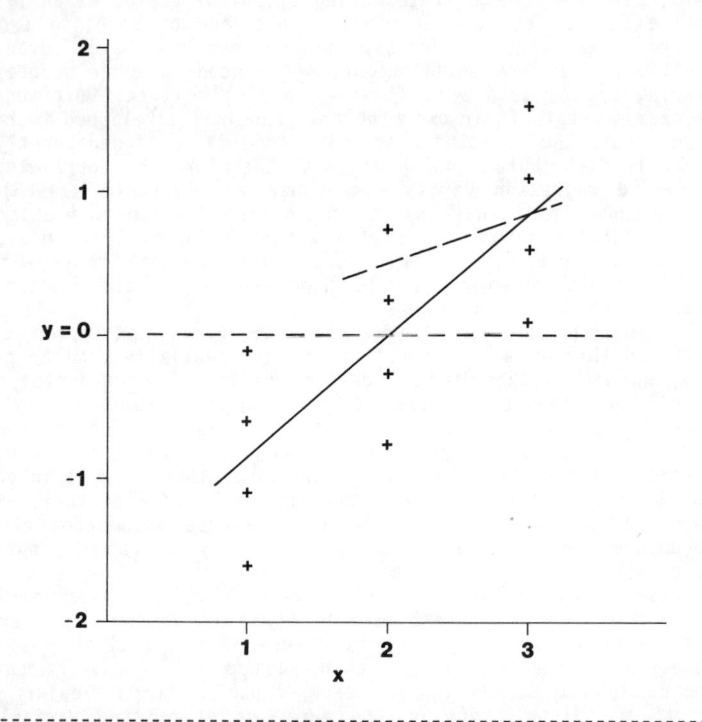

Two regression lines are shown of y on x. The continuous line is estimated using all the data; the discontinuous line is estimated only using cases for which y exceeds zero.

Fig. 2: Sample selection and slope estimation.

was independent in the whole population, it cannot be so distributed in each of these four sub-populations. However, unpublished simulations by the authors showed that under conditions typical of those of the data, the correlations induced by the sampling scheme were slight for renewal models with no or positive duration effects. In this instance the results we obtained from the Leeds data appear robust.

10. CONCLUSIONS

Geographers are interested in a large number of behavioural processes including local travel, inter-regional migration, unemployment, even disease incidence and accidents. Within each of these processes the population is unlikely to share identical propensities nor will these propensities be fully characterised by our measurements on explanatory variables alone. Unobserved explanatory variables will always exist. It is also to be expected that endogeneity will exist among some of the explanatory variables in which we are interested. We have shown that under such circumstances typical cross-sectional studies are unlikely to provide satisfactory results. Parameter estimates will be biased in an unknown fashion and the usual significance test can be wholly misleading. Our conclusions suggest that researchers must consider the theory underlying any behavioural study carefully, in terms of the presence of both omitted variables and endogeneity, and should treat results from cross-sectional studies with considerable caution.

This same theory must be exploited in deriving any appropriate longitudinal model. The residential mobility model derived in this paper is just one example of the many forms of longitudinal model possible. Not only do we have the usual issues in discrete choice modelling concerning the selection of variables and the overall functional form of the model (e.g. logit or probit), but we have to confront new and probably unfamiliar issues. The dynamics structure of the model, for example whether the model is to include duration or Markovian effects, determines how the initial conditions may be dealt with, which in turn has considerable implications for data requirements and survey design. Before the technical issues concerning conditional or marginal likelihood (Pickles and Davies, 1983), and parametric or non-parametric approaches to omitted variables can be solved, theory must be exploited to suggest how omitted variables should be included within the model, how they might vary over time, and whether they can be assumed independent of the included explanatory variables. These considerations of the dynamic structure and omitted variables must be undertaken with an eye on obtaining an identifiable model, given the data at hand. A simple example showed how not all empirical questions can be answered. Causal inference, even with longitudinal data, is neither assured nor always easy.

The concerns of the preceding paragraph should not divert us from the considerable promise that longitudinal studies offer. In many circumstances they will yield results in which we can trust, in contrast to those from a cross-sectional study. The availability and use of longitudinal data is rapidly increasing. The organisers of the

Michigan Panel Survey, now in its eighteenth year, are able to provide a no doubt incomplete bibliography of more than 300 references all of which have used data from just this one panel. We urge geographers to exploit this potential, but to do so whilst remaining mindsome of the problems that we have raised.

MCR Child Psychiatry Unit (A.R.P.)
Institute of Psychiatry
London
England

Centre for Applied Statistics (R.B.D.)
University of Lancaster
Lancaster
England

and

Department of Town Planning
University of Wales Institute of Science and Technology
Cardiff
Wales

ACKNOWLEDGEMENTS

The first author thanks the ESRC for the post-doctoral research Fellowship which hastened the completion of the work reported; the second author is indebted to the Nuffield Foundation for their financial support. Both authors are grateful to the Inter-University Consortium for Political and Social Research for the provision of data for the Michigan Panel Study of Income Dynamics. Neither the ESRC, the Nuffield Foundation, ICPSR nor the original collectors of the data are responsible for the analyses and conclusions presented.

BIBLIOGRAPHY

Anas, A.: 1982, *Residential Location Markets and Urban Transportation*, New York, Academic Press.

Andersen, E.B.: 1973, *Conditional Inference and Models for Measuring*, Copenhagen, Mentalhygiejniuk Forlag.

Barbolet, R.M.: 1969, 'Housing classes and the socio-ecological system', Centre for Environment Studies, Working Paper 4.

Basu, D.: 1977, 'On the elimination of nuisance parameters', *Journal of the American Statistical Association* **72**, 355-366.

Campbell, D.T. and J.C. Stanley: 1963, *Experimental and Quasi-Experimental Designs for Research*, Chicago, Rand McNally.

Chamberlain, G.: 1979, 'Heterogeneity, omitted variable bias and duration dependency', Harvard University for Economic Research, Discussion Paper 691.

Chesher, A. and T. Lancaster: 1983, 'The estimation of models of labour market behaviour', *Review of Economic Studies* **50**, 609-624.

Davies, R.B.: 1984, 'A generalized beta-logistic model for longitudinal data with an application to residential mobility', *Environment and Planning* **A 16**, 1375-1386.

Davies, R.B. and R. Crouchley: 1984, 'Calibrating longitudinal models of residential mobility and migration: an assessment of a non-parametric marginal likelihood approach', *Regional Science and Urban Economics* **14**, 231-247.

Davies, R.B. and A. Pickles: 1985a, 'A panel study of life cycle effects in residential mobility', *Geographical Analysis* **17**, 199-216.

Davies, R.B. and A. Pickles: 1985b, 'Longitudinal versus cross-sectional methods for behavioural research: a first round knock-out', *Environment and Planning* **A 17**, 1315-1329.

Dhrymes, P.J.: 1978, *Introductory Econometrics*, New York, Springer-Verlag.

Elbers, C. and G. Ridder: 1982, 'True and spurious duration dependence: the identifiability of the proportional hazards model', *Review of Economic Studies* **49**, 403-411.

Festinger, L.: 1957, *A Theory of Cognitive Dissonance*, Stanford, Stanford University Press.

Heckman, J.J.: 1981, 'The incidental parameters problem and the problem of initial conditions in estimating a discrete time discrete data stochastic process', in C.F. Manski and D. McFadden (eds.), *Structural Analysis of Discrete Data with Econometric Applications*, Cambridge, Mass, MIT Press.

Heckman, J.J. and B. Singer: 1984a, 'A method for minimising the impact of distributional assumptions in econometric models of duration', *Econometrica* **52**, 271-320.

Heckman, J.J. and B. Singer: 1984b, 'The identifiability of the proportional hazards model', *Review of Economic Studies* **165**, 231-242.

Heckman, J.J. and R.J. Willis: 1977, 'A beta-logistic model for the analysis of sequential labor force participation by married women', *Journal of Political Economy* **85**, 27-58.

Huff, J. and W.A.V. Clark: 1978, 'Cumulative stress and cumulative inertia: a behavioural model of the decision to move', *Environment and Planning* **A 10**, 1101-1119.

Ineichen, B.: 1981, 'The housing decisions of young people', *British Journal of Sociology* **32**, 252-258.

Kalbfleisch, J.D. and R.L. Prentice: 1980, *The Statistical Analysis of Failure Time Data*, New York, Wiley.

Laird, N.: 1978, 'Non-parametric likelihood estimates of a mixing distribution', *Journal of the American Statistical Association* **73**, 805-811.

Lancaster, T. and S. Nickell: 1980, 'The analysis of re-employment probabilities for the unemployed', *Journal of the Royal Statistical Society* **A 143**, 141-165.

Lindsay, B.G.: 1983a, 'The geometry of mixture likelihoods: a general theory', *The Annals of Statistics* **11**, 86-94.

Lindsay, B.G.: 1983b, 'The geometry of mixture likelihoods, Part II: the exponential family', *The Annals of Statistics* **11**, 783-792.

Moss, W.G.: 1979, 'A note on individual choice models of migration', *Regional Science and Urban Economics* **9**, 333-343.
Murie, A.: 1974, 'Household movement and housing choice', Centre for Urban and Regional Studies, University of Birmingham, Occasional Paper 28.
Neyman, J. and E. Scott: 1948, 'Consistent estimates based on partially inconsistent observations', *Econometrica* **16**, 1-32.
Pickles, A.R.: 1985, *An Introduction to Likelihood*, CATMOG Series, Norwich, Geo Abstracts.
Pickles, A.R. and R.B. Davies: 1983, 'Recent developments in the analysis of movement and recurrent choice', in G. Gaile and C. Willmott (eds.), *Spatial Models and Statistics*, Amsterdam, Reidel.
Pickles, A.R. and R.B. Davies: 1985, 'The longitudinal analysis of housing careers', *Journal of Regional Science* **25**, 85-101.
Ridder, G.: 1984, 'The distribution of single spell duration data', Mimeo, University of Amsterdam.
Simar, L.: 1976, 'Maximum likelihood estimation of a compound Poisson process', *The Annals of Statistics* **4**, 1200-1209.
Tardiff, T.J.: 1979, 'Definition of alternatives and representation of dynamic behaviour in spatial choice models', *Transportation Research Record* **723**, 25-30.
Winship, C.: 1984, 'Age dependence, heterogeneity and the interdependence of life-cycle transitions', Mimeo, Department of Sociology, Northwestern University, Evanston IL.

DAVID A. HENSHER

AN ASSESSMENT OF ATTRITION IN A MULTI-WAVE PANEL OF HOUSEHOLDS

1. INTRODUCTION

Attrition is an issue of great importance in empirical analysis using longitudinal data in which measurements are taken at two or more points in time on the same sample of units. Although attrition per se need not be a problem, any bias due to loss of sample size can have a profound effect on the usefulness of the empirical outputs of the study. For example, if in the current context of predicting automobile energy consumption the households that are lost at each recontact point are typically high-kilometre households then parameter estimates associated with a study in which vehicle use is endogenous could be significantly biased. If, however, there is no difference in the distribution of kilometres between the total sample and the continuing respondents, but there are some differences with respect to exogenous variables (e.g. number of workers, income, household size), it is not necessarily the case that attrition is a source of bias given the objectives of the study.

In recent years the attrition problem has been interpreted as a missing data issue, enabling one to consider the extensive number of approaches used in handling missing data in general to assist in both identifying attrition bias and accommodating it in modelling. Much of this literature however is specialised to either systematic 'global' procedures concerned with the detection of non-response bias as identified by comparisons of aggregate measures of central tendency on pairs of waves, or ad hoc imputation procedures for individual items in each sample observation. What is missing is a *systematic* procedure for identifying attrition bias and correcting for it at the individual unit/item level. A number of researchers have recognised that Heckman's sample selectivity work (in a cross-section context), which showed that missing items for a subset of respondents can be viewed as a specification problem, provides a suitable framework for investigating attrition bias in panel data (Heckman, 1979; Maddala, 1979; Hausman and Wise, 1979; Hensher and Wrigley, 1986; Hensher, 1986b; Winer, 1983).

In this paper we introduce the attrition problem as an issue in sample selectivity, and relate it to the widely used correction procedures of sample weighting. An econometric model is specified which models the participation decision and links this with the two key behavioural models, namely vehicle use and vehicle possession, in order to test for attrition bias. Attrition bias is one form of selectivity bias and is treated herein as the only source, enabling us to use both phrases for the same phenomenon. Likewise a model of participation choice may be interpreted equivalently as a model of attrition 'choice'. The data are a four-wave panel of Sydney households, interviewed annually from 1981 through to 1985. The small number of non-continuers between waves two and three (25 out of 1276 or 1.96%) and waves three and four (9 out of 1206 or .75%) eliminated

J. Hauer et al. (eds.), Urban Dynamics and Spatial Choice Behaviour, 105–124.

the need to test for attrition bias; although we make an attempt to consider the wave 2-wave 3 situation.

2. A DISAGGREGATE PRE-ANALYSIS APPROACH FOR SHORT PANELS

A panel of households is typically drawn from a closed population at an initial sampling point and reinterviewed on a fixed time cycle (in our study it is annually). Some analysts are primarily interested in studying behaviour over time of a sample of households and are less concerned with the 'representativeness' of the participants at each point in time (in relation to the sample population at that time point which may or may not be significantly different to the initially sampled population). Other analysts are interested in maintaining sample representativeness such that the continuing panellists are equivalent 'to a new sample drawn at the appropriate time' (the repeated cross-section). In the former circumstance it is common for analysts to exclude households that do not complete the panel cycle, whereas in the latter situation panel drop out can be an important source of bias in modelling the *population's* behaviour. In each subsequent wave we are interested in identifying the role that non-participants who are still in the closed sample population have on both the univariate descriptors of the data (mean, standard deviation, range) and the parameter estimates of the behavioural models. This requires us to exclude from consideration households which move out of the sampled population, and to include a representative set of "new" households in the sample population applicable to the wave under study. The "new" households are assumed to be the product of household formation out of the existing sample of households, which are formed in the population with known selection probabilities (Duncan *et al*, 1984). For a short panel of five years (1981-1985) and an interest in a topic which is unlikely to involve significant change in the behavioural variables over such a horizon, other considerations associated with a changing composition through time of the population are not relevant.

Let us view non-response as the equivalent of missing data for an item and/or a unit of information, and define it as the difference between the mean for the response strata and the population parameter being estimated. Formally,

$$\bar{Y} = W_R \bar{Y}_R + W_{NR} \bar{Y}_{NR} \tag{1}$$

where $\bar{Y}_R$, $\bar{Y}_{NR}$ are means for the response (R) and non-response (NR) strata; and W_R, W_{NR} are the respective proportions of the population in these two strata.

A survey wave produces an estimate of $\bar{Y}_R$. The difference between $\bar{Y}_R$ and the population parameter being estimated, $\bar{Y}$, is

$$\bar{Y}_R - \bar{Y} = W_{NR}(\bar{Y}_R - \bar{Y}_{NR}) \quad (2)$$

Minimisation of non-response bias is consistent with keeping W_{NR} small and/or keeping $\bar{Y}_{NR}$ close to $\bar{Y}_R$.

Missing data due to panel attrition are unlikely to occur *completely at random* (*CAR*) since the means of the observed variables for responders in subsequent waves differ usually from the means of non-responders. While this very strong assumption that data are missing at random *and* observed at random is likely to be unacceptable in practice, an alternative weaker assumption is that data are missing at random (*MAR*): given particular levels of the observed variables, the levels missing on other variables are missing at random. This implies that responders and non-responders with the same characteristics on the observed variables do not differ systematically on other variables. This is the underlying sampling strategy of most attrition-correction studies, be they ad hoc or systematic.

The *MAR* assumption requires a minimum amount of information on nonresponders, which for short-panels (at least) is typically drawn from the preceding wave. This information is typically non-behavioural background variables (mainly socio-demographic and economic), denoted by the x-vector, which are assumed to be linked to a y-vector of behavioural variables (e.g. vehicle use) by the *MAR* assumption. Because y-variables are not observed for the non-responders, models which adopt the *MAR* assumption are referred to as *ignorable* models.

The *MAR* specification is problematic if panel drop out probability is strongly influenced by behavioural variables. For example define Y as the annual kilometres of household vehicles (*HKM*) and X as household income (*HINC*) in $Y = \beta X$. Then if relatively high *HINC* households are not included in the response set, the coefficient of *HINC* is still an unbiased estimate of β. This is because the probability of participation is influenced only by the x-variable. This is the *MAR* assumption where knowledge of Y is irrelevant (it can be ignored), given the *correct specification* of the relationship between Y and X. However, if the probability of participation is related to *HKM* and hence the unobserved influences (ε), the estimated coefficient of *HINC* may be biased. In this situation we cannot ignore the role of Y (via ε), and hence the *MAR* assumption is violated. What we have then are two roles for a missing data-correction procedure:

role 1: for an ignorable (*MAR*) situation, the inclusion of attrition correction is a means of providing protection against nonresponse bias introduced by the misspecification of the relationship between Y and X.

role 2: for a non-ignorable situation where the unobserved components of the participation choice model and the Y-imputation model are correlated, attrition correction is introduced to allow for this correlation.

Rosenbaum and Rubin (1983) have formally shown that the inverse of the participation probability provides the most general sample weight for participants, because stratification on the participation probability eliminates participation bias without stratifying on the complete set of x-variables which normally should define a cell domain. Cells are formed by grouping participation probabilities into sufficiently fine increments, forming adjustment cells according to the intervals selected, and allocating units to each cell. The limiting cell would contain one household. Thus the weight for participating units in a cell is either the inverse of the observed participation rate in that cell or the inverse of the observed participation probability. This weighting approach produces the same estimated domain means as those obtained by imputing adjustment cells for missing y-vales. The weighted sample size based on participants only is the same as the unweighted sample size for participants plus non-participants.

We can now propose a model schema for attrition. We will assume that the participation choice model is of the binary probit form and the behavioural model for vehicle use (or vehicle possession) is OLS or GLS. Alternative forms for the participation choice model are discussed in Hensher and Wrigley (1986).

The behavioural model is given in Equation (3)

$$y_q = \boldsymbol{\beta}\mathbf{X}_q + \xi_q, \qquad q = 1,\ldots,Q \text{ households} \tag{3}$$

Define a binary-valued index, R_q, equal to 1 if y_q is observed and equal to zero if y_q is unobserved; and assume that $R_q = 1$ iff

$$r = \alpha y_r + \mathbf{X}_r\boldsymbol{\gamma} + \mathbf{Z}_r\boldsymbol{\theta} + \varepsilon_r \geq 0 \tag{4}$$

where y_r is a scalar behavioural variable;

$\mathbf{X}_r$ is a vector of explanatory variables influencing y_q, referred to herein as household-specific effects;

$\mathbf{Z}_r$ is a vector of variables that affect participation but do not directly affect y_q (hence affect y_q's probability of being observed), referred to herein as context-specific and respondent-specific effects (see below);

α, $\boldsymbol{\beta}$, $\boldsymbol{\gamma}$, $\boldsymbol{\theta}$, are parameter sets;

ξ_q, ε_r are unobserved components assumed each to be normally distributed with zero mean and respective variances of σ^2_ξ and σ^2_ε.

The probability of a household dropping out of or continuing in a sample wave can be due to behavioural effects (e.g. vehicle use), contextual effects (e.g. length of previous interview, quality of interviewer, month of interview), respondent effects (e.g. age, personal income, marital status), and household effects (e.g. household income, dwelling tenure status). Households that move out of the sample population between waves must be excluded since their participation choice is assumed to be unrelated to these effects ('our study is not sufficiently influential to cause residential relocation'), and more importantly they are no longer relevant to the representativeness of the continuing participants.

If behavioural effects influence attrition, then Equation (4) can be rewritten as Equation (5).

$$r = \mathbf{X}_r(\alpha\boldsymbol{\beta} + \boldsymbol{\gamma}) + \mathbf{Z}_r\boldsymbol{\theta} + (\alpha\xi_r + \varepsilon_r) \tag{5}$$

$$= \mathbf{X}_r\boldsymbol{\tau} + \mathbf{Z}_r\boldsymbol{\theta} + \mu_r \tag{6}$$

The participants in the continuing sample wave are a conditional set, conditional on y_q being observed (i.e. $R_q = 1$); thus Equation (3) must be restated as Equation (7).

$$E(y_q|\mathbf{X}_r, R_q = 1) = \boldsymbol{\beta}\,\mathbf{X}_r + E(\boldsymbol{\xi}_r|\mathbf{X}, R_q = 1) \tag{7}$$

The unobserved component in (7) is censored because some (not necessarily a random subset) of the disturbances are no longer present due to attrition. Thus in recognising that some levels of y_q and hence ξ_q only exist if $r \geq 0$ (Equation 4), we can write

$$E(\boldsymbol{\xi}_r|\mathbf{X}_r, r \geq 0) = E(\boldsymbol{\xi}_r|\mathbf{X}_r, \mu \geq -\mathbf{X}_r\boldsymbol{\tau} - \mathbf{Z}_r\boldsymbol{\theta}) \tag{8}$$

Assuming a standard normal distribution and a probit functional form for the participation choice model, Johnson and Kotz (1970, 81; 1972, 112) show that the relationship in Equation (9) exists.

$$E(\boldsymbol{\xi}_r|\mathbf{X}_r, \mu \geq -\mathbf{X}_r\boldsymbol{\tau} - \mathbf{Z}_r\boldsymbol{\theta}) = \left[\frac{\operatorname{cov}(\boldsymbol{\xi}_r,\mu_r)}{\sigma^2_{\mu_r}}\right] E(\mu_r)$$

$$= \left[\frac{\operatorname{cov}(\boldsymbol{\xi}_r,\mu_r)}{\sigma_{\mu_r}}\right]\left[\frac{f\left(\dfrac{\mathbf{X}_r\boldsymbol{\tau} + \mathbf{Z}_r\boldsymbol{\theta}}{\sigma_{\mu_r}}\right)}{F\left(\dfrac{\mathbf{X}_r\boldsymbol{\tau} + \mathbf{Z}_r\boldsymbol{\theta}}{\sigma_{\mu_r}}\right)}\right] \tag{9}$$

$f(\cdot)$ is the standard normal density function and $F(\cdot)$ the standard normal cumulative distribution function.

The essence of attrition bias is a recognition that the covariance between the unobserved influences on y_r (i.e. $\boldsymbol{\xi}_r$) and the unobserved influences on attrition (i.e. μ) is non-zero. If $\operatorname{cov}(\boldsymbol{\xi}_r,\mu_r)$ equals zero, then the parameter estimates in the behavioural Equation (3) for the participant set are unbiased. An operational specification of the behavioural model can be given as

$$Y_r = \mathbf{X}_r\boldsymbol{\beta} + \zeta AC_r + \nu_r \tag{10}$$

where $\zeta = [\mathrm{cov}(\boldsymbol{\xi}_r,\mu_r)]/\sigma_{\mu_r}$

$$AC_r = f(\cdot)/F(\cdot)$$

$$\nu_r \sim N(0,\sigma^2_{\nu_r}) \text{ with } E(\nu_r,\mu_r) = 0$$

If ζ is statistically significant then we have evidence of attrition bias, and it is necessary to include the attrition correction, AC_r, in the behavioural model. Even if ζ is not statistically significant, its presence may affect the point estimates of parameters in $\boldsymbol{\beta}$, and thus should be retained (in essence to satisfy role 1 above). Equations (6) and (10) define the attrition models for pre-analysis of panel data.

3. EMPIRICAL PRE-ANALYSIS OF THE SYDNEY PANEL

The interest in panel attrition stems from the collection, since mid-1981, of a panel data set developed from a sample of Sydney households. Each household was reinterviewed at twelve month intervals up to a maximum of four occasions. Each household was interviewed by a trained interviewer in a face-to-face situation. The instrument was a structured survey form with the emphasis on details about the households automobiles (number, type, use profile, cost profile) for the 12-month period prior to the interview data. An extensive set of socio-economic data were also compiled for each household member (see Hensher, 1985, 1986a for further detail). In most interview contexts a single individual supplied the information, some of which required the *respondent* to ascertain it from particular members of the household prior to the interview. Data were obtained on contextual effects such as duration of interview, month of interview, status of respondent within the household, cooperation of respondent, respondent interest in the subject matter etc. Together with respondent-specific socioeconomic data and household-specific data we can identify the sources of influence on household participation. The loss of households may be due in large measure to the contextual effects and/or respondent effects. If these are the main influences on attrition and they do not affect the relationship between household-specific effects and the household level behavioural variables, it is likely that bias is not attributable to attrition. In assessing attrition effects, we specialise our inquiry to the two critical issues of the broader study for which the data were collected - household automobile use and household vehicle possession.

The steps in estimation for the 4-wave panel are summarised in Figure 1, with the sample sizes reported therein given in detail in Figure 2. The attrition rates for adjacent waves are respectively 8.2%, 1.95%, and 0.75%. The 'global' attrition rate is 10%, slightly

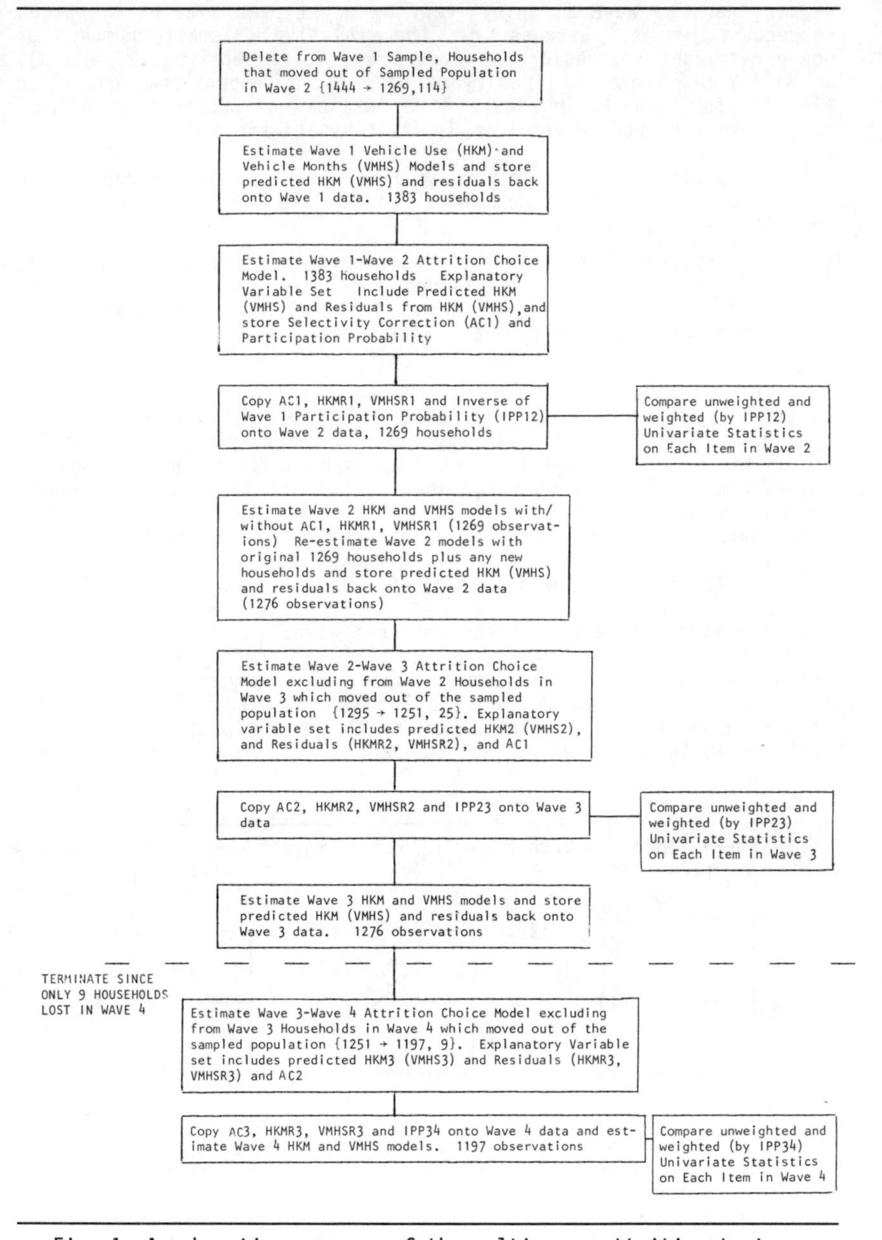

Fig. 1: A schematic sequence of the multi-wave attrition test.

higher than for wave 2. Intuitively we may expect some bias between subsequent waves. Because of the relatively small number of non-participant households in waves 3 and 4 (respectively 25 and 9), we will concentrate our inquiry on wave 2, but report some qualified findings for wave 3. In Figure 3 we outline the proposed structural relationship between waves 1 to 3. It is hypothesised that:

1. the probability of participating in wave 2 is influenced by the rate of vehicle use and level of vehicle possession in wave 1 $[\xi_{H1} \longrightarrow \mu_{12} \longleftarrow \xi_{V1}]$;
2. the expected level of vehicle use/vehicle possession in wave 2 is influenced by the unexplained rate of vehicle use/level of vehicle possession in wave 1(2) (i.e. correlated error structures) $[\xi_{H1} \longrightarrow \xi_{H2}; \xi_{V1} \longrightarrow \xi_{V2}]\cdot[\xi_{H2} \longrightarrow \xi_{H3}; \xi_{V2} \longrightarrow \xi_{V3}]$.

The assessment of attrition herein is confined to an interest in household-level behaviour, not that of individual members of each household. This is due to the emphasis in the sampling on households as unified decision units. Thus respondent effects are viewed as intervening influences on participation but not as direct influences on household behaviours such as vehicle use and possession. The behavioural models are specified on household-level variables.

3.1. *The Attrition or Participation Choice Models*

The attrition choice models for adjacent waves (1, 2) and (2, 3) are summarised in Table I. The final set of explanatory variables have been selected from an assessment of a large number of contextual, respondent and household variables[1]. The contextual and respondent effects have a dominant influence on the participation decision in contrast to the relatively few household-level variables. There is a

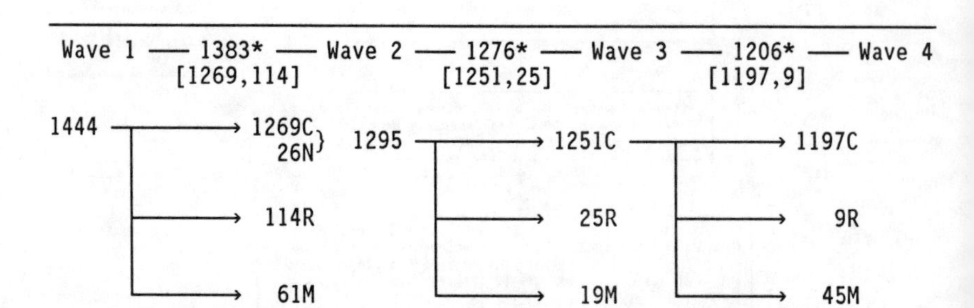

* Sample Size in Attrition Choice Model (Equation (6))
C = Continuing Sample, N = New Households, R = Refusals,
M = Moved out of Sampled Population

Fig. 2: A profile of the sample size across 4 waves.

noticeable absence of household socioeconomic effects, suggesting that the decision by a respondent on the household's continuation status tends not to have any biasing effect on the household-level socioeconomic profile of the sample. Interestingly, the respondent profile which is typically more likely to define a non-continuing household is a non-female head, non-possession of a driver's licence, low number of working hours and low employment skills. Retired male heads are an example of such respondents. This profile however only applies to the initial sample loss (wave 2); the loss in wave 3 is markedly different, as might be expected, and is characterised more by the marital status of the respondent. In particular, respondents with the marital status of "single" tended to have a higher than equal chance of discontinuing their household[2]. It is clear from Table I that respondent effects are very weak for wave 3.

The quality of the interviewer has a very important influence on the household's participation. The INTi dummy variables however also reflect the location of households; interviewer's A and PP had locations in wave 1 which were at the lower end of the socioeconomic status scale, but were selected for these areas because of their competence in communication with such households. The negative parameter estimates in a sense are the net effect of minimising attrition by allocating the "better" interviewers to difficult areas. The same logic applies in wave 2 to interviewer BL who was given the task of following up the initial refusal households to try and keep these households in the sample. The success of BL is reflected in a considerably lower attrition rate than otherwise would have eventuated. Interviewer F on the other hand was blessed with a middle-class area, however her especial talents of warmth and enthusiasm reflect a positive parameter estimate, which although not statistically significant is included as the most significant 'positive' interviewer effect. The implication is that interviewer effects are critical in influencing the attrition rate in locations where the socioeconomic composition is generally less cooperative. The

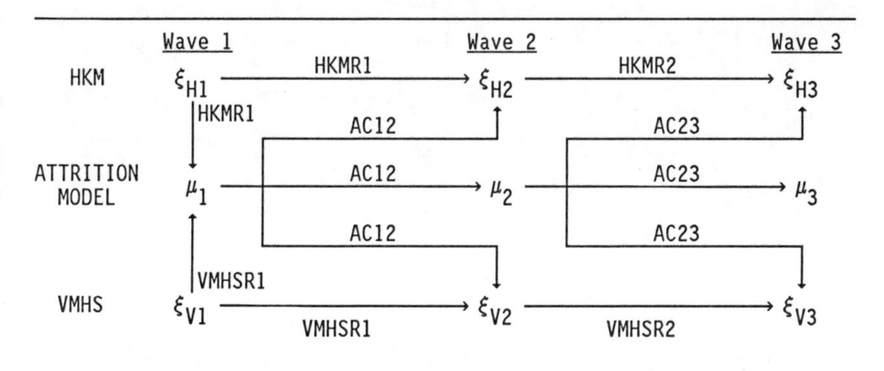

Fig. 3: Conditionality between unobserved components and attrition correction.

TABLE I
Attrition models - Wave 2 and Wave 3

Dependent variable: Probability of participating vs not participating in subsequent wave

Explanatory Variable	ACRONYM	WAVES (1, 2) Estimated Coefficient	(t-value)
CONTEXTUAL EFFECTS			
Interviewer A dummy	INT1	-0.5891	(-2.72)
Interviewer BL dummy	INT4	-	
Interviewer F dummy	INT6	-	
Interviewer JB dummy	INT8	-0.6765	(-1.77)
Interviewer PP dummy	INT11	-0.5799	(-3.44)
January interview dummy	MTHINT1	-0.2558	(-1.38)
February interview dummy	MTHINT2	-0.3168	(-2.34)
Detached residence dummy	DWELLNG1	-	
Semi-detached residence dummy	DWELLNG2	-	
RESPONDENT EFFECTS			
Respondent drivers licence dummy	DRVLIC	0.3827	(2.12)
Female head respondent dummy	ROLE2	0.2045	(1.66)
Income of respondent	INCMRES	-	
Respondent occupation status dummy :			
3539 = Labourer	ANU3539	-0.5148	(-2.20)
4044 = Tradesperson	ANU4044	-0.2568	(-1.26)
5054 = Technicians	ANU5054	-	
7579 = Professionals	ANU7579	-0.3843	(-1.56)
Respondent - married	MRSTAT1	-	
Respondent - single	MRSTAT2	-	
Respondent U.K. ethnic dummy	ETHNC2	-	
Respondent - Full-time hours	HRSFT	0.0059	(2.01)
HOUSEHOLD EFFECTS			
Household income ('00's)	HINCOME	-0.000010	(-0.02)
Rent residence dummy	RENTPLAC	-0.1929	(-1.21)
Household-business registered vehicle dummy	HBDUM	-	
Other-business registered vehicle dummy	OBDUM	-	
Extremely extrovert household dummy	HCHARA1	-	
Slightly extrovert household dummy	HCHARA2	-	
Total annual vehicle fuel costs	TFUELCS	-0.2361	(-2.85)
BEHAVIOURAL EFFECTS			
Predicted annual vehicle use	PHKM	0.000049	(3.28)
Residual annual vehicle use	HKMR1 (ξ_{H1})	0.0000081	(1.47)
Predicted annual vehicle months	PVMHS1	-0.0952	(-3.29)
Residual annual vehicle months	VMHSR1 (ξ_{V1})	-0.0349	(-2.14)
Attrition Correction Wave 12	AC1	-	
Intercept	CONSTANT	1.9671	(6.25)
Likelihood Ratio Index	ρ^2	.057	

AC1 = -0.9002 + 0.9665 (IPP12), R^2 = 0.88 (1269 observations)
(-80.93) (95.04)
IPP12 = inverse of probability of pariticpation in Wave 2

TABLE I (Continued)

ACRONYM	WAVES (2, 3)			
	With AC12 Estimated Coefficient (t-value)		Without AC12 Estimated Coefficient (t-value)	
INT1	-		-	
INT4	-0.9594	(-3.51)	-1.0102	(-3.77)
INT6	0.5613	(1.06)	0.5214	(1.00)
INT8	-		-	
INT11	-		-	
MTHINT1	-		-	
MTHINT2	-0.7515	(-3.36)	-0.8326	(-3.84)
DWELLNG1	0.5409	(1.86)	0.5052	(1.77)
DWELLNG2	0.2309	(0.45)	0.2093	(0.41)
DRVLIC	-		-	
ROLE2	-		-	
INCMRES	0.0948	(1.07)	0.1010	(1.15)
ANU3539	-		-	
ANU4044	-		-	
ANU5054	3.5437	(0.05)	3.9448	(0.03)
ANU7579	-		-	
MRSTAT1	-0.6929	(-1.55)	-0.6798	(-1.56)
MRSTAT2	-0.9621	(-2.01)	-0.9824	(-2.09)
ETHNC2	-0.3780	(-1.37)	-0.3557	(-1.30)
HRSFT	-		-	
HINCOME	-		-	
RENTPLAC	-0.1627	(-0.61)	-0.1320	(-0.50)
HBDUM	3.0481	(0.49)	3.5020	(0.29)
OBDUM	-0.0459	(-0.18)	-0.0431	(-0.17)
HCHARA1	-0.2333	(-0.49)	-0.2192	(-0.47)
HCHARA2	-0.3660	(-1.60)	-0.3840	(-1.69)
TFUELCS	-0.1629	(-1.98)	-0.1830	(-2.23)
PHKM	-		-	
HKMR1 (ξ_{H1})	-		-	
PVMHS1	-		-	
VMHSR1 (ξ_{V1})	-		-	
AC1	-1.3843	(-1.44)	-	
CONSTANT	3.1626	(5.47)	2.9855	(5.41)
ρ^2	.089		.087	

month-of-interview variables support the belief that interviewing during the main vacation period is likely to result in a higher incidence of refusal than at other periods.

The only household-variable which is consistently significant is annual fuel cost which has a mean of $973 (s.d. = 956) for wave 1 continuers and $1191 (s.d. = 1285) for wave 1 non-continuers. Other things equal, the propensity to drop out of the sample increases as the annual vehicle fuel bill increases. This helps to explain why in Table II the coefficient of unit fuel cost (*PTCSKM*) in the vehicle use model for wave 2 is the only household-level variable whose coefficient is changed significantly by the inclusion of selectivity. When *TFUELCS* is interpreted in relation to the behavioural effects it is tempting to suggest an inconsistency with the sign of the vehicle use variables *PHKM* and *HKMR*1 (ξ_{H1}). However, the link has to also consider the vehicle months variables (*PVMHS*, *VMHSR*1 (ξ_{V1})) which are significant and negative. When both vehicle use and vehicle months are considered in relation to fuel cost we find that there are clearly other influences on vehicle use than fuel cost which provide a counterbalancing positive participation to the negative participation effect of fuel cost.

The signs of *PHKM* and *PVMHS* provide evidence that positive correlation exists between household mobility and panel participation, a result also found by Kitamura and Bovy (1987), but a negative correlation between vehicle possession and participation. The signs of ξ_{H1} and ξ_{V1} further indicate that households which had higher than expected annual kilometres tended to continue into wave 2, whereas households with less than expected vehicle months tended to drop out of wave 2.

The attrition correction (*AC*1) derived from the wave (1, 2) attrition model has been included in the wave (2, 3) attrition model to test whether the unobserved propensity to participate in wave 2 (μ_1) is an influence on the propensity to participate in wave 3. The correction term is not strongly statistically significant, and its presence does not noticeably alter the magnitudes of the parameter estimates of the other variables. We can conclude from this that the very mild correlation between the unobserved propensity to participate in wave 2 and the propensity to participate in wave 3 supports the hypothesis that the reasons for non-participation are largely contemporaneous and linked to contextual and respondent effects. The noticeably absent link between attrition in wave 3 and the other behavioural effects in Table I further support this conclusion.

3.2. *The Behavioural Models*

The attrition correction term *ACW* together with the residual from the previous wave (ξ_{HW} or ξ_{VW}), and the contemporaneous set of explanatory variables are included in the estimation of the vehicle use and vehicle months models (Table II). Since the interest is on the possibility of attrition bias, models are estimated with and without the attrition correction term; and furthermore because of the added complexity of the correlation of errors between adjacent waves (with a one-way chronological structure) to enhance the model specification, we consider the models with and without the lagged residual term.

In both waves and both behavioural models the lagged residual terms (ξ_{H1}, ξ_{H2}, ξ_{V1}, ξ_{V2}) are positive and highly significant (t-statistics ranging from 8.75 to 32.3). As well these terms individually contribute non-marginally to the overall fit of the model, and notably so for the *VMHS* model in wave 2. As anticipated, the unexplained vehicle use (vehicle months) in wave W is strongly correlated with the unexplained vehicle use (vehicle months) in wave $W + 1$, and that excluding this effect is a source of misspecification.

The attrition correction term has a mixed impact. In the wave 2 vehicle use model it is not statistically significant (t value = -0.66) in the absence of ξ_{H1} and still retains non-significance but to a lesser extent in the presence of ξ_{H1} (t = -1.67). The standard errors have been corrected for the inter-temporality of the attrition term and the behavioural model, but not adjusted for the loss of efficiency due to the 'errors-in-variables' property of an index variable (as is *ACW*) which is derived from sample estimates of parameters. Thus the t-values should be interpreted as upward biased estimates. In the wave 2 *VMHS* model, the correction term is statistically significant (and negative) in the absence of ξ_{V1} but becomes nonsignificant (and positive) in the presence of ξ_{V1}. The sensitivity of the correction term to the lagged residuals must be expected, especially in wave 2, since these residuals are a significant influence on the participation probability (and hence μ_W).

In wave 3, the attrition correction term is consistently nonsignificant, maintaining its sign in the presence and absence of the lagged residual, but changing dramatically in magnitude. As anticipated there is no evidence of attrition bias in wave 3 and if we assume that the correct specification of the wave 2 models includes the lagged residual term, the same conclusion applies to wave 2. Any significant impact of *SCW* on the parameter estimates of the other contemporaneous explanatory variables is limited to the set of statistically nonsignificant effects (e.g. *LOCNA*, *LOCNB*, *LOCND*, *NUNI*, *LIFCYSB*, *LIFCYSD* in *VMHS*2; *RENTPLAC* in *HKM*3; and *LOCNG* in *VMHS*3). Thus the evidence suggests that the bias due to attrition can be rejected but that failure to recognise and allow for chronological correlated errors between behavioural models is a source of specification bias.

The evidence on the general absence of attrition is confirmed by a comparison of the univariate measures of central tendency for the unweighted and weighted variable sets in both waves 2 and 3 (Table III). The results are virtually indistinguishable. The sample weight applied to each continuing observation is the inverse of the participation probability. The mean of the 'attrition weight' is 1.09 (s.d. = .09) for wave 2 and 1.020 (s.d. = .050) for wave 3, which reinstate the sample size 'as if all units had continued in the panel'.

TABLE II
Behavioural models for vehicle use and vehicle possession (MLE)

Dep. Variables: HKMW = annual household vehicle kilometres in Wave W
VMHSW = annual household vehicle months in Wave W

Explanatory Variable	ACRONYM	WAVE 2 (1269)							
		HKM 2				VMHS 2			
		Base	Base + AC 1	Base + HKMR 1	Base + AC 1 + HKMR 1	Base	Base + AC 1	Base + VMHSR 1	Base + AC 1 + VMHSR 1
		$\hat{\beta}$ (t)	$\hat{\beta}$ (t)	$\hat{\beta}$ (t)	$\hat{\beta}$ (t)	$\hat{\beta}$ (t)	$\hat{\beta}$ (t)	$\hat{\beta}$ (t)	$\hat{\beta}$ (t)
Residential-location specific dummy variables:									
North-East (Manly-Warringah)	LOCNA	5144.6	4969.0	5169.8	4773.0	-0.4166	-0.9918	-0.4925	-0.3083
		(4.11)	(3.77)	(4.60)	(4.20)	(-0.52)	(-1.19)	(-0.85)	(-0.51)
North Shore	LOCNB	-	-	-	-	0.5065	0.0315	0.2683	0.3880
						(0.71)	(0.50)	(0.68)	(0.96)
North West (Hills District)	LOCNC	5971.1	6052.6	5616.0	5797.5	4.6467	4.5924	4.5465	4.5634
		(3.97)	(3.17)	(3.23)	(3.33)	(5.03)	(5.30)	(7.14)	(7.20)
West, (Penrith, Blacktown)	LOCND	6437.4	6203.5	7078.0	6554.5	0.5757	-0.1124	0.3078	0.5272
		(4.71)	(3.82)	(4.88)	(4.41)	(0.68)	(-0.12)	(0.52)	(0.86)
South West (Liverpool, Campbelltown)	LOCNE	2719.1	2896.1	3046.5	3449.1	2.3319	2.5181	1.9560	1.8944
		(1.63)	(1.83)	(2.17)	(2.41)	(2.29)	(2.94)	(3.21)	(3.08)
South (Sutherland)	LOCNF	5143.1	5423.9	5438.7	6074.8	3.6436	3.9841	3.6986	3.5902
		(1.80)	(2.04)	(2.07)	(2.28)	(2.17)	(2.28)	(3.45)	(3.35)
Central-Eastern Suburbs	LOCNG	-	-	-	-	-	-	-	-
Technical College education at least one household member	TEC	-	-	-	-	-	-	-	-
Household income ('00's)	HINCOME	7.9461	8.0722	7.2546	7.5342	0.0110	0.0112	0.0089	0.0088
		(2.49)	(2.19)	(2.25)	(2.32)	(6.03)	(5.84)	(6.72)	(6.60)
Number of full-time workers	NFTW	2492.2	2487.2	2476.9	2608.4	2.6660	2.7430	2.6409	2.6161
		(4.06)	(3.77)	(4.24)	(4.48)	(7.45)	(6.30)	(9.74)	(9.65)
Number of part-time workers	NPTW	1870.3	1900.3	1952.3	2020.6	1.9832	2.0251	2.1555	2.1429
		(2.17)	(2.16)	(2.48)	(2.56)	(3.97)	(3.39)	(5.15)	(5.11)
Rent residence dummy	RENTPLAC	3169.6	3097.3	3014.2	2849.4	-2.1806	-2.2431	-1.9138	-1.8925
		(3.02)	(2.50)	(2.75)	(2.62)	(-3.38)	(-3.41)	(-4.32)	(-4.27)

TABLE II (Continued)

Explanatory Variable	ACRONYM	WAVE 3 (1251)							
		HKM 3				VMHS 3			
		Base	Base + AC 2	Base + HKMR 2	Base + AC 2 + HKMR 2	Base	Base + AC 2	Base + VMHSR2	Base + AC 2 + VMHSR2
		$\hat{\beta}$ (t)	$\hat{\beta}$ (t)	$\hat{\beta}$ (t)	$\hat{\beta}$ (t)	$\hat{\beta}$ (t)	$\hat{\beta}$ (t)	$\hat{\beta}$ (t)	$\hat{\beta}$ (t)
Residential-location specific dummy variables:									
North-East (Manly-Warringah)	LOCNA	3327.2	3404.1	3391.9	3412.6	-	-	-	-
		(2.72)	(2.81)	(3.17)	(3.18)				
North Shore	LOCNB	-	-	-	-	-	-	-	-
North West (Hills District)	LOCNC	6710.5	6710.4	7209.0	7208.7	3.4924	3.4912	3.4313	3.4309
		(4.47)	(3.45)	(4.69)	(4.69)	(2.79)	(3.93)	(4.38)	(4.37)
West, (Penrith, Blacktown)	LOCND	6088.9	6139.2	6170.6	6184.1	1.6676	1.6391	1.6991	1.6827
		(4.65)	(4.34)	(4.77)	(4.74)	(1.52)	(1.24)	(1.30)	(1.30)
South West (Liverpool, Campbelltown)	LOCNE	5199.9	5216.6	5100.4	5105.0	-	-	-	-
		(3.36)	(3.78)	(4.16)	(4.16)				
South (Sutherland)	LOCNF	4616.7	4698.9	4502.0	4524.3	3.6122	3.5719	3.5792	3.5233
		(1.58)	(2.11)	(1.91)	(1.92)	(1.49)	(2.22)	(2.71)	(2.71)
Central-Eastern Suburbs	LOCNG	-	-	-	-	-	-	-	-
Technical College education at least one household member	TEC	2255.9	2286.3	2532.9	2540.9	-0.0557	-0.0644	-0.0179	-0.0215
		(2.91)	(2.86)	(3.54)	(3.58)	(-0.07)	(-0.08)	(-0.02)	(-0.30)
Household income ('00's)	HINCOME	6.6626	6.6834	8.8208	8.8251	0.0095	0.0095	0.0084	0.0083
		(2.35)	(2.30)	(3.29)	(3.33)	(4.06)	(4.63)	(4.22)	(4.23)
Number of full-time workers	NFTW	4392.1	4352.1	3948.5	3938.1	2.8435	2.8660	2.8584	2.8673
		(7.17)	(5.77)	(6.44)	(6.36)	(5.63)	(6.29)	(6.90)	(6.89)
Number of part-time workers	NPTW	2500.6	2535.6	2258.8	2268.5	2.4139	2.3900	2.4734	2.4639
		(2.77)	(2.82)	(2.71)	(2.73)	(3.24)	(2.57)	(3.01)	(2.99)
Rent residence dummy	RENTPLAC	-69.29	-297.52	-807.3	-868.28	-1.9414	-1.8078	-1.9299	-1.8769
		(-0.06)	(-0.23)	(-0.74)	(-0.80)	(-2.17)	(-1.56)	(-1.87)	(-1.71)

TABLE II (Continued)

Explanatory Variable	ACRONYM	WAVE 2 (1269)							
		HKM 2				VMHS 2			
		Base	Base + AC 1	Base + HKMR 1	Base + AC 1 + HKMR 1	Base	Base + AC 1	Base + VMHSR 1	Base + AC 1 + VMHSR 1
		$\hat{\beta}$ (t)	$\hat{\beta}$ (t)	$\hat{\beta}$ (t)	$\hat{\beta}$ (t)	$\hat{\beta}$ (t)	$\hat{\beta}$ (t)	$\hat{\beta}$ (t)	$\hat{\beta}$ (t)
Lifecycle stage dummy variables:									
Young adults, no children	LIFCYSA	-	-	-	-	-1.8016	-1.9710	-0.7653	-0.7063
						(-1.68)	(-1.75)	(-1.03)	(-0.94)
Two heads, preschool children	LIFCYSB	2511.9	2399.1	2236.9	1979.7	0.9150	0.7273	1.4293	1.4921
		(1.95)	(1.78)	(1.71)	(1.51)	(0.89)	(0.80)	(2.40)	(2.49)
Two heads, preschool & young children	LIFCYSC	1637.9	1565.5	1377.9	1211.9	1.7992	1.7373	2.1553	2.1768
		(1.30)	(1.45)	(1.36)	(1.19)	(1.78)	(1.90)	(3.52)	(3.56)
Two heads, young school children	LIFCYSD	2498.3	2415.5	2071.6	1881.1	0.9870	0.9320	1.7544	1.7756
		(2.23)	(1.83)	(1.78)	(1.62)	(1.01)	(0.95)	(2.92)	(2.97)
Two heads, older school children	LIFCYSE	2667.3	2595.9	1904.9	1737.3	1.7406	1.7557	2.4532	2.4517
		(1.89)	(1.54)	(1.27)	(1.16)	(1.51)	(1.45)	(3.41)	(3.42)
One or two heads, all ≥ 16	LIFCYSF	-	-	-	-	4.9540	5.3837	5.7709	5.6369
						(4.53)	(4.31)	(7.81)	(7.63)
Older adults, ≥65 <65, no children	LIFCYSG	-	-	-	-	-0.0991	-0.0778	1.0055	1.0037
						(-0.10)	(-0.09)	(2.06)	(2.07)
Cost per vehicle km (petrol) cents	PTCSKM	-40.36	-11.36	-212.36	-148.19	-	-	-	-
		(-0.16)	(0.04)	(-0.81)	(-0.56)				
Annual vehicle months	VMHS	1077.8	1075.7		1075.8	-	-	-	-
		(22.8)	(18.8)		(20.5)				
Household lifestyle 2nd quartile	LSH2650	-	-	-	-	1.8919	1.8425	1.4431	1.4568
						(5.09)	(4.12)	(4.64)	(4.70)
Community lifestyle 2nd quartile	LSC2650	-	-	-	-	0.1765	0.1492	0.2261	0.2350
						(0.74)	(0.58)	(1.38)	(1.44)
Residual vehicle months (previous wave)	VMHSRW	-	-	-	-	-	-	0.7697	5.8822
								(32.2)	(32.04)
Residual vehicle use (previous wave)	HKMRW	-	-	0.3937	0.3969	-	-	-	-
				(8.75)	(8.96)				
Attrition Correction	ACW	-	-3392.2	-	-7667.8	-	-7.4333	-	2.3848
			(-0.66)		(-1.67)		(-2.15)		(1.22)
Intercept	CONSTANT	13681.9	-5032.5	-4759.6	-3817.3	9.6633	10.970	9.8776	9.4595
		(50.4)	(-2.85)	(-3.49)	(-2.51)	(10.9)	(11.42)	(19.34)	(15.65)
R-squared	R^2	0.475	0.475	0.559	0.561	0.359	0.363	0.704	0.704

TABLE II (Continued)

Explanatory Variable	ACRONYM	WAVE 3 (1251)							
		HKM 3				VMHS 3			
		Base	Base + AC 2	Base + HKMR 2	Base + AC 2 + HKMR 2	Base	Base + AC 2	Base + VMHSR2	Base + AC 2 + VMHSR2
		$\hat{B}$ (t)	$\hat{B}$ (t)	$\hat{B}$ (t)	$\hat{B}$ (t)	$\hat{B}$ (t)	$\hat{B}$ (t)	$\hat{B}$ (t)	$\hat{B}$ (t)
Lifecycle stage dummy variables:									
Young adults, no children	LIFCYSA	-	-	-	-	-	-	-	-
Two heads, preschool children	LIFCYSB	-	-	-	-	3.6819	3.6715	3.5714	3.5674
						(3.30)	(2.63)	(2.63)	(2.64)
Two heads, preschool & young children	LIFCYSC	3568.3	3540.3	3471.3	3463.8	2.3365	2.3372	2.3353	2.3356
		(2.77)	(3.28)	(3.29)	(3.28)	(2.21)	(2.58)	(2.83)	(2.84)
Two heads, young school children	LIFCYSD	5142.96	5107.7	4531.3	4522.3	2.4968	2.5042	2.5099	2.5129
		(4.26)	(3.67)	(3.81)	(3.82)	(2.49)	(2.64)	(2.95)	(2.95)
Two heads, older school children	LIFCYSE	4670.6	4654.4	4801.9	4797.5	4.1412	4.1386	3.9835	3.9826
		(3.04)	(2.80)	(3.35)	(3.36)	(3.32)	(3.09)	(3.20)	(3.21)
One or two heads, all ≥ 16	LIFCYSF	5261.3	5172.9	5300.4	5276.6	6.2630	6.3032	6.3496	6.3656
		(3.88)	(3.56)	(3.93)	(3.92)	(5.58)	(5.65)	(6.43)	(6.43)
Older adults, ≥65 <65, no children	LIFCYSG	-51.945	-103.45	-366.9	-380.63	-	-	-	-
		(-0.04)	(-0.11)	(-0.46)	(-0.48)				
Cost per vehicle km (petrol) cents	PTCSKM	-516.98	-524.30	-420.5	-422.57	-	-	-	-
		(-2.84)	(-2.85)	(-2.60)	(-2.65)				
Annual vehicle months	VMHS	516.4	517.04	500.9	510.13	-	-	-	-
		(15.14)	(6.35)	(6.27)	(6.29)				
Household lifestyle 2nd quartile	LSH2650	-	-	-	-	0.8608	0.8553	0.9558	0.9536
						(1.56)	(1.58)	(1.94)	(1.94)
Community lifestyle 2nd quartile	LSC2650	-	-	-	-	-	-	-	-
Residual vehicle months (previous wave)	VMHSRW	-	-	-	-	-	-	-	0.7852
									(12.47)
Residual vehicle use (previous wave)	HKMRW	-	-	0.508	0.5077	-	-	0.7856	-
				(10.40)	(10.40)			(12.50)	
Attrition Correction	ACW	-	5551.6	-	1494.2	-	-3.3603	-	-1.3356
			(0.58)		(0.22)		(-0.81)		(-0.35)
Intercept	CONSTANT	1638.77	1525.6	1522.1	1491.78	10.0919	10.1785	10.4049	10.4391
		(1.14)	(0.94)	(1.09)	(1.06)	(12.25)	(14.52)	(15.91)	(15.91)
R-squared	R^2	0.425	0.425	0.552	0.552	0.228	0.228	0.341	0.341

TABLE III
Univariate sample statistics for weighted and unweigted waves
(mean and st. dev.; only for those variables cited in Tables I and II)

ACRONYM (See Tables1, 2 for definition)	WAVE 2				WAVE 3			
	UNWEIGHTED (1269)		WEIGHTED (1383)		UNWEIGHTED (1251)		WEIGHTED (1276)	
INT 1	.064	(.245)	.066	(.248)	.061	(.239)	.061	(.239)
INT 4	.069	(.253)	.069	(.253)	.085	(.279)	.088	(.283)
INT 6	.091	(.288)	.092	(.289)	.086	(.280)	.084	(.278)
INT 8	.011	(.104)	.012	(.110)	.011	(.105)	.011	(.105)
MTHINT 1	.086	(.280)	.087	(.281)	.064	(.245)	.064	(.245)
MTHINT 2	.162	(.369)	.167	(.373)	.149	(.356)	.153	(.360)
DWELLING 1	.847	(.360)	.850	(.358)	.851	(.356)	.848	(.359)
DWELLING 2	.032	(.177)	.032	(.176)	.034	(.180)	.034	(.181)
HCHARA 1	.052	(.222)	.051	(.220)	.047	(.212)	.047	(.212)
HCHARA 2	.312	(.464)	.312	(.463)	.306	(.461)	.307	(.461)
DRVLIC	.942	(.233)	.940	(.237)	.945	(.229)	.944	(.231)
ROLE 2	.463	(.499)	.460	(.499)	.462	(.499)	.461	(.499)
INCRMES	14831	(12341)	14808	(12348)	16063	(13142)	16015	(13088)
ANU 3539	.020	(.139)	.020	(.142)	.018	(.135)	.018	(.134)
ANU 4044	.102	(.303)	.104	(.305)	.090	(.266)	.091	(.288)
ANU 5054	.089	(.285)	.089	(.285)	.084	(.277)	.083	(.276)
ANU 7579	.032	(.177)	.033	(.178)	.029	(.168)	.029	(.168)
MRSTAT 1	.739	(.439)	.739	(.439)	.742	(.438)	.742	(.438)
MRSTAT 2	.137	(.344)	.138	(.345)	.123	(.328)	.124	(.330)
ETHNC 2	.122	(.328)	.121	(.326)	.118	(.323)	.119	(.324)
HRFST	21.71	(21.09)	21.66	(21.08)	21.91	(21.24)	21.90	(21.23)
LOCNA	.112	(.315)	.108	(.331)	.115	(.319)	.115	(.319)
LOCNB	.358	(.480)	.354	(.478)	.352	(.478)	.353	(.478)
LOCNC	.077	(.267)	.079	(.269)	.074	(.261)	.073	(.261)
LOCND	.099	(.298)	.095	(.293)	.102	(.303)	.102	(.303)
LOCNE	.065	(.246)	.066	(.249)	.071	(.257)	.071	(.257)
LOCNF	.019	(.136)	.020	(.141)	.018	(.131)	.017	(.131)
LOCNG	.271	(.455)	.278	(.448)	.269	(.443)	.269	(.443)
LIFCYSA	.119	(.324)	.117	(.321)	.102	(.303)	.103	(.304)
LIFCYSB	.118	(.323)	.116	(.320)	.107	(.309)	.107	(.309)
LIFCYSC	.122	(.328)	.121	(.326)	.125	(.331)	.125	(.331)
LIFCYSD	.163	(.370)	.162	(.369)	.159	(.366)	.159	(.366)
LIFCYSE	.102	(.302)	.103	(.304)	.099	(.299)	.099	(.299)
LIFCYSF	.132	(.338)	.138	(.345)	.144	(.351)	.145	(.352)
LIFCYSG	.153	(.360)	.152	(.360)	.161	(.367)	.161	(.367)
HINCOME	31860	(17685)	32029	(17783)	33100	(18707)	33134	(18699)
NFTW	1.377	(.923)	1.389	(.935)	1.326	(.927)	1.330	(.931)
NPTW	.224	(.472)	.222	(.471)	.205	(.436)	.203	(.435)
RENTPLAC	.181	(.385)	.179	(.384)	.160	(.367)	.163	(.370)
NUNI	.640	(.813)	.634	(.811)	.651	(.839)	.651	(.840)
LSH 2650	.284	(.641)	.284	(.641)	.278	(.610)	.279	(.611)
LSC 2650	.870	(.973)	.870	(.977)	.852	(.980)	.854	(.983)
HBDUM	.156	(.363)	.155	(.362)	.168	(.374)	.166	(.372)
OBDUM	.149	(.256)	.149	(.356)	.129	(.335)	.129	(.355)
TFUELCS	1066	(1017)	1080	(1061)	1127	(1041)	1135	(1075)
VMHS	19.60	(9.88)	19.67	(9.98)	20.30	(12.65)	20.30	(12.64)
HKM	24866	(18889)	24959	(19020)	23512	(17662)	23563	(17755)

4. CONCLUSION

The econometric approach outlined and applied herein to a multi-wave panel data set is an important pre-analysis procedure. Since it uses standard econometric models but combines them in an interesting way to assess the influence of missing data on sample representativeness, all users of panel data are encouraged to implement these procedures. We would add that since many panel data sets are multi-purpose, it is necessary for each analyst to determine the existence of attrition bias with respect to their particular objectives as embodied in the appropriate behavioural equations.

In the current context of the application of a short panel to predict household vehicle use and possession, there is no evidence of attrition bias; however there is very strong evidence to support a model specification which accommodates the correlated error structure between time periods. This is an adjunct justification for panel data per se, and will be built on in the estimation of dynamic vehicle use and possession models in the main analysis phase.

If panels such as that considered herein were continued over a longer period (e.g. resurvey on a 5-year cycle) it is likely that attrition bias will become an even more critical issue. The practical capability of producing suitable sample weights will then reinforce the value of the pre-analysis approach outlined in this paper.

ACKNOWLEDGEMENT

This paper is one in a series prepared for a study on the dimensions of automobile demand, supported financially by The Australian National Energy Research Development and Demonstration Program and a Macquarie University Research Grant. I am indebted to Nora Bodkin, Frank Milthorpe, Nariida Smith, Ryuichi Kitamura, Peter Barnard and a referee for comments.

NOTES

[1] The set of contextual effects investigated were: interviewer, number of interviews per interviewer assigned to a unit, day, month and year of interview, duration of interview, days difference between due date and interview date, nature of residential dwelling, change of interviewer between waves, number of interviewer changes, land use in street, conditions of buildings, grounds and roads in street, number of household participants in the interview, level of cooperation during interview, character and outlook of household as perceived by interviewer.

The set of respondent effects were: role in household, age, drivers licence holder, hours worked full/part time, occupation status, retired/home duties, personal income, education level attained, ethnic background, flexibility of work schedule, life style in relation to work, household, community, leisure. Household effects investigated were: number of vehicle months, business-registered

vehicles in mix, household size, income, annual fuel costs, number of full-time and part-time workers, tenure status, annual kilometres, number of income recipients, stage in lifecycle, location of residence, lifestyle of household, number of students.
[2] In this situation we suspect that the problem is linked to the interviewer tying the active respondent down to a suitable interview time, and pressure from other household members who failed to cooperate when the interviewer made the initial recontact telephone call and found the respondent 'not at home'.

School of Economic and Financial Studies
Macquarie University
Australia

BIBLIOGRAPHY

Duncan, G.J., D.H. Hill, and M. Ponzo: 1984, 'How representative is the PSID?: A response to some questions raised in the UNICON report', Survey Research Center, University of Michigan (mimeo).

Hausman, J.A. and D.A. Wise: 1979, 'Attrition bias in experimental and panel data: the Gary income experiment', *Econometrica* **47**, 455-473.

Heckman, J.J.: 1979, 'Sample selection bias as a specification error', *Econometrica* **47**, 153-161.

Hensher, D.A.: 1985, 'Longitudinal surveys in transport: an assessment', in E.S. Ampt, E.J. Richardson, and W. Brog (eds.), *New Survey Methods in Transport*, VNU Science Press, Utrecht.

Hensher, D.A.: 1986a, 'Dimensions of automobile demand: an overview of an Australian research project', *Environment and Planning* **A 18**, 1339-1374.

Hensher, D.A.: 1986b, 'Issues in the pre-analysis of panel data', *Transportation Research* **A 21**, 265-286.

Hensher, D.A. and N. Wrigley: 1986, 'Statistical modelling of discrete choices in discrete time with panel data', in Ministry of Transport and Public Works (ed.) *Behavioural Research for Transport Policy*, VNU Science Press, Utrecht.

Johnson, N.L. and S. Kotz: 1970, *Distribution in Statistics: Continuous Univariate Distributions - - 1*, John Wiley and Sons, New York.

Johnson, N.L. and S. Kotz: 1972, *Distribution in Statistics: Continuous Multivariate Distributions*, John Wiley and Sons, New York.

Kitamura, R. and P.H.L. Bovy: 1987, 'Analysis of attrition biases and trip reporting errors for panel data', *Transportation Research* **A 21**, 287-302.

Maddala, G.S.: 1979, 'Selectivity problems in longitudinal data', *Annales de L'Insee* **30-31**, 423-450.

Rosenbaum, P.R. and D.E. Rubin: 1983, 'The central role of the propensity score in observational studies for causal effects', *Biometrika* **70**, 41-55.

Winer, R.S.: 1983, 'Attrition bias in econometric models estimated with panel data', *Journal of Marketing Research* **20**, 177-186.

PETER NIJKAMP AND AURA REGGIANI

SPATIAL INTERACTION AND DISCRETE CHOICE: STATICS AND DYNAMICS

1. INTRODUCTION

Interaction analysis has become an important topic in social science research. Especially in spatial interaction analysis - dealing with flows of commodities, persons, information etc. between regions in a spatial system - much attention has been devoted to models incorporating the push and pull effects as well as the distance friction effects between regions. In this context, the traditional gravity model has become a very popular analytical tool. From the seventies onwards it has increasingly been realized that entropy theory - originating from statistical mechanics or from information theory - might provide a new foundation for the use of the gravity model (see, for instance, Wilson, 1970; Batten, 1983; Haynes and Fotheringham, 1984).

In this respect two important methodological questions arose, viz. the *macro-behavioural interpretation* of the entropy model (e.g. in terms of social utility) and the *micro-behavioural basis* of the (aggregate) entropy model (see also Reggiani, 1985).

The first question has been dealt with in various ways in the literature. It has been shown that the dual specification of the entropy model (implying a geometric programming model) leads to a certain (aggregate) generalized (non-linear) cost function to be minimized subject to some plausible constraints on spatial push and pull effects (see, for instance, Nijkamp, 1975, 1979; Jefferson and Scott, 1977; Charnes *et al.* 1977). Another interpretation was based on a reformulation of the entropy model by showing that the transportation model of linear programming is a special limit case, so that the entropy approach could also be justified as a general formulation of a conventional linear assignment model related to activity analysis (see, for instance, Evans, 1973; Coelho and Wilson, 1977; Erlander, 1977).

The second question pertains to the correspondence between the (aggregate) spatial interaction model emerging from entropy theory and micro-behavioural utility theory (especially discrete probabilistic choice theory).

The present paper aims at exploring the second question, viz. the extent to which spatial interaction models are in agreement with discrete utility theory (especially the multinomial logit model). Section 2 of this paper will give a brief overview of results from the recent literature and will also point out some unsatisfactory elements in these results. Then Section 3 will deal with a more general class of spatial interaction analyses based on Alonso's general theory of movement (see Alonso, 1978), and it will be shown that it is possible to derive an analytical correspondence between the conventional Alonso (macro) model and a (micro) discrete choice model. Next, the paper will develop an optimal control approach in order to interpret a dynamic spatial interaction model (based either on entropy theory or

J. Hauer et al. (eds.), Urban Dynamics and Spatial Choice Behaviour, 125–151.

on Alonso's general theory) in terms of a dynamic disaggregate choice model. It will be shown that here indeed under certain conditions a correspondence between (macro) spatial interaction models and (micro) discrete choice models can be found. Thus it is concluded that the correspondence between discrete choice models and spatial interaction models appears to hold for both static and dynamic spatial choice analysis. The structure of this paper is shown in Figure 1.

2. RELATIONSHIPS BETWEEN STATIC SPATIAL INTERACTION AND LOGIT MODELS

Discrete choice models - especially of the Multinomial Logit (MNL) type - have in the past decade become increasingly popular in many research areas (marketing, transportation, migration etc.). In the area of regional and urban research a wide variety of applications can be found, inter alia in residential choice analysis, modal split analysis, route choice models, job search analysis, and so forth.

The increasing popularity of such disaggregate models of choice evokes also important methodological questions: what is the relationship between discrete spatial choice models and (aggregate) spatial interaction (SI) models of the entropy (or gravity) type? Does the mathematical specification of the underlying utility function or the statistical properties of the error terms of a probabilistic discrete choice model exert a significant impact on the consistency of results from aggregate or disaggregate choice models?

In this section, a brief overview of various results in the recent literature will be presented. It will be demonstrated here that - despite the commonly accepted views on the consistency of MNL models and SI models - some weaknesses have not yet been solved and that there are still some open research questions.

Van Lierop and Nijkamp (1979) have tried to analyze the relationships and similarities between SI models and MNL models by using some

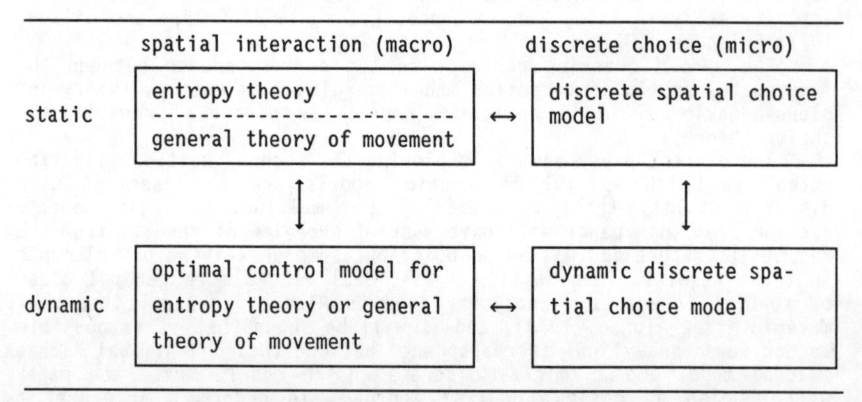

Fig. 1: Typology of linkages between spatial interaction and discrete choice models.

results from information theory. It is well known that entropy models (which may be regarded as the foundation of SI models) can easily be interpreted in terms of statistical information theory (see Batten, 1983). Assume a binary choice situation with p_i the probability that alternative i will be chosen. Then the standard expression for the entropy E is in the context of information theory:

$$E = - p_i \ln p_i - (1-p_i) \ln(1-p_i) \tag{1}$$

Now the first-order derivative of (1) with respect to p_i can be written as:

$$\frac{\partial E}{\partial p_i} = -\ln \left(\frac{p_i}{1-p_i}\right) \tag{2}$$

This is a standard expression for (minus) the logit (see Theil, 1972), so that the logit measures - in information-theoretic terms - the sensitivity of uncertainty (implied by the entropy) for a variation in the probability of choosing an alternative. In a more general sense, one may interpret this result as follows: entropy measures the probability of occurrence of a macro state of a system which may have a great many configurations depending on the individual events; the logit - being the first-order derivative of the entropy - reflects the marginal change in this entropy due to changes in individual choice probabilities. This interpretation suggests at least a certain correspondence between (discrete) MNL models and (aggregate) SI models (based on entropy).

A more rigorous analysis of relationships between MNL models and SI models has been provided by Coelho (1977) and Coelho and Williams (1978). These authors started off from the standard expression for entropy in (either a singly or a doubly constrained) SI model and were able to show that the results were in agreement with the standard expression for an MNL model.

The standard expression for the doubly constrained entropy model is:

$$\left.\begin{aligned}
&\max \omega = - \sum_{i=1}^{I} \sum_{j=1}^{J} T_{ij}(\ln T_{ij} - 1) \\
&\text{subject to} \\
&\sum_{j=1}^{J} T_{ij} = O_i, \qquad \forall\, i \\
&\sum_{i=1}^{I} T_{ij} = D_j, \qquad \forall\, j \\
&\sum_{i=1}^{I} \sum_{j=1}^{J} c_{ij} T_{ij} = C
\end{aligned}\right\} \tag{3}$$

where T_{ij} = volume of flows from i to j
O_i = aggregate volume of flows at point of origin i
D_j = aggregate volume of flows at point of destination j
c_{ij} = unit transportation costs from i to j
C = total transportation cost.

The maximization of the entropy function subject to all constraints leads to the following standard solution for T_{ij}:

$$T_{ij} = A_i\, B_j\, O_i\, D_j \exp(-\beta c_{ij}) \tag{4}$$

where the balancing factors A_i and B_j (guaranteeing the fulfillment of the additivity conditions) are defined as:

$$\left.\begin{aligned} A_i &= \exp(-\lambda_i)/O_i \\ B_j &= \exp(-\mu_j)/D_j \end{aligned}\right\} \tag{5}$$

The parameters λ_i, μ_j and β represent the Lagrange multipliers associated with the constraints O_i, D_j and C.

Now it can easily be derived that the share of flows from origin i to any point of destination j (i.e., the probability p_{ij} of choice from i to j) is equal to:

$$p_{ij} = T_{ij}/O_i = B_j D_j \exp(-\beta c_{ij}) / \sum_{j=1}^{J} B_j D_j \exp(-\beta c_{ij}) \tag{6}$$

This functional form corresponds to the well-known multinomial logit model (see among others McFadden, 1974; Domencich and McFadden, 1975). Thus, an entropy model as an aggregate representation has an analytical correspondence to a logit model from disaggregate choice theory. The only difference here is that (6) contains a weighting factor reflecting the attractiveness of a point of destination j (see also Wegener *et al*., 1985).

Various alternative derivations of similar or analogous results have been made in the past years. For instance, Coelho (1977) and Coelho and Wilson (1977) used a slightly different specification of the entropy function by including the cost constraint directly into the maximand, viz.

$$\max \omega = -1/\beta \sum_{i=1}^{I} \sum_{j=1}^{J} T_{ij}(\ln T_{ij} - 1 + \beta c_{ij}) \tag{7}$$

In Coelho's view, objective function (7) reflects the maximization of total group surplus, subject to appropriate constraints; it reproduces essentially the dispersion inherent in individual choice in the random utility approach. The result is however exactly the same as (6).

Leonardi (1985) has demonstrated the asymptotic equivalence between random utility theory and maximization of entropy (for both the single constrained and the double constrained spatial interaction problem), by using a slightly different entropy formulation:

$$\max \omega = -1/\beta \sum_{i=1}^{I} \sum_{j=1}^{J} T_{ij}(\ln T_{ij}/O_i D_j + \beta c_{ij}) \tag{8}$$

This formulation leads also to the same solution for p_{ij}.

Finally, in the context of residential location, Mattson (1983) employs an (aggregate) welfare-theoretic framework based on a random utility approach to individual utility-maximizing behaviour. Under certain conditions he was able to demonstrate the equivalence between an aggregate expected utility approach and entropy maximization.

The latter contribution leads to the question which kind of discrete utility maximizing model is compatible with the entropy model. In this context, interesting contributions have been made by various authors, and some of them will briefly be discussed here.

Anas (1983) starts from a stochastic utility maximization approach, based on individual choice behaviour. He lists the following conditions for a formal derivation of the MNL model from stochastic utility theory:

- equal linear utility functions for all individuals
- no taste variations among individual choices
- an additive stochastic part of the utility function, in which the error terms are independently and identically distributed (IID) over the population and for each individual according to a Gumbel distribution
- individual utility maximization based on the choice of the most preferred alternative.

Given these conditions, the derivation of the MNL model is straightforward. Next, Anas shows that the same specification for the MNL model can be obtained by maximizing an entropy function. The author then claims that information theory is more general than utility maximization. In our view this is a statement which has to be considered with caution. First, the MNL model is only a very specific case of random utility maximizing models; other more general stochastic discrete choice models (like probit analysis; see Daganzo, 1979; Van Lierop, 1985) are not necessarily compatible with entropy models. Secondly, if a one-to-one correspondence does exist between SI and MNL models, this also means that SI models are hampered by the same highly restrictive conditions as the MNL models.

Leonardi (1985) has extended the analysis of Anas by demonstrating that - instead of a Gumbel distribution - also any alternative probability distribution for the stochastic terms in a linear utility function may lead to the same MNL model, provided the tail of one minus the cumulative distribution function of the stochastic terms has asymptotically an exponential form. In addition, by replacing the term βc_{ij} in Equation (8) by $-v_{ij}$ (representing the deterministic part in the individual utility function, composed of the utility attached to j by individuals in i, corrected for transportation costs), he was able to demonstrate the equivalence between an MNL approach based on random utility theory and SI analysis.

An alternative approach was proposed by Bröcker (1980) who used in a trade allocation model an individual utility function that did not only include the consumption x_m of commodity type m (m=1,...,M), but also a spatial diversity indicator ($\sum_{m=1}^{M} x_m \ln x_m$). This probabilistic utility framework (based on a doubly exponentially distributed stochastic preference indicator) appears to be in agreement with the well known doubly constrained SI model.

Finally, Erlander (1980) has also established a link between expected utility theory and entropy theory. On the basis of individual expected utility, he was able to derive a cost function to be minimized subject to some constraints, including an entropy constraint. Thus a dual specification of an entropy model could be shown to be in agreement with expected utility maximization. A similar approach was also followed by Boyce *et al.* (1983) and Fisk and Boyce (1984). Boyce *et al.* employed the concept of hierarchical dispersion constraints and were able to derive a nested logit function determining mode and destination choice. Fisk and Boyce (1984) used an entropy expression based on prior information, thus also introducing a modified composite cost function.

The conclusions from this section seem to be rather straightforward: entropy maximization and stochastic utility maximization are not in contrast with each other, but are two sides of the same coin. Some caution is however needed here.

First, stochastic utility maximization is a fairly general approach; only under very restrictive assumptions may the MNL model be derived. This means that entropy maximization is at best in agreement with a very specific class of discrete probabilistic choice models. Alternative and more general models (like Probit or Dogit models) - which also arise from stochastic utility maximization - are not necessarily in agreement with entropy maximization. Alternatively, starting from a general stochastic utility approach does not imply that one ends up with an SI model of the standard gravity type.

An additional complication is that various discrete choice models are hampered by severe implicit restrictions, such as the independence of irrelevant alternatives (IIA) axiom. This also implies that the standard SI model is suffering from the same shortcomings, in as far as this model is based on an MNL approach. It is noteworthy that, in order to circumnavigate the IIA problem, one uses in practical research often a binary choice model (see also Van Lierop and Nijkamp, 1982; Pindyck and Rubinfeld, 1981). Then one usually applies in case of multiple alternatives a binary MNL model sequentially, but this is evidently a fairly time-consuming and theoretically less elegant choice analysis. In this context, the nested logit formulations given by Ben-Akiva and Lerman (1979), Boyce *et al.* (1983), Fisk and Boyce

(1984), Reggiani (1985), Roy and Lesse (1985) and Wilson (1982) are highly interesting.

The consistency problem related to entropy and logit models is caused by the uni-directional approach adopted by almost all authors. They start from the entropy model and demonstrate that the result of entropy maximization may lead to a logit formulation which is compatible with the MNL model from stochastic utility maximization. So far however, attempts to start with a specific type of discrete choice model (e.g., a probit model) and to derive the aggregate SI pattern that is consistent with the underlying individual choice model have not been successful due to the mathematical and statistical complexities involved. In this context, it may be worthwile to pay attention to a general theory of movement developed by Alonso (1978), which may lead to a general class of SI models encompassing the gravity-type model as a special case. This will be the subject of Section 3.

3. ALONSO'S GENERAL THEORY OF MOVEMENT AND DISCRETE CHOICE THEORY

Alonso (1978) has developed a general theory of movement which provides a comprehensive framework for a broad class of SI models including unconstrained, singly (production or attraction) constrained and doubly constrained interaction models as special cases. His approach has two interesting features, viz. (a) a simultaneous and consistent treatment of origin-destination flows and of total in- and outflows in each place, and (b) the inclusion of variables and parameters indicating the impact of the spatial system as a whole on place-to-place flows. Thus the flows T_{ij} from i to j are not only determined by push variables in i and pull variables in j, but also by the attributes of alternative origins and destinations. All these considerations lead to an appealing general framework for SI analysis.

The Alonso model has been further interpreted and extended among others by Anselin and Isard (1979), Hua (1980), Ledent (1980, 1981), Wilson (1980), and Nijkamp and Poot (1987), but its essence has remained the same. Here a brief presentation of the main characteristics of the Alonso model will be given, using the notation from Section 2.

First, the total flows out of each region of origin i are proportional to the unfavourable characteristics of region i (denoted by the repulsion factor V_i) and depend also on the relative attractiveness exerted by the rest of the system as seen from i (denoted by the pull-in factor W_i) with a rate of response α_i, i.e.,

$$\sum_{j=1}^{J} T_{ij} = O_i = V_i W_i^{\alpha_i}, \qquad \forall\ i \tag{9}$$

Similarly, we may assume that flows towards the region of destination j depend on the intrinsic attractive features of j (denoted by the attraction factor Y_j) and are also dependent on the relative unattractiveness of the rest of the system as seen from j (denoted by the push-out factor Z_j) with a response rate ε_j, i.e.,

$$\sum_{i=1}^{I} T_{ij} = D_j = Y_j \, Z_j^{\varepsilon_j}, \qquad \forall\, j \tag{10}$$

It should be noted that - in contrast to the standard entropy model - O_i and D_j are not necessarily known in advance.

Next, one may define a (standardized) repulsion factor O_i^* related to the push-out effects of region i as:

$$O_i^* = \frac{O_i}{W_i} = \frac{V_i \, W_i^{\alpha_i}}{W_i} = V_i \, W_i^{\alpha_i - 1} \tag{11}$$

Analogously, one may define a (standardized) attraction factor D_j^* related to the total pull effects of j as:

$$D_j^* = \frac{D_j}{Z_j} = \frac{Y_j \, Z_j^{\varepsilon_j}}{Z_j} = Y_j \, Z_j^{\varepsilon_j - 1} \tag{12}$$

Next, one may also define a general distance friction function F_{ij} reflecting the ease of movement between i and j. This function may have the standard exponential form from SI models, but this is not necessary. Furthermore, F_{ij} is not necessarily symmetric.

Following now the basic principle from SI analysis that flows from i to j depend on origin and destination characteristics as well as on the distance friction, one may specify the following relationship for T_{ij}:

$$T_{ij} = O_i^* \, D_j^* \, F_{ij} = V_i Y_j W_i^{\alpha_i - 1} \, Z_j^{\varepsilon_j - 1} \, F_{ij} \tag{13}$$

In order to ensure consistency between (13) on the one hand and (9) and (10) on the other, it is clear that the following conditions have to be fulfilled:

$$O_i = V_i W_i^{\alpha_i - 1} \sum_{j=1}^{J} Y_j Z_j^{\varepsilon_j - 1} F_{ij} \tag{14}$$

and:

$$D_j = Y_j Z_j^{\varepsilon_j - 1} \sum_{i=1}^{I} V_i W_i^{\alpha_i - 1} F_{ij} \tag{15}$$

Now it is easily seen that:

$$W_i = \sum_{j=1}^{J} Y_j Z_j^{\varepsilon_j - 1} F_{ij} \tag{16}$$

and:

$$Z_j = \sum_{i=1}^{I} V_i W_i^{\alpha_i - 1} F_{ij} \tag{17}$$

Alonso's model, based on a general theory of movement, appears to lead to a general type of SI model. Now the question arises whether this model can be made consistent with an entropy model in a more general context. In Appendix 1 it is demonstrated that indeed a more general version of the conventional entropy model leads to the same results as the Alonso model.

Consequently, the Alonso model is able to generate a wide variety of alternative specific SI models (see Wilson, 1980). An unconstrained SI model is generated, if $\alpha_i = 1$, $\forall\ i$, and $\varepsilon_j = 1$, $\forall\ j$. A production-constrained SI model arises, if $\alpha_i = 0$, $\forall\ i$, and $\varepsilon_j = 1$, $\forall\ j$, while an attraction-constrained SI model arises, if $\alpha_i = 1$, $\forall\ i$, and $\varepsilon_j = 0$, $\forall\ j$. Finally, a doubly constrained SI model is attained, if $\alpha_i = 0$, $\forall\ i$, and $\varepsilon_j = 0$, $\forall\ j$. It is also evident that by using alternative values of α_i ($\alpha_i \geq 0$) and ε_j ($\varepsilon_j \geq 0$) a broad class of SI models which are not covered by the conventional entropy approach may emerge. Furthermore, mixed values of α_i and ε_j may be used, as it has *not* been assumed that $\alpha_i = \alpha$, $\forall\ i$, and $\varepsilon_j = \varepsilon$, $\forall\ j$, so that the Alonso model offers indeed a general and flexible framework for SI analysis.

The Alonso model has been applied in various cases (see among others Anselin, 1982; Fisch, 1981; Nijkamp and Poot, 1987; Poot, 1984; Porell, 1982; Porell and Hua, 1981; Tabuchi, 1984).

The question to be dealt with now is whether the Alonso model is also consistent with discrete choice theory, particularly with an MNL approach. The question can be analyzed by writing p_{ij}, the probability that a trip starting in i will end up in j, as follows (see (6)):

$$p_{ij} = \frac{T_{ij}}{O_i} = \frac{Y_j Z_j^{\varepsilon_j - 1} F_{ij}}{\sum_{j=1}^{J} Y_j Z_j^{\varepsilon_j - 1} F_{ij}} \tag{18}$$

This model is evidently a logit expression, though it is more general than the one described in (6). In case of a production-constrained model (with $\varepsilon_j = 1$), it is clear that we obtain an expression analogous to (6), which is equivalent to an MNL model. In all other cases, a more general logit model arises, as can easily be seen by substituting (12) into (18), i.e.,

$$p_{ij} = \frac{D_j Z_j^{-1} F_{ij}}{\sum_{j=1}^{J} D_j Z_j^{-1} F_{ij}} \tag{19}$$

where Z_j^{-1} may be interpreted as a general balancing factor related to the accessibility to j. If F_{ij} is a standard exponential distance decay function, then (19) is an MNL model that is directly related to a standard stochastic discrete choice model based on a linear utility function (including a linear trip cost component) and all other assumptions mentioned in Section 2. However, if F_{ij} has a different (i.e., non-exponential) form, the underlying utility function from stochastic choice analysis might have a non-linear cost component (see also Diappi and Reggiani, 1985).

Furthermore, it is interesting to observe that the Alonso model can be related to a *nested logit model*. In fact, it is worth noting that, since Z_j and W_i are interrelated (see Eqs. (16), (17)), the resulting p_{ij} will depend not only on the characteristics of destination j and origin i, but also on the characteristics of other origins. This result confirms our interpretation of the compatibility of the Alonso model with a nested logit model (see for a formal derivation also Nijkamp and Reggiani, 1986), in which the inter-individual utility independence is not violated. This has the important implication that the Alonso model does not suffer from the restrictive IIA-axiom which hampers the application of the conventional MNL model.

In conclusion, Alonso's general theory of movement offers two important extensions compared to the results discussed in Section 2. In the first place, the Alonso model can be shown to emerge from a more general type of entropy model, and hence leads to a more general type of SI model. And secondly, the Alonso model is also in harmony with a general class of disaggregate choice models, so that this model is (implicitly) based on more general spatial interaction principles. Especially the consistency between the Alonso model and the nested logit model is important in this context.

4. DYNAMIC SPATIAL INTERACTION MODELS

4.1. *A Brief Review*

In the few past years increasing attention has been devoted to the analysis of dynamic spatial systems (see e.g. Griffith and Lea, 1983; Kahn, 1981). This is also reflected in the analysis of SI systems (see Williams and Wilson, 1980), by using among others Markov transition probabilities between different states of a spatial system (cf. Byler and Gale, 1978).

Various types of dynamic analysis of SI systems may be used. Examples are:

a). An analysis of the *demand* side of SI behaviour, by using inter alia a dynamic entropy specification (for example, the dynamic shopping model developed by Coelho, 1977). This is an interesting approach, but it is easily seen that such straightforward extensions do not drastically affect the structure of SI models. Adjusted methods, based on Bayesian approaches to multiperiod entropy formulations, have been put forward, among others by Batty and March (1976). A more general approach based on Gokhale and Kullback's minimum information discrimination was developed by Haynes and Phillips (1982) who were also able to establish a link with individual choice behaviour.

An interesting approach may emerge, if the decisions of individuals lead to congestion effects, so that in a next period the flows from i to j are affected by capacity constraints (see, for instance, Batten and Boyce, 1986; Boyce and Southworth, 1979). This might be taken into account by replacing the constant unit transportation costs c_{ij} by a function including the volume of flows T_{ij} or the capacity of the road network. If we replace the term $\sum_{i=1}^{I} \sum_{j=1}^{J} c_{ij} T_{ij}$ by the term $\sum_{i=1}^{I} \sum_{j=1}^{J} T_{ij} \int_{0}^{T_{ij}} c_{ij}(z)dz$, we have to solve the following adjusted entropy model:

$$\max \omega = - \sum_{i=1}^{I} \sum_{j=1}^{J} T_{ij}(\ln T_{ij}-1)$$

subject to

$$\left.\begin{aligned} &\sum_{j=1}^{J} T_{ij} = O_i, \qquad \forall\, i \\ &\sum_{i=1}^{I} T_{ij} = D_j, \qquad \forall\, j \\ &\sum_{i=1}^{I} \sum_{j=1}^{J} \int_{0}^{T_{ij}} c_{ij}(z)dz = C \end{aligned}\right\} \qquad (20)$$

The solution to this non-linear entropy model can be found in Appendix 2.

b). An analysis of the *supply* side of SI behaviour by linking the supply side to user behaviour. A wide variety of interesting dynamic models have been generated in this context (see e.g., Lombardo and Rabino, 1983; Wilson, 1981). This may lead to the design of interesting differential equations based on (4). An example is the following dynamic model:

$$\dot{R}_j = \varepsilon \, [\sum_{i=1}^{I} Q_i \, (R_j^{\alpha} \, e^{-\beta c_{ij}} / \sum_{j=1}^{J} R_j^{\alpha} \, e^{-\beta c_{ij}}) - \kappa_j R_j] \tag{21}$$

where R_j is the value of an endogenous variable in area j ($\dot{R}_j$ is its time derivative, i.e., the velocity of change); Q_i is the value of an exogenous variable in area i (e.g., population); c_{ij} is the transport cost from i to j, while κ_j, α, β, ε are relevant parameters. Model (21) comprises, of course, the equilibrium conditions given by:

$$\sum_{i=1}^{I} Q_i (R_j^{\alpha} \, e^{-\beta c_{ij}} / \sum_{j=1}^{J} R_j^{\alpha} \, e^{-\beta c_{ij}}) = \kappa_j R_j \tag{22}$$

Since Equation (22) is non-linear, it is clear that interesting bifurcation patterns may emerge, when the parameters α and β vary. Particular specifications of such dynamic models including attraction forces of retail facilities in a spatial system are (see Allen *et al.*, 1978; Wilson, 1981):

$$\dot{R}_j = \varepsilon R_j [N(F_j / \sum_{j'=1}^{J} F_{j'}) - R_j] \tag{23}$$

where R_j represents the scale of retail facilities in area j, F_j an attraction variable (e.g., population), N the total demand for a certain service, and $F_j \, / \sum_{j'=1}^{J} F_{j'}$ the share of total demand attracted to zone j.

Other dynamic models based on structural changes in space and time can be found among others in Allen *et al.* (1978), Clarke and Wilson (1983) and Dejon (1983), who also studied the equilibrium conditions of the related social system. Models of the type described above may exhibit various kinds of dynamic behaviour including catastrophic behaviour and bifurcations. (For a review of such models see also Rabino, 1985; Birkin and Wilson, 1985.) A specific application based on ecological dynamics of the Volterra-Lotka model, put in the framework of a dogit model, can be found in Sonis (1984).
c). An *integration of supply and demand* in SI models. In this context only a few attempts have been made (see e.g. Leonardi, 1983), mainly because this leads to fairly complex dynamic systems. To demonstrate the difficulties involved, we will employ in the following Section 4.2 a simple dynamic entropy model which incorporates feedback effects from the supply side via an adjustment of O_i.

4.2. *An Optimal Control Approach to Spatial Interaction Models*

Having briefly reviewed some earlier research in the field of dynamic SI modelling, we will now present an optimal control version of a dynamic entropy model. Once again we will assume a transportation system in which all variables T_{ij} are time-dependent. The T_{ij} variables will be regarded as control variables in the Pontryagin sense. In a dynamic context, the objective function can be regarded as cumulative entropy function (see also Dendrinos and Sonis, 1986; Sonis, 1986) and reflects the homogeneity of the system in a certain time period. In this case, the question which has to be raised concerns which of the other variables that are normally given in the conventional static entropy model are time-dependent state variables. In our view, it is plausible to regard the O_i variables as such state variables, as their evolution may be dependent on net population growth. The reason for this is that a rise in population may generate more transportation, so that the total push-out effects from origin i may be a linear function of the population size in i, denoted by P_i. Consequently, we obtain:

$$O_i = \delta_i P_i \tag{24}$$

or:

$$\dot{O}_i = \delta_i \dot{P}_i \tag{25}$$

where a dot represents a time rate of growth of the variable concerned.

Population in place i may change due to natural increase and net inmigration M_i. The rate of change in population is then equal to:

$$\dot{P}_i = \rho_i + \dot{M}_i \tag{26}$$

where we have assumed for the sake of simplicity (but without loss of generality) a linear time-dependent growth curve, which leads to a constant population growth factor ρ_i. Net inmigration is regarded here in a simple way as the result of net residential push and pull effects of place i. It is assumed that a high net transportation inflow into the city reflects a high attractiveness and will generate a demand for housing, so that after some time the time rate of net inmigration growth is increasing due to the urban pull-in effects reflected by the net transportation inflows, i.e.,

$$\dot{M}_i = \gamma_i\left(\sum_{j=1}^{J} T_{ji} - \sum_{j=1}^{J} T_{ij}\right) \tag{27}$$

Clearly, congestion effects in the sense of negative attraction effects might also be taken into account, for instance, by making γ_i a decreasing function of the net transportation flows.

By next substituting (27) and (26) into (25) one obtains the following result:

$$\dot{O}_i = \delta_i \rho_i + \delta_i \gamma_i (\sum_{j=1}^{J} T_{ji} - \sum_{j=1}^{J} T_{ij})$$

$$= \delta_i \rho_i + \delta_i \gamma_i (D_i - O_i) \quad (28)$$

Altogether we obtain the following adjusted optimal control entropy model:

$$\max \omega = \int_0^T -\sum_{i=1}^{I} \sum_{j=1}^{J} T_{ij} (\ln T_{ij} - 1) dt$$

subject to

$$\left.\begin{aligned} & \sum_{j=1}^{J} T_{ij} = O_i, \quad \forall i \\ & \sum_{i=1}^{I} T_{ij} = D_j, \quad \forall j \\ & \sum_{i=1}^{I} \sum_{j=1}^{J} c_{ij} T_{ij} = C \\ & \dot{O}_i = \delta_i \rho_i + \delta_i \gamma_i (\sum_{j=1}^{J} T_{ji} - \sum_{j=1}^{J} T_{ij}) \end{aligned}\right\} \quad (29)$$

Two remarks are in order here. First, it might be in greater agreement with dynamic economic theory to include a discount factor in the integrand of the entropy function, but this would not fundamentally change the results. Secondly, it might also, in principle, be possible to include a similar relationship for D_j, though this is less plausible for O_i. If D_j is not regarded as fixed and if no dynamic relationship for O_i were to be included, one would of course end up with a production-constrained SI model.

The derivation of the SI model associated with (29) is somewhat tedious and can be found in Appendix 3. The final result appears to be:

$$T_{ij} = A_i^* B_j^* O_i D_j \exp(-\beta c_{ij}) \quad (30)$$

where the generalized balancing factors A_i^* and B_j^* are defined in Appendix 3. Consequently, the final result is again a SI model which can be linked to discrete choice theory based on a (general) MNL model (see for further details Appendix 3).

If the same approach is applied to the Alonso model discussed in Section 3, it is clear that no drastic changes will occur. A dynamic equation for O_i can be introduced in an analogous way. In this case, we will, of course, again obtain a more general model of which Equation (30) is a particular case (see for full details Appendix 4). The consequences for the model results may again be calculated by means of simulation experiments (see Nijkamp and Poot, 1987). The main difference with the previous approach is that the Alonso model is not *explicitly* based on an optimal control model for a generalized entropy function, so that, in dealing with the dynamic behaviour of this system by means of simulation experiments, there is less need to worry about the choice of control and state variables or about complicated necessary conditions for an optimum evolution of the spatial system. In this respect, the Alonso model is easier to deal with.

5. CONCLUDING REMARKS

The class of aggregate (macro-oriented) spatial interaction models (of either the entropy type or the Alonso type) is - in comparison with disaggregate (micro-oriented) discrete choice models (especially multinomial logit models) - not a completely different way of representing spatial choice behaviour. Both types of models are essentially two sides of the same coin. This paper has shown that an analytical correspondence exists between multinomial logit models and spatial interaction models, both in a static and in a dynamic context. It has also been shown in the present paper that an optimal control approach to a dynamic entropy model complies with the above-mentioned result. A problem which remains to be studied in this context is the derivation of stability and equilibrium conditions in case of a more general objective function for a dynamic spatial interaction model. This might require a closer analysis of Brownian process movements in a non-linear dynamic spatial framework.

Dept. of Economics
Free University
Amsterdam
The Netherlands

Dept. of Social and Economic Analysis
University of Venice
Venice
Italy

BIBLIOGRAPHY

Allen, P.M., J.L. Deneubourg, M. Sanglier, F. Boon, and A. de Palma: 1978, 'The dynamics of urban evolution', Final Report of the U.S. Department of Transportation, Washington D.C..

Alonso, W.: 1978, 'A theory of movements', in N.M. Hansen (ed.), *Human Settlement Systems*, Ballinger, Cambridge, Mass..

Anas, A.: 1983 'Discrete choice theory, information theory and the multinomial logit and gravity models', *Transportation Research* **B 17**, 13-23.

Anselin, L.: 1982, 'Implicit functional relationships between systemic effects in a general model of movement', *Regional Science and Urban Economics* **12**, 365-380.

Anselin, L. and W. Isard: 1979 'On Alonso's general theory of movement', *Man, Environment, Space and Time* **1**, 52-63.

Batten, D.F.: 1983, *Spatial Analysis of Interacting Economies*, Martinus Nijhoff, Dordrecht, The Netherlands.

Batten, D.F. and D.E. Boyce: 1986, 'Spatial interaction, transportation and interregional commodity flow models', in P. Nijkamp (ed.), *Handbook in Regional Economics*, North-Holland Publ. Co., Amsterdam.

Batty, M. and L. March: 1976, 'The method of residues in urban modelling', *Environment and Planning* **A 8**, 189-214.

Ben-Akiva, M. and S.R. Lerman: 1979, 'Disaggregate travel and mobility choice models and measures of accessibility', in D.A. Hensher and D.R. Stopher (eds.), *Behavioral Theory Modelling*, Croom Helm, London.

Berechman, J.: 1981, 'Analytical problems in linking an activity model with a transportation network model', *Environment and Planning* **A 13**, 449-462.

Birkin, M. and A.G. Wilson: 1985, 'Some properties of spatial-structural-economic-dynamic models', Paper presented at the 4th European Colloquium on Quantitative and Theoretical Geography, Veldhoven.

Boyce, D.E.: 1984, 'Urban transportation network-equilibrium and design models: recent achievements and future projects', *Environment and Planning* **A 16**, 1445-1474.

Boyce, D.E., K.S. Chon, Y.J. Lee, K.T. Lin, and L.J. Le Blanc: 1983, 'Implementation and computational issues for combined models of location, destination, mode and route choice', *Environment and Planning* **A 15**, 1214-1230.

Boyce, D.E. and F. Southworth: 1979, 'Quasi-dynamic urban location models with endogenously determined travel costs', *Environment and Planning* **A 11**, 575-584.

Bröcker, J.: 1980, 'An application of economic interaction models to the analysis of spatial effects of economic integration', *Environment and Planning* **A 12**, 321-338.

Byler, J.W. and S. Gale: 1978, 'Social account and planning for changes in urban housing markets', *Environment and Planning* **A 10**, 247-266.

Charnes, A., K.E. Haynes, F.Y. Phillips, and G. White: 1977, 'New equivalencies and qualities from the extremal solution of the gravity model of spatial interaction', *Journal of Regional Science* **17**, 71-76.

Clarke, M. and A.G. Wilson: 1983, 'The dynamics of urban spatial structure: progress and problems', *Journal of Regional Science* **23**, 1-18.

Coelho, J.D.: 1977, 'The use of mathematical methods in model based land use planning', Ph.D. Diss., Dept. of Geography, University of Leeds, Leeds.

Coelho, J.D. and H.C.W.L. Williams: 1978, 'On the design of land use plans through locational surplus maximization', *Papers of the Regional Science Association* **40**, 71-85.

Coelho, J.D. and A.G. Wilson: 1977, 'Some equivalence theorems to integrate entropy maximizing sub-models within overall mathematical programming frameworks', *Geographical Analysis* **9**, 160-173.

Daganzo, C.: 1979, *Multinomial Probit*, Academic Press, New York.

Dejon, B.: 1983, 'Attractivity-regulated dynamic equilibrium models of migration of the multiplicative type', in D. Griffith and A.C. Lea (eds.), *Evolving Geographical Structures*, Martinus Nijhoff, The Hague.

Dendrinos, S. and M. Sonis: 1986, 'Variational principles and conservation conditions in Volterra's ecology and in urban relative dynamics', *Journal of Regional Science* **26**, 359-377.

Diappi, L. and A. Reggiani: 1985, 'Modelli di trasporto e geografia del territorio', in A. Reggiani (ed.), *Territorio e Trasporti*, Franco Angeli, Milano.

Domencich, T.A. and D. McFadden: 1975, *Urban Travel Demand. A Behavioural Analysis*, North-Holland Publ. Co., Amsterdam.

Erlander, S.: 1977, 'Entropy in linear programming - an approach to planning', Linköping Institute of Technology, Report LITH-Mat-R-77-3, Linköping.

Erlander, S.: 1980, *Optimal Spatial Interaction and the Gravity Model*, Springer Verlag, Berlin.

Evans, S.P.: 1973, 'A relationship between the gravity model for trip distribution and the transportation problem in linear programming', *Transportation Research* **7**, 39-61.

Evans, S.P.: 1976, 'Derivation and analysis of some models for combining trip distribution and assignment', *Transportation Research* **10**, 37-57.

Fisch, O.: 1981, 'Contributions to the general theory of movement', *Regional Science and Urban Economics* **11**, 157-173.

Fisk, C.S. and D.E. Boyce: 1984, 'A modified composite cost measure for probabilistic choice modelling', *Environment and Planning* **A 16**, 241-248.

Florian, M.: 1983, 'Nonlinear cost network models in transportation analysis', Centre de Recherche sur les Transports, publication 287 - Département d'Informatique et de la Recherche Operationelle, publication 469 - Université de Montreal, Montreal, Quebec.

Florian, M. and S. Nguyen: 1978, 'A combined trip distribution, modal split and trip assignment model', *Transportation Research* **12**, 241-246.

Griffith, D. and A.C. Lea (eds.): 1983, *Evolving Geographical Structures*, Martinus Nijhoff, The Hague.

Haynes, K.E. and A.S. Fotheringham: 1984, *Gravity and Spatial Interaction Models*, Sage Publ., Beverly Hills.

Haynes, K.E. and F.Y. Phillips: 1982, 'Constrained minimum discrimination information: a unifying tool for modelling spatial and individual choice behaviour', *Environment and Planning* **A 14**, 1341-1354.

Hua, C.: 1980, 'An exploration of the nature and rationale of a systemic model', *Environment and Planning* **A 12**, 713-726.

Jefferson, T.R. and C.H. Scott: 1977, 'Entropy maximizing models or residential location via geometric programming', *Geographical Analysis* **9**, 181-187.
Kahn, D. (ed.): 1981, *Essays in Societal System Dynamics and Transportation: Report of the Third Annual Workshop in Urban and Regional Systems Analysis*, Research and Special Programs Administration, U.S. Department of Transportation, Washington D.C..
Kamien, M.I. and N.L. Schwartz: 1981, *Dynamic Optimization: The Calculus of Variations and Optimal Control in Economics and Management*, North-Holland Publ. Co., Amsterdam.
Ledent, J.: 1980, 'Calibrating Alonso's general theory of movement: the case of interprovincial migration flows in Canada', *Sistemi Urbani* **2**, 327-358.
Ledent, J.: 1981, 'On the relationship between Alonso's theory of movement and Wilson's family of spatial interaction models', *Environment and Planning* **A 13**, 217-224.
Leonardi, G.: 1983, 'An optimal control representation of stochastic multistage multiactor choice process', in D.A. Griffith and A.C. Lea (eds.), *Evolving Geographical Structures*, Martinus Nijhoff, The Hague.
Leonardi, G.: 1985, 'Equivalenza asintotica tra la teoria delle utilità casuali e la massimizzazione dell'entropia, in A. Reggiani (ed.), *Territorio e Trasporti*, Franco Angeli, Milano.
Lierop, W.F.J. van: 1985, 'Spatial interaction modelling and residential choice analysis', Ph.D. thesis, Dept. of Economics, Free University, Amsterdam.
Lierop, W. van and P. Nijkamp: 1979, 'A utility framework for interaction models', *Sistemi Urbani* **1**, 41-64.
Lierop, W. van and P. Nijkamp: 1982, 'Disaggregate models of choice in a spatial context', *Sistemi Urbani* **4**, 331-369.
Lombardo, S.T. and G.A. Rabino: 1983, 'Some simulations of a central place theory model', *Sistemi Urbani* **5**, 315- 332.
Los, H. and S. Nguyen: 1983, 'Solution algorithms for a combined residential location and transportation model', *Environment and Planning* **A 15**, 515-529.
Mattson, L.G.: 1983, 'Equivalence between welfare and entropy approaches to residential location', Research Paper TRITA-MAT-1983-18, Royal Institute of Technology, Stockholm.
McFadden, D.: 1974, 'Conditional logit analysis of qualitative choice behaviour', in P. Zarembka (ed.), *Frontiers in Econometrics*, Academic Press, New York.
Miller, R.E.: 1979, *Dynamic Optimization and Economic Applications*, McGraw-Hill, New York.
Nijkamp, P.: 1975, 'Reflections on gravity and entropy models', *Regional Science and Urban Economics* **5**, 203-225.
Nijkamp, P.: 1979, 'Gravity and entropy models: the state of the art', in G.R.M. Jansen, P.H.L. Bovy, J.P.J.M. van Est, and F. le Clerq (eds.), *New Developments in Modeling Travel Demand and Urban Systems*, Gower, Aldershot, UK.
Nijkamp, P. and H.J. Poot: 1987, 'Dynamics of generalized spatial interaction models', *Regional Science and Urban Economics* **17**, 367-390.

Nijkamp, P. and A. Reggiani: 1986, 'A synthesis between macro and micro models in spatial interaction analysis with special reference to dynamics', Research Memorandum, Dept. of Economics, Free University, Amsterdam.
Pindyck, R.S. and D.L. Rubinfeld: 1981, *Econometric Models and Economic Forecasts*, McGraw-Hill, London.
Poot, H.J.: 1984, 'Models of New Zealand internal migration and residential mobility'. Ph.D. Diss., Victoria University, Wellington.
Porell, F.W.: 1982, 'Intermetropolitan migration and quality of life', *Journal of Regional Science* **22**, 137-158.
Porell, F.W. and C.M. Hua: 1981, 'An econometric procedure for estimation of a generalized system gravity model under incomplete information about the system', *Regional Science and Urban Economics* **11**, 585-606.
Rabino, G.: 1985, 'Modelli dinamici del sistema integrato territorio e trasporti', in A. Reggiani (ed.), *Territorio e Trasporti*, Franco Angeli, Milano.
Reggiani, A. (ed.): 1985, *Territorio e Trasporti; Modelli Matematici per l'Analisi e la Planificazione*, Franco Angeli, Milano.
Roy, J.R. and P.L. Lesse: 1985, 'Entropy models with entropy constraints on aggregated events', *Environment and Planning* **A 17**, 1669-1674.
Sonis, M.: 1984, 'Dynamic choice of alternatives, innovation diffusion, and ecological dynamics of the Volterra-Lotka model', in D.E. Pitfield (ed.), *Discrete Choice Models in Regional Science*, Pion, London.
Sonis, M.: 1986, 'A unified theory of innovation diffusion, dynamic choice of alternatives, ecological dynamics and urban/regional growth and decline', Paper presented at the Conference on Innovation Diffusion, Venice.
Tabuchi, T.: 1984, 'The systemic variables and elasticities in Alonso's general theory of movement', *Regional Science and Urban Economics* **14**, 249-264.
Tan, K.C. and R.J. Bennett: 1984, *Optimal Control of Spatial Systems*, George Allen & Unwin, London.
Theil, H.: 1972, *Statistical Decomposition Analysis*, North-Holland Publ. Co., Amsterdam.
Wegener, M., F. Gnad, and M. Vannahme: 1985, 'The time scale of urban change', in B. Hutchinson and M. Batty (eds.), *Advances in Urban Systems Modeling*, North-Holland Publ. Co., Amsterdam.
Williams, H.C.W.L. and A.G. Wilson: 1980, 'Some comments on the theoretical and analytic structure of urban and regional models', *Sistemi Urbani* **2**, 203-242.
Wilson, A.G.: 1970, *Entropy in Urban and Regional Modelling*, Pion, London.
Wilson, A.G.: 1980, 'Comments on Alonso's 'theory of movement'', *Environment and Planning* **A 12**, 727-732.
Wilson, A.G.: 1981, *Catastrophe Theory and Bifurcation*, Croom Helm, London.
Wilson, A.G.: 1982, 'Transport, location and spatial systems: planning with spatial interaction models and related approaches', School of Geography, University of Leeds, mimeographed.

APPENDIX 1. CORRESPONDENCE BETWEEN THE ALONSO MODEL AND A GENERAL ENTROPY MODEL

In this appendix it will be shown that Alonso's general theory of movement is in agreement with a generalized entropy model (see also Poot, 1984). Assume the following entropy-maximizing model:

$$\max \omega = - \sum_{i=1}^{I} \sum_{j=1}^{J} T_{ij}(\ln T_{ij}-1)$$

subject to

$$\left.\begin{aligned} &\sum_{j=1}^{J} T_{ij} = O_i = V_i W_i^{\alpha_i} \\ &\sum_{i=1}^{I} T_{ij} = D_j = Y_j Z_j^{\varepsilon_j} \\ &\sum_{i=1}^{I} \sum_{j=1}^{J} c_{ij} T_{ij} = C \end{aligned}\right\} \tag{31}$$

where relationships (9) and (10) have been substituted into the conventional entropy model.

Then it can be shown in a straightforward way that the solution of (31) is:

$$T_{ij} = V_i Y_j W_i^{\alpha_i - 1} \; Z_j^{\varepsilon_j - 1} \; \exp(-\beta c_{ij}) \tag{32}$$

This result can be proved by differentiating the Lagrangian expression L for (31) as:

$$\frac{\partial L}{\partial T_{ij}} = -\ln T_{ij} - \lambda_i - \mu_j - \beta c_{ij} = 0 \tag{33}$$

so that:

$$T_{ij} = \exp(-\lambda_i - \mu_j - \beta c_{ij}) \tag{34}$$

In addition, the following conditions hold:

$$\sum_{j=1}^{J} T_{ij} = V_i W_i^{\alpha_i} = \exp(-\lambda_i) \sum_{j=1}^{J} \exp(-\mu_j - \beta c_{ij}) \tag{35}$$

and

$$\sum_{i=1}^{I} T_{ij} = Y_j Z_j^{\varepsilon_j} = \exp(-\mu_j) \sum_{i=1}^{I} \exp(-\lambda_i - \beta c_{ij}) \tag{36}$$

so that

$$\exp(-\lambda_i) = V_i W_i^{\alpha_i} \{ \sum_{j=1}^{J} \exp(-\mu_j - \beta c_{ij}) \}^{-1} \tag{37}$$

and

$$\exp(-\mu_j) = Y_j Z_j^{\varepsilon_j} \{ \sum_{i=1}^{I} \exp(-\lambda_i - \beta c_{ij}) \}^{-1} \tag{38}$$

By now writing:

$$\sum_{j=1}^{J} \exp(-\mu_j - \beta c_{ij}) = W_i \tag{39}$$

and

$$\sum_{i=1}^{I} \exp(-\lambda_i - \beta c_{ij}) = Z_j \tag{40}$$

we obtain:

$$W_i^{\alpha_i - 1} = \exp(-\lambda_i)/V_i \tag{41}$$

and

$$Z_j^{\varepsilon_j - 1} = \exp(-\mu_j)/Y_j \tag{42}$$

Substition of (41)-(42) into (34) yields the final result for the flows T_{ij} presented in (32). Q.E.D.

Consequently, in the Alonso model the expressions for the Lagrange multipliers λ_i and μ_j are different from those in the standard entropy model, which is of course caused by the differences in the corresponding constraints. It is also easily seen that by using Equations (39)-(42), we obtain the following result for W_i and Z_j:

$$W_i = \sum_{j=1}^{J} Y_j Z_j^{\varepsilon_j - 1} \exp(-\beta c_{ij}) \tag{43}$$

and

$$Z_j = \sum_{i=1}^{I} V_i W_i^{\alpha_i - 1} \exp(-Bc_{ij}) \tag{44}$$

The latter two expressions correspond to Equations (16) and (17) in Section 3.

APPENDIX 2. SOLUTION TO AN ENTROPY MODEL WITH NON-LINEAR COST CONSTRAINTS

In this section the following entropy model with a non-linear cost constraint will be dealt with (see (20)):

$$\left.\begin{aligned}
&\max \omega = - \sum_{i=1}^{I} \sum_{j=1}^{J} T_{ij}(\ln T_{ij} - 1) \\
&\text{subject to} \\
&\sum_{j=1}^{J} T_{ij} = O_i \\
&\sum_{i=1}^{I} T_{ij} = D_j \\
&\sum_{i=1}^{I} \sum_{j=1}^{J} \int_0^{T_{ij}} c_{ij}(z)dz = C
\end{aligned}\right\} \tag{45}$$

The latter problem can also be interpreted as a combined trip distribution-trip assignment problem (see for an interesting historical review Boyce, 1984; as well as Berechman, 1981; Florian, 1983). Various operational solution techniques for this model have been developed, such as the Frank-Wolfe approach, Evans' algorithm, the diagonalization method etc. (see, among others, Evans, 1976; Florian and Nguyen, 1978; Los and Nguyen, 1983).

The analytical solution to (45) can be derived by differentiating the Lagrangian expression L:

$$\begin{aligned}
L = &- \sum_{i=1}^{I} \sum_{j=1}^{J} T_{ij}(\ln T_{ij} - 1) + \sum_{i=1}^{I} \lambda_i (O_i - \sum_{j=1}^{J} T_{ij}) + \\
&+ \sum_{j=1}^{J} \mu_j (D_j - \sum_{i=1}^{I} T_{ij}) + \beta \, (C - \sum_{i=1}^{I} \sum_{j=1}^{J} \int_0^{T_{ij}} c_{ij}(z)dz)
\end{aligned} \tag{46}$$

The optimality conditions are:

$$\frac{\partial L}{\partial T_{ij}} = - \ln T_{ij} - \beta c_{ij}(T_{ij}) - \lambda_i - \mu_j = 0 \tag{47}$$

If we now assume the following non-linear approximation for $c_{ij}(T_{ij})$:

$$c_{ij}(T_{ij}) = \eta + \theta \ln T_{ij} \tag{48}$$

(47) can be written as follows:

$$- \ln T_{ij} - \beta\theta \ln T_{ij} - \beta\eta - \lambda_i - \mu_j = 0 \tag{49}$$

so that

$$T_{ij} = T_{ij}^{-\beta\theta} \exp(-\beta\eta) \exp(-\lambda_i) \exp(-\mu_j) \tag{50}$$

By substituting next:

$$\exp(-\lambda_i) = A_i \, O_i \tag{51}$$

and

$$\exp(-\mu_j) = B_j \, D_j \tag{52}$$

into (50), we obtain:

$$T_{ij} = T_{ij}^{-\beta\theta} \exp(-\beta\eta) \, A_i O_i B_j D_j \tag{53}$$

This is an adjusted expression for the standard solution of a conventional entropy model and can also be written as:

$$T_{ij} = \{\exp(-\beta\eta) \, A_i O_i B_j D_j\}^{\frac{1}{1+\beta\theta}} \tag{54}$$

Clearly, alternative specifications for (48) can also be used, for instance, an exponential expression.

APPENDIX 3. DERIVATION OF AN SI MODEL BASED ON A DYNAMIC ENTROPY APPROACH

The general optimal control entropy model to be solved here is found in (29). This is apparently a bounded optimal control model (see for instance Miller, 1979; Kamien and Schwartz, 1981; Tan and Bennett, 1984), so that both the Hamiltonian and the Lagrangian function have to be used for the derivation of the optimal control variables T_{ij} and the optimal state variables O_i.

The Hamiltonian H (with ψ_i being a co-state variable) is equal to:

$$H(T_{ij}, O_i, t) = -\sum_{i=1}^{I} \sum_{j=1}^{J} T_{ij}(\ln T_{ij} - 1) + \sum_{i=1}^{I} \psi_i \dot{O}_i \tag{55}$$

while the Lagrangian L is equal to:

$$\left. \begin{aligned} L(T_{ij}, O_i, t) = {} & H(T_{ij}, O_i, t) \\ & + \sum_{i=1}^{I} \lambda_i (O_i - \sum_{j=1}^{J} T_{ij}) \\ & + \sum_{j=1}^{J} \mu_j (D_j - \sum_{i=1}^{I} T_{ij}) \\ & + \beta (C - \sum_{i=1}^{I} \sum_{j=1}^{J} c_{ij} T_{ij}) \end{aligned} \right\} \tag{56}$$

Now the necessary (first-order) conditions are:

$$\left.\begin{aligned} &\frac{\partial L}{\partial T_{ij}} = 0 \qquad \forall\ i,j \\ &\frac{\partial L}{\partial O_i} = -\dot{\psi}_i \qquad \forall\ i \\ &\frac{\partial L}{\partial \psi_i} = \dot{O}_i \qquad \forall\ i \end{aligned}\right\} \tag{57}$$

The second-order conditions are also satisfied, as:

$$\frac{\partial}{\partial T_{ij}}\left(\frac{\partial L}{\partial T_{ij}}\right) = -\frac{1}{T_{ij}} < 0 \tag{58}$$

Now it can be demonstrated after some mathematical manipulations that:

$$\frac{\partial L}{\partial T_{ij}} = -\ln T_{ij} - \lambda_i - \mu_j - \beta c_{ij} - \delta_i \gamma_i \psi_i + \delta_j \gamma_j \psi_j \tag{59}$$

or, by making use of (57):

$$T_{ij} = \exp(-\lambda_i - \delta_i \gamma_i \psi_j) \cdot \exp(-\mu_j - \beta c_{ij} + \delta_j \gamma_j \psi_j) \tag{60}$$

or

$$\sum_{j=1}^{J} T_{ij} = O_i = \exp(-\lambda_i - \delta_i \gamma_i \psi_i) \sum_{j=1}^{J} \exp(-\mu_j - \beta c_{ij} + \delta_j \gamma_j \psi_j) \tag{61}$$

so that

$$p_{ij} = \frac{T_{ij}}{O_i} = \exp(-\mu_j - \beta c_{ij} + \delta_j \gamma_j \psi_j) / \sum_{j=1}^{J} \exp(-\mu_j - \beta c_{ij} + \delta_j \gamma_j \psi_j) \tag{62}$$

The latter expression is again a logit model, but it has a more general form than the standard MNL model emerging from the conventional stochastic discrete choice theory. This implies that our dynamic SI model is equivalent to a more comprehensive disaggregate stochastic utility model. By defining now:

$$\left.\begin{aligned} &\exp(-\lambda_i)/O_i = A_i \\ &\exp(-\mu_i)/D_j = B_j \\ &\exp(\delta_j \gamma_j \psi_j) = G_j \end{aligned}\right\} \tag{63}$$

we may write p_{ij} as follows:

$$p_{ij} = \frac{T_{ij}}{O_i} = B_j D_j G_j \exp(-\beta c_{ij}) / \sum_{j=1}^{J} B_j D_j G_j \exp(-\beta c_{ij}) \qquad (64)$$

If we compare (64) with (6), it is easily seen that (64) includes an additional weighting factor G_j caused by the attraction factor (in terms of a dynamic accessibility ψ_j) associated with each place of destination j.

The expression for T_{ij} is obviously also more general than the one presented in (4). It can also be demonstrated that in our case the following result arises:

$$T_{ij} = A_i G_i^{-1} O_i B_j D_j G_j \exp(-\beta c_{ij}) \qquad (65)$$

By defining the following balancing factors:

$$A_i^* = A_i G_i^{-1} \qquad (66)$$

and

$$B_j^* = B_j G_j \qquad (67)$$

it is easily seen that these factors are equal to:

$$A_i^* = \left\{ \sum_{j=1}^{J} B_j^* \, D_j \exp\,(-\beta c_{ij}) \right\}^{-1} \qquad (68)$$

and

$$B_j^* = \left\{ \sum_{i=1}^{I} A_i^* \, O_i \exp\,(-\beta c_{ij}) \right\}^{-1} \qquad (69)$$

Consequently, the final result of our dynamic entropy model leads to a generalized SI model.

After the discussion of the optimality conditions for the control variables, we have to analyze also the conditions for the state and co-state variables, as it is clear that the optimal control solution for T_{ij} still contains the (as yet unknown) state variables O_i. This is the next step of the analysis.

The optimality conditions for the state variables are:

$$\frac{\partial L}{\partial O_i} = -\dot{\psi}_i \qquad (70)$$

or:

$$\frac{\partial \left\{ -\sum_{i=1}^{I} \sum_{j=1}^{J} T_{ij}(\ln T_{ij} - 1) + \sum_{i=1}^{I} \psi_i \dot{O}_i \right\}}{\partial O_i} = -\dot{\psi}_i - \lambda_i \qquad (71)$$

By next substituting (65) and (28) into (71), some tedious mathematical operations are necessary in order to arive at a result for the state and co-state variables. For the sake of simplicity, we will only give the result here for the state and co-state variables in the *unconstrained* SI model:

$$-\left\{\sum_{j=1}^{J} G_i^{-1} D_j G_j \exp(-\beta c_{ij}) \ \ln\ [O_i G_i^{-1} D_j G_j \exp(-\beta c_{ij})]\right\}$$

$$- \psi_i \delta_i \gamma_i = -\dot{\psi}_i - \lambda_i \tag{72}$$

or

$$\dot{\psi}_i - \psi_i \delta_i \gamma_i - \sum_{j=1}^{J} (T_{ij}/O_i) \ \ln\ T_{ij} + \lambda_i = 0 \tag{73}$$

The result cannot be elaborated in an analytical sense, as is the usual situation in SI models. However, it can be solved in a recursive numerical way. One may expect that the final result is unique, as we are dealing with a concave programming model subject to linear constraints.

APPENDIX 4. AN OPTIMAL CONTROL FOR THE ALONSO MODEL

In the light of Appendix 3, it is clear that in the case of Alonso's model only the expressions for the marginals O_i and D_j will change (see (9) and (10)).Thus the result analogus to (62) is:

$$p_j = \frac{T_{ij}}{V_i W_i^{\alpha_i}} \tag{74}$$

$$= \exp(-\mu_j - \beta c_{ij} + \delta_j \gamma_j \psi_j) / \sum_{j=1}^{J} \exp(-\mu_j - \beta c_{ij} + \delta_j \gamma_j \psi_j)$$

By now defining (see expressions (41), (42) and (63)):

$$\left.\begin{aligned} \exp(-\lambda_i)/V_i &= W_i^{\alpha_i - 1} \\ \exp(-\mu_j)/Y_j &= Z_j^{\varepsilon_j - 1} \\ \exp(\delta_j \gamma_j \psi_j) &= G_j \end{aligned}\right\} \tag{75}$$

we may write p_{ij} as follows:

$$p_{ij} = \frac{T_{ij}}{V_i W_i^{\alpha_i}} \tag{76}$$

$$= Y_j Z_j^{\varepsilon_j - 1} G_j \exp(-\beta c_{ij}) / \sum_{j=1}^{J} Y_j Z_j^{\varepsilon_j - 1} G_j \exp(-\beta c_{ij})$$

Expression (76) is analogous to (18) the only difference being that (76) contains the usual attraction factor G_j. Therefore:

$$T_{ij} = V_i W_i^{\alpha_i - 1} G_i^{-1} Y_j Z_j^{\varepsilon_j - 1} G_j \exp(-\beta c_{ij}) \tag{77}$$

If we define (see expression (11))

$$V_i W_i^{\alpha_i - 1} = O_i^* \tag{78}$$

and

$$Y_j Z_j^{\varepsilon_j - 1} = D_j^* \tag{79}$$

we get:

$$T_{ij}^* = O_i^* \, G_i^{-1} D_j^* \, G_j \exp(-\beta c_{ij}) \tag{80}$$

We can easily see that when α and ε are zero, we obtain as a particular case expression (65) (where $A_i = W_i^{-1}$ and $B_j = Z_j^{-1}$). Thus the dynamic Alonso model is also compatible with an optimal control entropy model.

For the second optimality condition from (57), i.e. $\frac{\partial L}{\partial O_i} = -\dot{\psi}_i$, we can easily see that no drastic changes occur. The only difference is that the original expression (71) contains T_{ij}^* instead of T_{ij}. Finally, we will find the following results analogous to (72) and (73):

$$-\left\{ \sum_{j=1}^{J} G_i^{-1} D_j^* \, G_j \exp(-\beta c_{ij}) \, \ln[O_i^* \, G_i^{-1} D_j^* \, G_j \exp(-\beta c_{ij})] \right\}$$

$$-\psi_i \delta_i \gamma_i = -\dot{\psi}_i - \lambda_i \tag{81}$$

and:

$$\dot{\psi}_i - \gamma_i \delta_i \psi_i - \sum_{j=1}^{J} (T_{ij}^* / O_i^*) \ln T_{ij}^* + \lambda_i = 0 \tag{82}$$

Compared to (72) and (73) the latter result from the Alonso model is more interesting, as (81) and (82) are based on a doubly constrained spatial interaction model, whereas (72) and (73) are based on an unconstrained model.

PART II

DYNAMIC URBAN MODELS

DENISE PUMAIN

SPATIAL DYNAMICS AND URBAN MODELS

1. THE SELF-ORGANIZATION IDEAS: NEW WINE OR NEW BOTTLES?

Analogies and model transfer from one branch of research to another are often fruitful. They encourage new development in knowledge by using already elaborated formalisms or models and by suggesting new questions which may lead to new fields of research. But they do not avoid the two opposite risks: vague and empty analogy or the use of ill-adapted methodology which severely distorts the subject under study.

1.1. *Characteristics of Self-Organizing Systems*

One aim in designing dynamic models in geography is to understand and anticipate the evolution of socio-spatial systems. New dynamic models mainly draw upon analogies from physics, for instance chemical kinetics (Allen, 1978) or laser theory (Haag and Weidlich, 1984). The systems under study are described on at least two levels: system-wide (macroscopic variables) and elementary (microscopic level). There is a very large number of elements in the system. Sometimes a third level of description is added, with subsystems comprising unfixed numbers of elements. Differential equations describe the variation of a macroscopic variable over time. They are obtained by applying definitions of interactions at the microscopic level, i.e. between the elements of the system. The passage from the microscopic level to the macroscopic description of the system is the most arduous problem in such model building. It may lead to stochastic formulations when interactions between the elements do not have a deterministic form but their results are known only as probability distributions.

According to physical theories such as the "dissipative structures theory", phenomena of self-organization and of bifurcation may occur in open systems when they are maintained under an influx of energy. Those systems may organize themselves in structures which are created or destroyed during the evolution of the system. This evolution is both deterministic, according to a trajectory which can be predicted using the equations of a model describing interdependencies between variables; and random, or undetermined, during periods of instability when a change in structure (or a phase transition) can occur. The equations of the model may therefore admit several solutions or multiple equilibria, several possible trajectories corresponding to qualitatively different structures, between which the system may be driven towards one branch or another, towards a given form of organization, by the amplification of a small fluctuation (after Prigogine and Stengers, 1979 and Allen and Sanglier, 1981).

Many sources of instability intervene in the evolution of such open systems when they are situated far from equilibrium. On the one hand, they continually undergo internal fluctuations, variations in

J. Hauer et al. (eds.), Urban Dynamics and Spatial Choice Behaviour, 155–173.

the level of their characteristic variables (which may result from changes in the micro-states of the elements of the system); on the other hand, they are always subject to external perturbations stemming from their environment. An open system is then continually adjusting the level of its variables or the size of its subsystems. It maintains a relative structural persistence only when this structure constitutes, under the given conditions, a stable state for the system; that is a state towards which the system comes back after having gone slightly apart from it. The structure is then viewed as an attractor on the system's trajectory. The dynamic instability may induce a passage from one trajectory to another, from one structure to another structure, that is from a given qualitative behaviour of the system to another, through a bifurcation point.

This kind of model structure provides the tools for analyzing the system's dynamics in connection with the following questions:

- is there any equilibrium state? are there many? (the reference here is to a dynamic equilibrium, that is a stable structure, a qualitative behaviour of the system having a certain persistence);
- what is the kind of the equilibrium state: stationary, unstable, periodical?
- what is the consequence of a modification in the values of the parameters of the model upon the evolution of the system and upon its possible equilibrium states?

1.2. *Grounds for Analogy*

To what extent does it appear useful to develop an analogy between such physical systems and the systems which are usually studied by geographers?

When formalizing a geographical object in this framework as a dynamic spatial system, one has to define it as a set of elements which are either located elements or geographical zones. Location (at least relative location) must appear as a basic property of the elements of the system; the interactions between those elements must be at least partly spatial interactions, which are linked to expressions of the absolute (site) or relative (situation) location of the elements; and the state of the system is defined as the geographical configuration of its characteristic variables. A geographical structure is then given by a particular relative size and evolution for the state variables and/or located subsystems which are used for the description of the geographical system.

At the most disaggregated level, a spatial system can be formalized as a set of localized and interacting actors (individuals or groups as persons, households, firms, associations, etc.) who are using and continuously re-creating geographical differences and spatial configurations. These interactions are diverse in nature and form: competition for space, propensity to agglomerate, segregative tendencies, imitation, etc. According to the case, they generate the homogeneity or the diversity of contents, the increase or the decrease in gradients, the concentration or the dispersal in distributions. One has to hypothesize that the dynamics of these interactions is creating the spatial structures, for which they are together both an expression and a precondition for their existence. Such a formalization has many appealing features for geographers:

- It allows stress to be placed upon the linkages between the individual behaviour of the actors and interactions defined at this microscopic level on the one hand, and the macroscopic descriptors of the system on the other hand. Much research has been conducted and empirical regularities have been established for each scale of study separately, though a clear connection between the two levels of observation is not always established. Here, self-organization phenomena, meso- or macro-scale structures, are described as consequences of an interaction game between individuals, each animated by their own objectives. These consequences are not always intuitive to the observer, they are often not concerted and most of the time are not perceived by the actors.
- This approach may also provide an interpretative framework for observed regularities in spatial systems. Due to the instability of elements being considered[1], it is only because bifurcations occur in the evolution of dynamic systems, because their trajectories "jump" from one attractor to another, that identifiable and separable categories, or large-scale regularities, may be observed. Temporal series appear then as "possible" sequences of complex dynamics, and very often the problem for social sciences, where experimentation is impossible, consists in identifying the dynamics which produced a particular temporal series of observable structures; i.e. a specific trajectory (Prigogine and Stengers, 1979). In geography, according to R. Brunet (1980), the number of simple spatial structures (which he calls: "chorèmes") is reducible to around forty. They could be considered as a series of attractors among the set of all theoretically imaginable trajectories which represent the evolution of spatial configurations where random fluctuations are occurring.
- Another interesting feature of this approach is that the historical dimension of social systems is taken into account via the concept of irreversibility. On the one hand, the explanation of the state of a system at a given date integrates its previous trajectory (i.e. its history) and the contemporary structure includes the "memory" of previous bifurcations. On the other hand, the characteristic fluctuations of dynamic systems imply that it is impossible to prepare initial conditions which would lead to identical futures. The impossibility of exact prediction is then given as a theoretical "a priori". However, the analysis of the dynamic behaviour of the systems and of their sensitivity to changes in parameter values allows the exploration of a limited number of possible futures, according to the assumptions made about the evolution of the parameters (Allen and Sanglier, 1981).

These ideas may appear seductive to geographers. But it is not because of their novelty: all of them were originally expressed a considerable time ago. Rather it is because they allow the relaxation of some of the more oppressive restrictions imposed by the previous methods of model building, and because new experimental tools related to these old ideas are therefore available.

2. SOME DIRECTIONS IN URBAN AND REGIONAL MODELLING

Instead of trying to present a complete review of existing models, which could never be exhaustive and would soon become obsolete, this chapter will outline some examples which illustrate the variety of approaches used in this approach to model building[2].

At least in human geography, most of the available examples are found in fields where rather stable, general and well-known spatial structures can be observed and for which well-developed and efficient descriptive and sometimes predictive models already existed:

- intra-urban or interregional migration flows;
- intra-urban distribution of population densities and land prices or rents;
- intra-urban distribution of residential population and service activities (extensions of the Garin-Lowry types of models);
- inter-urban distribution of population sizes and functional hierarchical levels (development of central places networks).

However, few applications of these new dynamic models to real-world data have yet been attempted. As such it is still difficult to give them as much credit for their empirical usefulness as it is for their potential theoretical value.

It has been shown that, when using nonlinear equations and/or interdependencies between variables, even very simple mathematical models can produce very complicated dynamics (May, 1976). Some of the models which have been proposed by geographers are too complex to be analytically tractable and must be solved and calibrated by means of simulation. Another potential method of classifying dynamic models of spatial structures is therefore to consider how they choose between the advantages of complexity (allowing possibly more realistic representations and also more complete descriptions of spatial features) and the gain in generality and soundness of results offered by the existence of more or less well-known analytical solutions.

2.1. *'Simple' Models*

In these models, formulations in use in urban and regional analysis are more or less directly drawn from findings in other fields: catastrophe theory in mathematics, Volterra-Lotka models of interacting species in biology, and master-equation techniques from physics. All these models can be solved analytically with the implication that one knows how to identify the equilibrium states of the system, how many there are, if they are stable or not. One is then able to situate the observed trajectory with respect to these equilibria and to make predictions about the stability of the structure according to hypothetical variations in the values of parameters.

2.1.1. *Catastrophe Theory*

Direct applications of R. Thom's (1974) catastrophe theory to the study of spatial structures seem to be limited. Despite that fact that this theory provides very good analytical tools and theorems about the number and precise qualitative shape of the functions linking the state-variables to the elementary catastrophes, it allows too small a number of variables and parameters to describe the system (respectively a maximum of two and four), so that it cannot be of great help in most geographical problems. Lung (1985) underlines the difficulties in constructing models which satisfy the basic assumptions of the theory: how to define a potential function, how to incorporate competition for space, and how to find relevant state-variables which may undergo sudden changes yet produce another qualitative structure for smooth variations in the values of parameters.

Most of the studies in this field are still exploratory. A good review is given by Wilson (1981). Usually the spatial structure is summarized by a global measure of size, or density, or intensity of spatial interaction, whose variations are related to a few control parameters. For instance, Amson (1975) considers urban density as a state-variable, which is dependent upon two parameters: it is proportional to rental and inversely proportional to "opulence", which is a measure of the benefit from urban interaction. He then shows that urban density follows variations according to the fold model if a saturation effect is introduced, with rental becoming a logistic function of density. If a minimal threshold of available space per person is added, the model for the variations of the density becomes a cusp. The condition for a smooth and continuous behaviour of the density is that one of the parameters, rental or opulence, will keep high values.

Catastrophe theory is used by many authors to describe discontinuities in urban growth. For example, Mees (1975) gives an interpretation of the revival of cities in Europe between the 11th and 13th centuries, following a long period of urban decline, as a discontinuity in the relationship between the urbanization rate and the possibility of external trade. This discontinuity appears at a critical value in decreasing transportation prices. A more comprehensive model involving a butterfly catastrophe takes as a state-variable the population working in manufacturing in towns, which is a function of four control variables: density of regional population, average productivity, urban-rural productivity differential and difficulty of transport. Papageorgiou (1980) tries to explain the cases of sudden urban growth observed in the 19th and 20th centuries by the existence of scale economies in the utility function of cities for individuals. Technological changes, even when utility increases in a continuous manner, may at a given level introduce a discontinuity in the urbanization rate, which then jumps from one equilibrium point to another, in accordance with a fold catastrophe.

A third approach consists in explaining the existence of thresholds and hierarchical levels in urban systems by the jumps and hysteresis phenomena of catastrophe theory. In this fashion Casti and Swain (1975) suggest that the functional level of a central place may be seen as a cusp function of two control variables: population and per capita income, whereas Wilson (1981) relates the size of various urban facilities and urban centers to the benefits of facility size and to the disbenefits of travel. If the first control variable is a logistic function of size, and the second a linear one, the size may vary with the slope of this last function according to a kind of fold catastrophe.

The most interesting feature of this family of models seems then that they provide good theoretical insights in the discontinuous behaviour of some aggregate variables. But they are still of little help in the modelling of spatial structure or spatial interaction.

2.1.2. *Volterra-Lotka's Model*

This formulation depicts the evolution of the number of individuals belonging to two interacting biological species. Each species is also characterized by a birth-death process of the logistic type. Translations of this formulation to geographical analysis have been attempted for instance by Dendrinos and Mullaly (1981), but without true spatial dynamics. In their urban model the variables used in place of species are the city's population and its mean per capita income. The model then simulates the evolution of the share of an SMSA's population in the total national population x and of its relative level of per capita income y. So spatial interaction is not easily included in that kind of model structure. Equations of the model are of the following form:

$$\frac{dx}{dt} = \alpha \ (x(y - 1) - \beta x)$$

$$\frac{dy}{dt} = y \ \gamma \ (\bar{x} - x)$$

where $\bar{x}$ is a carrying capacity of the SMSA, α and γ are speeds of adjustment and β a coefficient of "urban friction". This model was chosen because it allows replication of one of the most frequent behaviours observed among the 90 SMSAs of the United States between 1940 and 1977. Such behaviour is the sink-spiral type, with oscillations around successive equilibria. However, the scarcity of available data does not allow a strong empirical support for this model (Figure 1).

Spatial interaction is not easily included in this kind of model structure: it appears only in a weak and implicit manner, since the size of one element in the system is expressed by a relative measure, as a proportion of the total size of the elements of the system. The same way of solving the spatial disaggregation problem has also been used for a dynamic model of regional development (Dendrinos, 1984) which describes "interaction" between the population and income of nine regions of the United States. In this application, the precise analogy with the original model is by no means clear; moreover the meaning of some of the parameters used in the equations is not well specified.

2.1.3. *Master-Equation Approach*

This technique relies upon theoretical principles used in the field of synergetics. It is potentially of great interest to geographers, because it links explicitly the state transition probabilities of individuals at a micro-level, and the evolution of some variables describing a macroscopic structure. The master equation gives the variation in time of the probability of the possible configurations in the space of the state-variables. This probability of transition from one configuration to another depends on the assumptions made about the number and nature of parameters affecting the individual transition probabilities. This stochastic formulation is then used to derive a deterministic equation for the evolution of the mean values which, in turn, allows the estimation of the parameters.

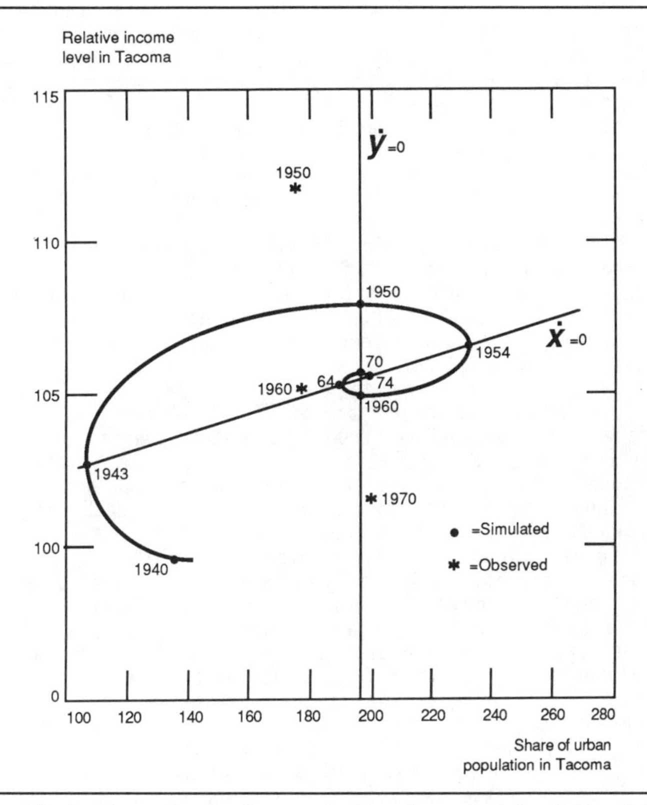

Fig. 1: Simulation of an urban evolution by a model of the Volterra-Lotka type (after Dendrinos and Mullaly, 1980).

This procedure has been used for deriving a two-populations-two zones model of the Volterra-Lotka type (Haag, 1984). The migration rate of each population group from one zone to another depends on its preference for a given part of the city, of its propensity to cluster ("internal sympathy"), and to join members of the other group ("external sympathy"), and of a general mobility level (flexibility). The authors explore analytically all possible types of spatial configuration and levels of urban segregation according to the sets of values taken by these four parameters. In this simple version, such a model is a good pedagogic presentation of elementary spatial dynamics.

The master equation approach has also been used for reformulating an intra-urban model of residential rent and density interactions (Haag and Dendrinos, 1983) and has been applied to twelve SMSAs of the United States (Dendrinos and Haag, 1984). The SMSAs are divided into two parts: the city centre, and the suburbs. Utility functions depending on rent and density levels through land availability are used to describe the individual behaviour of typical land-buyers (who move) and sellers (who transfer rental value from one zone to another). Aggregation of these individual behaviours produces a stochastic master equation, whose deterministic mean value describes the evolution of the population share and of the relative rent price in the central part of the urban unit. From simulations with empirical data, the authors predict a probable reversal in the suburbanization trend for the 1990s, if the parameter values which are standing for common general conditions of an SMSA's environment do not change over time.

A model of migration for a whole set of regions has been constructed under the same approach (Haag and Weidlich, 1984). Individual transition probabilities from one spatial subdivision to another are defined as functions of the difference in utility of regions. They are aggregated to define an equation of movement of the probability of states of the system (e.g.: all possible configurations of population repartition among regions), whose mean values equation is used for the estimation of parameters. The dynamics of the spatial system are then related to trend parameters whose values may be compared from one system to another, or over time (global mobility rate, "cooperation", that is an agglomeration effect, and a saturation effect), and to a set of preferences for each region. The socio-economic variables which could explain these preferences are not included in the dynamic model. However, when fitted by a regression model upon a temporal series of regional preferences as established from the model (after for instance annual migration tables), they allow to a certain extent a prediction of future migration patterns and of the further evolution of the regional spatial structure.

An interesting feature of this approach is that it provides a clear relationship between the spatial behaviour of individuals and the global dynamics of a spatial structure as measured by aggregated variables. All parameters, being defined at the micro-level, receive a straightforward interpretation. This is always the case in models dealing only with state-variables for the whole system.

2.2. *'Simple-Complex' Models*

In this category the models of the "Leeds school" developed under A. Wilson's direction can be considered. They are mainly models of intra-urban structure, concerning the location of activities and populations and the interaction flows that they generate. As information about those models is developed elsewhere, only the novel features of this approach will be emphasized here.

The "Leeds' approach" is related to the types of modelling previously quoted in the sense that analytical solutions are explored, at least for the more general formulations of the models (Wilson, 1981; Beaumont *et al.*, 1981a, 1981b). However, the approach always refers to a spatial structure (allowing a large number of zones in consideration) and attempts to model the interdependencies between the supply-side and the demand-side in the spatial dynamics. The aim is to produce an integrated model of urban structure, with relevant disaggregation of population and activities and a connection with determinants of individual behaviour.

The basic equations of these models stem from the simple interaction model:

$$F_{ij} = \frac{P_i W_j^{\alpha} e^{-\beta c_{ij}}}{\sum_j W_j^{\alpha} e^{-\beta c_{ij}}}$$

where the flow F_{ij} between a zone of residence i and a zone of destination j where a service is offered depends on W_j, a measure of attractivity of zone j for this service, P_i being the number of residents in zone i. α is a parameter of sensitivity to scale economies for consumers, and β a measure of their sensitivity to travel costs. In the dynamic versions of the model the supply variable W_j is determined endogenously as it adjusts itself to the total demand in each zone $D_j = \sum_i F_{ij}$:

$$\frac{dW_j}{dt} = \varepsilon\ (D_j - kW_j)$$

where ε is a speed of adjustment.

According to the values of the parameters, bifurcations may occur in the determination of the size and number of service centres. Such bifurcations and equilibrium points may be studied analytically for variations of one parameter, the other being held constant (see Figure 2), but the global dynamics of the system can only be examined by means of simulation.

An analogous form is retained for a second model, connected with the previous one, which locates the residental population of the city after the location of basic employment. Disaggregated versions considering various income groups, types of housing and kinds of

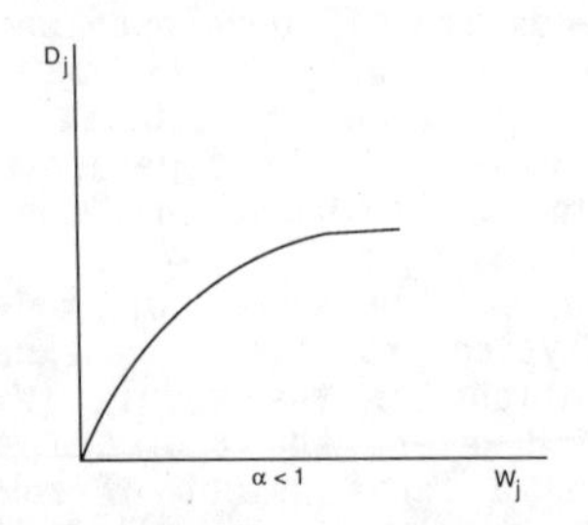

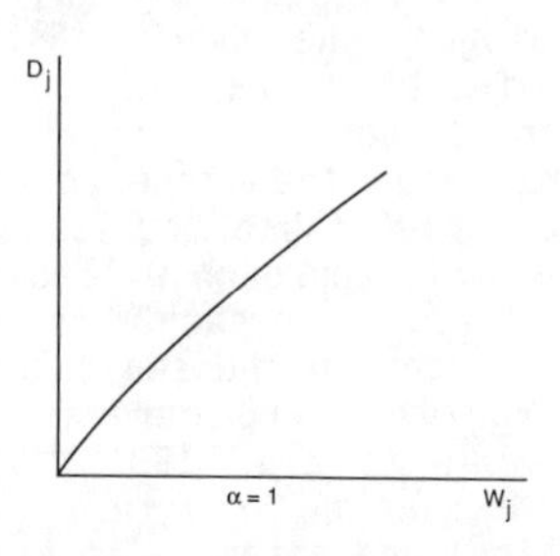

Effect of the taste of consumers regarding the size of centres (α) on the relationship between supply (W_j) and demand (D_j) for services.

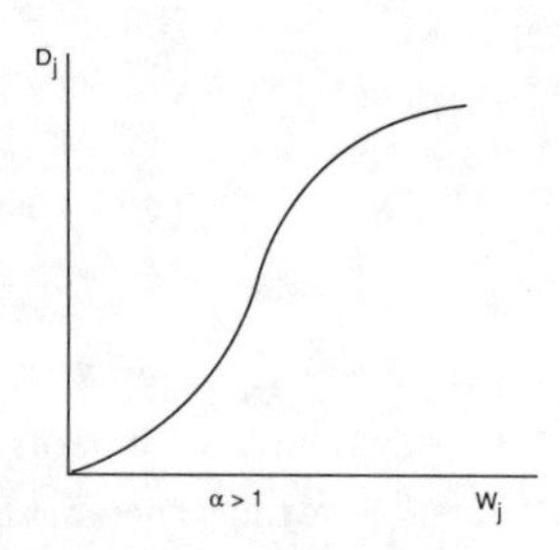

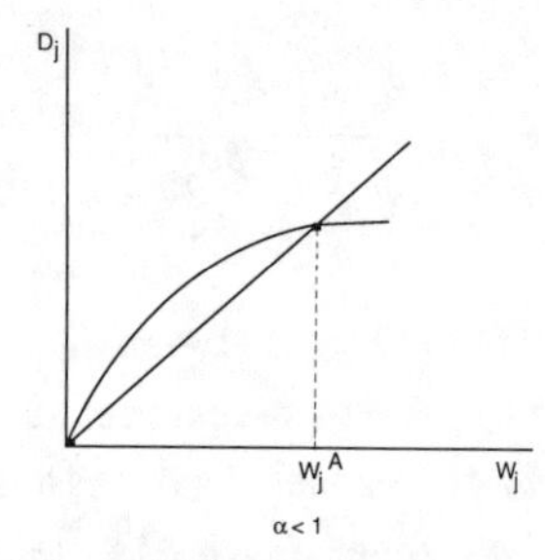

Points of stable (B) and instable (A) equilibria between supply (W_j) and demand (D_j) for services.

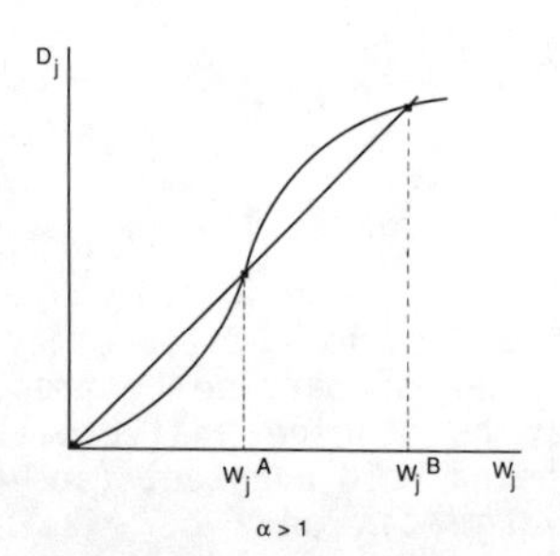

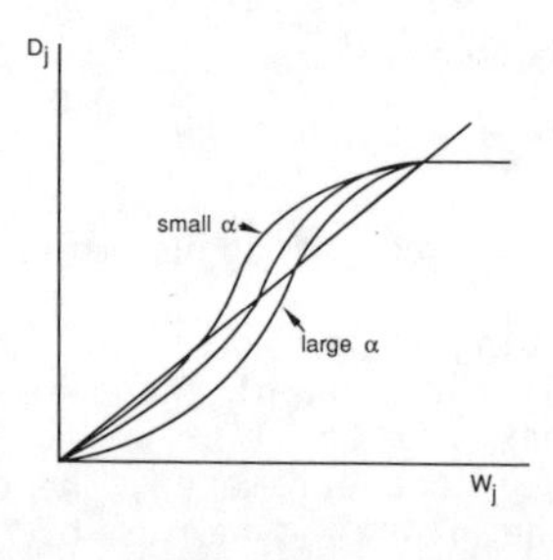

Effect of the taste of consumers regarding the size of centres (α) on size thresholds.

Fig. 2: Bifurcations in the development of service centers (after Wilson, 1981).

economic activities have been studied theoretically but not yet applied empirically to the evolution of particular urban areas.

2.3. *Complex Models*

Models of the Brussels school are undoubtedly more complex than the models previously considered. They deal mainly with central-place development (Allen and Sanglier, 1979b) interregional shifts in population and tertiary employment (with one application at the state-level in the United States: Allen and Engelen, 1984 and one at the province-level in the Netherlands: Allen et al., 1984a) and to intra-urban redistribution in labour force and residential population (Allen et al., 1981), with an application to the Brussels region (Allen et al., 1984b).

In these models, the connection with well-known mathematical processes is less clear and the equations cannot be solved analytically. Spatial interactions are modelled mainly by the means of attractivity functions, which characterize the advantages of one location in comparison with all other possible locations for each variable in the system. The mathematical expression of such attractivity functions is very complex and the dynamic behaviour of the variables can be studied only by means of simulation. However, in these equations integration is achieved of a very large number of the theoretical proposals which were formulated to explain the dynamics of geographical structures. For instance, in the intra-urban model, mathematical relations or parameters stand for: multiplier mechanisms of the economic base, spatial interactions of an intervening opportunities-type, growth upon limited space according to a logistic curve, mechanisms of competition for space and of social segregation, unequal speed of reaction, and similarity in behaviour according to the information available to the urban actors.

This kind of model provides then a useful tool to test the effects of processes which act simultaneously in the evolution of a spatial structure. They allow us to observe the evolution of the system in a global way, which is in a sense closer to real-world conditions than is the case with other types of models. However, when using these very complex models for applications, one has to be very confident of their quality of representation, in their selection of relevant descriptors as well as the meaning of the parameters that they include. Some of the specific problems associated with the *use* of dynamic models of complex spatial structures will be discussed next.

3. PROBLEMS IN APPLYING COMPLEX DYNAMIC MODELS OF SPATIAL STRUCTURE

The discussion in this section relies heavily upon an experimental use of P. Allen's intra-urban model to study the evolution of four French agglomerations (Pumain et al., 1985), but evidence is also derived from the use of a Wilson-type model for Rome (Lombardo and Rabino, 1983). Bearing in mind the complexity of these dynamic models, application to real-world situations might seem a premature exercise. Certainly such applications involve a plurality of difficulties, primarily because they use mathematical devices whose behaviour is still not well-known, and because they call upon numerous and not always readily available data.

Large urban models have suffered from many criticisms. Nowadays, however, stress is placed more upon qualitative features of spatial urban development than on absolute prediction of growth. The ability of dynamic models to simulate such qualitative changes is one of their characteristics which give them more realistic chances of success. However, a valuable way to improve the design of future models is to consider the main difficulties encountered in their application. Problems are of diverse nature, being either conceptual or more practical.

3.1. *Conceptual Problems*

The objectives of such applications are usually threefold: first is to check that the model is a good description of the kind of system under study; second is to test that the model is able to replicate an observed evolution of a particular system; third is to assess whether it may under given assumptions lead to plausible predicted trajectories for the future of the system. The imprecision which may arise when borrowing concepts from another field have then to be clarified. For instance, when taken theoretically, the use of the bifurcation concept looks rather simple: one slight change in the value of a parameter can drive the system on another trajectory with a different kind of spatial configuration, i.e. a structural change can occur. However, when using those models for practical purposes, we face the reverse problem. Observing a qualitative change in spatial configuration we need to assess if it corresponds to a bifurcation. Would one slight modification in one parameter value have been sufficient to avoid it, or to alter it, in this region of instability, or is this change more deeply tied to the functioning of the system? In that case, it would be naturally produced by the same trajectory defined by the nonlinear equations and the corresponding set of values for the parameters would have to be estimated. So, observing the evolution of an urban structure, we face the question of discovering a plausible dynamic path which may have generated it. But to what extent is it reliable enough to support prediction?

As a practical illustration, we might consider two different kinds of interpretations which have been proposed to explain the reversal in migratory flows between the core and periphery of urban areas which has appeared recently in most industrialized countries:

1. a "clean break with the past", a discontinuity or even the end of the urbanization process; or
2. just part of a cyclical behaviour as regards the agglomeration and dispersion tendencies, due to continuous variations in the values of parameters like income, ease of transportation, or average spatial needs.

In these cases, alternative trajectories which could be constructed in order to illustrate the bifurcation would not be the same. This most irritating and fascinating aspect of transferring concepts is in the end linked to the classical problem of model specification and selection of variables - which in turn is closely tied to the problem of the availability of data.

3.2. *Data Problems*

These are not specific to this kind of model building but are of decisive importance in the quality of applications. The scarcity of localized temporal statistical series at a detailed geographical level is well-known, even for aggregate variables. For instance, when testing his model of urban development, Dendrinos and Mullaly (1981) had only three points in time to check the possible interaction between urban income and population (Fig. 1). For models needing information at a micro-level, the lack of data is even worse, as surveys establishing conditional probabilities in spatial behaviour are still far too rare.

The lack of empirical knowledge about the relevant range of values for the parameters of the models is another worrying problem; for instance in applying the intra-urban model to French cities, no published estimations of parameters relating to propensity of activities to agglomerate, differential sensitivity of populations to the length of commuting, intensity of segregative tendencies between social groups, induction rates between service employment and resident population, agglomeration economies, speed of adjustment of entrepreneurs to the demand, etcetera could be found. All these values had to be estimated by means of calibration of the observed evolution of employment and residental population in the communes of the four French agglomerations being studied. Dynamic modelling could improve this situation by helping to select the parameters which are decisive in the construction of spatial structures and thus to encourage the development of the corresponding measurements.

3.3. *Calibration Problems*

Considering the extreme complexity of this kind of model and its sensitivity to small variations in parameters, the calibrating operations, trying to reproduce by simulation an observed evolution, are always long and difficult. The usual automatic procedures for calibration, even when designed for a large number of simultaneously varying parameters, are ineffective. Because of bifurcation properties of such systems, many sets of parameter values may give local optima for the criteria which evaluate the global deviation between observed and simulated evolution. The values of the parameters then have to be determined by a trial and error method. This type of calibration is arduous and costly, since the effect of one slight variation in the value of one parameter may differ according to the configuration of values for the other parameters: each time a bifurcation occurs, the calibration process has to come back to a previous stage.

The evaluation of a goodness-of-fit which is good enough to stop the calibration process is not straightforward either. As usual in geographical modelling, there is a contradiction between the hypothesis that general processes are valuable for all spatial units, and the lack of precision in estimation is due to specific local conditions. For instance in both P. Allen's intra-urban model and in Wilson's model, the city is represented as an open system where internal interactions determine the urban structure, starting from initial conditions which incorporate the whole history of the system.

In addition, a more or less implicit assumption is that these interactions operate in the context of a liberal economy, where adjustments between supply and demand and a relatively free competition for space are the main mechanisms governing the urban dynamics. As a result, the few applications which have been conducted have shown that calibration was more difficult and that the residuals between observed and simulated evolutions were larger when planning operations had been into effect. This was the case, for example, in regions of the Netherlands where large housing programs had been developed (Allen *et al.*, 1984a) and in those areas of French agglomerations where large building zones for social housing had been designed. This leads to the paradoxical situation (known as the "polder paradox") that the more efficient previous planning has been, the less the model can operate; therefore the better you planned the less you can predict a future evolution.

3.4. *Interpretation Problems*

In most of the applications discussed above it appears that the models were able to reproduce a large proportion of the wide variety of changes observed in most of the spatial units, and that the largest residuals between simulated and observed evolution reflected specific local conditions which were not taken into account by the model. In other words, the general mechanisms incorporated into the model seem to describe rather well the evolution of urban structure.

However, a problem remains with the interpretation of the values calibrated for the parameters. The estimated values are not independent of the number of spatial divisions being considered, of the size of these zones, or of the measurement unit of the variables. As a result, it is almost impossible to arrive at a standardization of the parameters. For example, in order to avoid the inequalities in areas of administrative urban units for which data are collected, one may think of converting absolute numbers of jobs and residents in Allen's model to densities. But the theoretical considerations which are included in the specification of the model are actually related to quantities. Moreover, the general logistic form of the equations renders parameter values dependent on the absolute numbers which express the size order of variables.

In addition, it is important to remember that the parameter values resulting from calibration reflect the joint effects of both the actual dynamics of the system and the way it has been disaggregated. When constructing or applying these types of dynamic models, special attention must therefore be given to the disaggregation problem: the design of spatial units, the selection of the state-variables at the micro and macro levels, according to the range of their spatial action and also to the time-scale of their intervention (i.e. whether they are fast or slow dynamics).

These considerations are of particular importance when such models are used for comparative studies. If the parameters of the models cannot be standardized, procedures which permit the comparison of the complex dynamics of spatial structures have to be developed. For instance, in the case of urban models where spatial dynamics are created by differential values in the attractivity functions, one may compute ratios between values taken by those functions for two different sub-areas. In the author's own application of Allen's model (Pumain *et al.*, 1985), ratios between the highest and second highest values for each component of the attractivity functions were used to compare the intensity of agglomerative tendencies in four urban areas where concentration of activities in the center was a permanent but inequally evolving feature of the spatial structure. By comparing attractivity ratios for spatial units which were either contiguous or separated by a given distance, it was possible to measure differentials in the effects of accessibility upon the development of service activities in the four urban areas.

The above represents an example of what has to be done if such models are to be used in a more operational context. If the objective is not only to make a qualitative exploration of various possible spatial configurations under given rules of individual behaviour, but also to get a deeper insight in the evolution of urban structures where no spectacular bifurcation occurs, it is vital to improve still further the experimental use of dynamic models.

4. CONCLUSION

This paper has focussed interest on dynamic models which try to explain the evolution of spatial structures at a meso or macro level from interactions operating between the elements of the system. Of course, this can only lead to a certain level of explanation. It can throw light on systems where the dynamics of elementary spatial interactions produces a morphogenesis; a spatial structure as observed at an upper level, the properties of which were not intended by the elements. For example, the intentions of the individual actors are not from the outset to produce an urban system, with a given spatial organization (but when it exists they use it and may try to improve some of its properties).

This particular level of understanding will also be of greater interest if many systems of different types appear to show the same kind of dynamic behaviour; if, as Winiwarter (1984) has suggested, some "isodynamics" (a word forged like "isomorphisms") can be observed. Much work has still to be done in linking spatial properties of individual behaviour, interaction processes and the resulting spatial configurations, as expressions of a limited number of typical dynamics.

Bifurcation models in the field of urban analysis seem for the moment better designed for a theoretical or even pedagogical understanding of the qualitative features of urban structures and of their development than for operation use. However, even if applications are still difficult, mainly because of data problems, they are essential for a better evaluation of the practical interest of each type of model. In particular, comparative applications of models with different specifications to the same urban area are essential.

I.N.E.D.
Paris
France

NOTES

[1] We have shown elsewhere (Pumain et al., 1986) examples of instability in the evolution of the elements belonging to urban systems.
[2] Models where spatial interaction is not explicitly taken into account will *not* be considered (e.g. many applications in geography try to explain discontinuities in the temporal evolution in the size of spatial entities. They may use several places to estimate the values of the parameters but they do not incorporate the effects of spatial competition in the construction of a spatial structure).

BIBLIOGRAPHY

Allen, P.: 1978, 'Dynamique des centres urbains', *Sciences et Techniques* **50**, 15-19.
Allen, P. and G. Engelen: 1984, 'Modélisation de l'évolution des Etats-Unis 1950-1970', in Y. Guermond (ed.) *L'Analyse de Système en Géographie*, Lyon, PUL.
Allen, P., G. Engelen, and M. Sanglier: 1984a, 'Computer handled efficiency stimuli exploration', Final Report, Noord Holland, Contract for Provinciaal Bureau Energiebesparing.
Allen, P., G. Engelen, and M. Sanglier: 1984b, 'Self-organizing systems and the laws of socio-economic geography', Brussels working papers on spatial analysis, series A, no. 4.
Allen, P. and M. Sanglier: 1978, 'Dynamic models of urban growth', *Journal of Social and Biological Structures* **1**, 265-280.
Allen, P. and M. Sanglier: 1979a, 'Dynamic models of urban growth', *Journal of Social and Biological Structures* **2**, 269-298.
Allen, P. and M. Sanglier: 1979b, 'A dynamic model of growth in a central place system', *Geographical Analysis* **11**, 256-272.
Allen, P. and M. Sanglier: 1981, 'Urban evolution, self-organization and decision-making', *Environment and Planning* **A 13**, 169-183.
Allen, P., M. Sanglier, F. Boon, J.L. Deneubourg, and A. de Palma: 1981, *Models of Urban Settlement and Structure as Dynamic Self-Organizing Systems*, Washington D.C., Department of Transportation.
Amoral: 1984, 'La dynamique de système, une méthode de modélisation des unités spatiales', *L'Espace Géographique* **2**, 81-93.
Amson, J.C.: 1975, 'Catastrophe theory: a contribution to the study of urban systems', *Environment and Planning* **A 2**, 175-221.
Batty, M.: 1976, *Urban Modelling*, Cambridge, Cambridge University Press.
Batty, M. and B. Hutchinson (eds.): 1983, *Systems Analysis in Urban Policy-Making and Planning*, Oxford, Plenum Press.
Batty, M. and Karmeshu: 1983, 'A strategy for generating and testing models of migration and urban growth', *Regional Studies* **17**, 223-236.
Beaumont, J.R., M. Clarke, and A.G. Wilson: 1981a, 'The dynamics of urban spatial structure: some exploratory results using difference equations and bifurcation theory', *Environment and Planning* **A 13**, 1473-1483.

Beaumont, J.R., M. Clarke, and A.G. Wilson: 1981b, 'Changing energy parameters and the evolution of urban spatial structure', *Regional Science and Urban Economics* **11**, 287-315.
Bennett, R.J., R.P. Haining, and A.G. Wilson: 1985, 'Spatial structure, spatial interaction, and their integration: a review of alternative models', *Environment and Planning* **A 19**, 625-645.
Bertuglia, C.S., S. Ocelli, G. Rabino, C. Salomone, and R. Tadei: 1983, 'The dynamics of Turin metropolitan area: a model for the analysis of the processes and for the policy evaluation', Paper presented at the 23rd Congress of the Regional Science Association, Poitiers.
Bertuglia, C.S., S. Ocelli, G. Rabino, and R. Tadei: 1980, 'A model of urban structure and development of Turin: theoretical aspects', *Sistemi Urbani* **2**, 59-90.
Brunet, R.: 1980, 'La composition des modèles dans l'analyse spatiale', *L'Espace Géographique* **4**, 253-264.
Bussiere, R. and T. Stovall: 1981, *Systèmes Evolutifs Urbains et Régionaux à l'Etat d'Equilibre*, Paris, C.R.U..
Casti, J. and H. Swain: 1975, 'Catastrophe theory and urban processes', RM-75-14, I.I.A.S.A., Laxenburg, Austria.
Clarke, M. and A.G. Wilson: 1983a, 'The dynamics of urban spatial structure: progress and problems', *Journal of Regional Science* **23**, 1-18.
Clarke, M. and A.G. Wilson: 1983b, 'Exploring the dynamics of urban housing structures: a 56 parameter residential location and housing model', Paper presented at the 23rd Congress of the Regional Science Association, Poitiers.
Couclelis, H.: 1985, 'Cellular worlds: a framework for modeling micro-macro dynamics', *Environment and Planning* **A 17**, 585-596.
Dendrinos, D.S.: 1980, 'A basic model of urban dynamics expressed as a set of Volterra-Lotka equations', in *Catastrophe Theory in Urban and Transport Analysis*, Washington D.C., Department of Transportation.
Dendrinos, D.S.: 1984, 'The structural stability of the US regions: evidence and theoretical underpinnings', *Environment and Planning* **A 16**, 1433-1443.
Dendrinos, D.S. and G. Haag: 1984, 'Toward a stochastic dynamical theory of location: empirical evidence', *Geographical Analysis* **16**, 287-300.
Dendrinos, D.S. and H. Mullaly: 1981, 'Evolutionary patterns of urban populations', *Geographical Analysis* **13**, 328-344.
Domanski, R. and A.P. Wierzbicki: 1983, 'Self organization in dynamic settlement systems', *Papers and Proceedings of the Regional Science Association* **51**, 141-160.
Ernst, I.: 1978, 'Formalisation du processus d'urbanisation: une application de la théorie des catastrophes', *Cahiers de la SEMA* **2**, 75-90.
Forrester, J.W.: 1969, *Urban Dynamics*, Cambridge (Mass.) M.I.T. Press.
Griffith, D.A. and A.C. Lea (eds.): 1984, *Evolving Geographical Structures*, NATO Advanced Institute Series, Martinus Nijhoff, Den Haag.
Griffith, D.A. and R. MacKinnon (eds.): 1981, *Dynamic Spatial Models*, Sijthoff and Noordhoff, Alphen aan den Rijn, Netherlands.
Guermond, Y. (ed.): 1984, *Analyse de Système en Géographie*, Presses Universitaires de Lyon.

Haag, G.: 1984, 'A non linear dynamic model for the migration of human population', in D.A. Griffith and A.C. Lea (eds.) *Evolving Geographical Structures*, NATO Advanced Institute Series, Martinus Nijhoff, Den Haag.

Haag, G. and D.S. Dendrinos: 1983, 'Toward a stochastic dynamical theory of location: a nonlinear migration process', *Geographical Analysis* **15**, 269-286.

Haag. G. and W. Weidlich: 1984, 'A stochastic theory of interregional migration', *Geographical Analysis* **16**, 331-357.

Isard, W. and P. Liossatos: 1979, *Spatial Dynamics and Optimal Space-Time Development*, North-Holland, New York.

Lombardo, S. and G. Rabino: 1983, 'Non linear dynamic models for spatial interaction: the result of some empirical experiments', Paper presented at the 23rd European Congress of the Regional Science Association, Poitiers.

Lung, Y: 1985, 'A la recherche de nouvelles techniques ou d'un nouveau paradigme: à propos d'approches récentes de l'espace économique', *Cahiers d'Econométrie Appliqueé* **1**, 69-77.

May, R.M.: 1976, 'Simple mathematical models with very complicated dynamics', *Nature* **261**, 459-467.

Mees, A.I.: 1975, 'The revival of cities in medieval Europe. An application of catastrophe theory', *Regional Science and Urban Economics* **5**, 403-425.

Paelinck, J.: 1980, 'Dynamic urban growth models', *Papers of the Regional Science Association* **24**, 25-37.

Papageorgiou, G.J.: 1980, 'On sudden urban growth', *Environment and Planning* **A 12**, 1035-1050.

Peschel, M. and W. Mende: 1983, *Leben wir in einer Volterra Welt?*, Akademie Verlag, Berlin.

Pred, A.: 1984, 'Place as historically contingent process: structuration and the time-geography of becoming places', *Annals of the Association of American Geographers* **74**, 279-297.

Prigogine, I. and I. Stengers: 1979, *La Nouvelle Alliance*, Gallimard, Paris.

Pumain, D.: 1982, *La Dynamique des Villes*, Economica, Paris.

Pumain, D. and T. Saint-Julien: 1978, 'Les dimensions du changement urbain', Mémoires et Documents de Géographie, C.N.R.S., Paris.

Pumain, D., T. Saint-Julien, and L. Sanders: 1984, 'Dynamics of spatial structure in French urban agglomerations', *Papers of the Regional Science Association* **55**, 71-82.

Pumain, D., T. Saint-Julien, and L. Sanders: 1984, 'Vers une modélisation de la dynamique intra-urbaine', *L'Espace Géographique* **2**, 125-135.

Pumain, D., T. Saint-Julien, and L. Sanders: 1986, 'Urban dynamics of some French cities', *European Journal of Operational Research* **25**, 3-10.

Pumain, D., L. Sanders, and T. Saint-Julien: 1985, 'Dynamique des systèmes et milieu urbain', rapport pour le PIREN, C.N.R.S., Paris.

Puu, T.: 1981, 'Catastrophic structural changes in a continuous regional model', *Regional Science and Urban Economics* **11**, 317-333.

Rechenmann, F.: 1982, 'La dynamique des systèmes et son double', Technique mathématiques pour l'analyse de système en géographie, stage C.N.R.S., Rouen.

Reymond, H.: 1981, 'Une problématique théorique de la géographie: plaidoyer pour une chorotaxie expérimentale', in H. Isnard, J.B. Racine, and H. Reymond (eds.) *Problématique de la Géographie*, PUF, Paris.

Reymond, H.: 1981, 'L'ouverture informatique en géographie urbaine: de l'analyse multivariée socio-économique à la simulation organique des systèmes urbaine', *Informatique et Sciences Humaines* **50**, 9-20.

Roehner, B.: 1983, 'Contribution à l'étude des processus stocastiques de naissance et mort: aspects théoriques et application', Université de Paris IV thèse de doctorat d'Etat.

Roehner, B. and K. Wiese: 1982, 'A dynamic generalization of Zipf's rank-size rule', *Environment and Planning* **A 14**, 1449-1467.

Sanders, L.: 1984, 'Interaction spatiale et modèlisation dynamique, application aux systèmes urbains', Université Paris VII, thèse de 3ème cycle.

Sheppard, E.: 1985, 'Urban system population dynamics: incorporating nonlinearities', *Geographical Analysis* **17**, 47-73.

Thom, R.: 1974, *Modèles Mathématiques de la Morphogénèse*, Christian Bourgeois, Paris.

Wagstaff, J.M.: 1977, 'A possible interpretation of settlement pattern development in terms of "catastrophe theory"', *Transactions, Institute of British Geographers* **3**, 165-178.

Wegener, M.: 1983, 'A simulation study of movement in the Dortmund housing market', *Tijdschrift voor Economische en Sociale Geografie* **74**, 267-281.

White, R.W.: 1977, 'Dynamical central place theory', *Geographical Analysis* **9**, 226-243.

White, R.W.: 1978, 'The simulation of central place dynamics: two sector systems and the rank size rule', *Geographical Analysis* **10**, 201-208.

White, R.W.: 1983, 'Chaotic behaviour and the self-organization of urban retail systems', Brussels working papers on spatial analysis, series A, no. 3.

Wilson, A.G.: 1981, *Catastrophe Theory and Bifurcation: Application to Urban and Regional Systems*, Croom Helm, London.

Wilson, A.G.: 1984, 'Making urban models more realistic: some strategies for future research', *Environment and Planning* **A 16**, 1419-1432.

Winiwarter, P.: 1984, 'Iso-dynamics of population-size distribution in hierarchical systems', Paper presented at the annual meeting of the Society for General Systems Research, Los Angeles.

MARK BIRKIN AND ALAN G. WILSON

SOME PROPERTIES OF SPATIAL-STRUCTURAL-ECONOMIC-DYNAMIC MODELS

1. INTRODUCTION

In this paper, we explore the properties of an increasingly complex family of dynamic urban models. We demonstrate the importance of being as explicit and as clear as possible about the particular hypotheses which are to be represented in particular situations given the variety of possibilities available. Our approach is applicable to many different urban models, but we employ the ever-useful retail model as an archetypal example.

The argument is presented as follows. In Section 2, we outline the retail model which includes hypotheses on stock dynamics in both comparative static and disequilibrium frameworks. We show in Section 3 how this can be expanded to incorporate locational price indexes of retail goods and land rents for retail developers. In Section 4, we then show that there are many ways in which a particular dynamic model can be specified in relation to additional hypotheses about the mechanisms of change. And this variety of possibilities is the central focus of the paper. We explore the properties of models based on different hypotheses, both to show the consequences of adopting particular hypotheses and to make some preliminary deductions from the results on the appropriateness of particular hypotheses in different circumstances. In Section 5, we briefly list a range of further extensions which can be made to the models. A selection of results is presented in Section 6 and some concluding comments in Section 7.

2. THE BACKGROUND: RETAIL STOCK DYNAMICS

Our starting point is the model stated in Harris and Wilson (1978). Let S_{ij} be the flow of expenditure from residents of i to shops in j; E_i the expenditure by residents of i; W_j, a measure of the attractiveness of j (usually taken as proportional to size for illustrative purposes); and c_{ij} as the cost of travel from i to j. Then the flow model is (for suitable parameters α and β)

$$S_{ij} = A_i \, E_i \, W_j^{\alpha} \, e^{-\beta c_{ij}} \tag{1}$$

$$\text{where } A_i = 1/\sum_k W_k^{\alpha} \, e^{-\beta c_{ik}} \tag{2}$$

The total revenue attracted to j is

J. Hauer et al. (eds.), Urban Dynamics and Spatial Choice Behaviour, 175–201.

$$D_j = \sum_i S_{ij} \tag{3}$$

and the cost of providing facilities of 'scale' W_j is

$$C_j = k_j W_j \tag{4}$$

for suitable constants k_j. Given α, β, $\{E_i\}$ and $\{c_{ij}\}$ and initial values of $\{W_j\}$, say $\{W_j^o\}$, then an obvious hypothesis for the dynamics of $\{W_j\}$ is:

$$\dot{W}_j = \varepsilon(D_j - C_j)W_j \tag{5}$$

for a suitable ε, with equilibrium condition

$$D_j = C_j \tag{6}$$

By making the appropriate substitutions, Equations (5) and (6) can be written explicitly in terms of $\{W_j\}$ as

$$\dot{W}_j = \varepsilon \left[\sum_i \frac{E_i W_j^\alpha e^{-\beta c_{ij}}}{\sum_k W_k^\alpha e^{-\beta c_{ik}}} - k_j W_j \right] \tag{7}$$

and

$$\sum_i \frac{E_i W_j^\alpha e^{-\beta c_{ij}}}{\sum_k W_k^\alpha e^{-\beta c_{ik}}} = k_j W_j \tag{8}$$

It was shown in Harris and Wilson (1978) that the nonlinear Equations (8) had interesting bifurcation patterns when parameters like α and β vary; and in Wilson (1981a), building on May (1976), that (7) has interesting bifurcation properties in relation to ε. The variety of patterns and trajectories which arise for different parameter values (together with a discussion of some of the computational difficulties) can be found in Wilson and Clarke (1979) and Clarke and Wilson (1983). The model has also been explored in a number of other papers: see, for example, Phiri (1980), Leonardi (1981a, 1981b), Harris, Choukroun and Wilson (1982), Lombardo and Rabino (1983), Rijk and Vorst (1983a, 1983b), Kaashoek and Vorst (1984), Chudzynska and Slodkowski (1984), Clarke (1984), Pumain, Saint-Julien and Sanders (1984), Roy and Brotchie (1984), Fotheringham (1985) and Clarke (1986).

3. EXTENSIONS TO INCLUDE PRICES AND LAND RENTS

The model presented above is a simple representation of the process of competition across space by retailers. Those with a particular locational advantage are likely to press this home with price reductions; though this will be tempered by land owners at those sites being able to demand higher land 'rents'. (We assume that all monetary variables are adjusted to refer to a suitable common time or time period.) In this section, we follow the argument of Wilson (1985) and extend the 1978 model to include prices and rents.

As a background to this, it can be argued that it is best to set out hypotheses in relation to the main agents involved: consumers, retailers, developers, land owners and manufacturers of retail goods - treating them all as separate for convenience, though in many cases a single person may wear several hats. The manufacturers play a role in the prices they charge to retailers in balancing consumer demand and supply (mediated by retailers) at an aggregate scale. However, we neglect this relationship as it is not our central concern here. We also assume that consumers move instantly to a new equilibrium following a change. Again, this assumption is easy to relax (as in Haag, 1989, for instance). Then, for simplicity, we assume that developers determine the scale and patterns of provision, $\{W_j\}$; retailers, the prices, $\{p_j\}$ and land owners, the land rents, $\{r_j\}$. We assume for our present purposes that demand $\{E_i\}$ is fixed, though again, this assumption can be easily relaxed.

An extended model is then given by amended versions of Equations (1) and (2) for consumers' behaviour as follows:

$$S_{ij} = A_i \, E_i \, W_j^{\alpha} \, p_j^{-\gamma} \, e^{-\beta c_{ij}} \tag{9}$$

$$\text{where } A_i = 1/\sum_k W_k^{\alpha} \, p_k^{-\gamma} \, e^{-\beta c_{ik}} \tag{10}$$

The main change is that a factor $p_k^{-\gamma}$ is added to the attractiveness term, so that destinations are more attractive to consumers when the price index is lower. γ is a suitable parameter.

Revenue is now partly determined by the price index:

$$D_j = p_j \sum_i S_{ij} \tag{11}$$

If $p_j < 1$, this can be interpreted as a benefit to the consumer which can be spent in other ways; and vice versa. The cost of provision is

$$C_j = (k_j + r_j)W_j \tag{12}$$

where k_j are the assumed non-varying costs, including buildings (annualised, say), labour, purchase of goods from manufacturers, while r_j is unit land rent.

The dynamics are then given by a version of Equation (6) extended to include price and land rent adjustments:

$$\dot{W}_j = \varepsilon_1(D_j - C_j)W_j \tag{13}$$

$$\dot{p}_j = \varepsilon_2(C_j - D_j)p_j \tag{14}$$

$$\dot{r}_j = \varepsilon_3(D_j - C_j)r_j \tag{15}$$

4. THE MECHANISMS OF CHANGE

In practice, Equations (13)-(15) are solved as difference equations. For reasons which will become clear shortly, we need a fairly complicated notation to describe the solution procedure. Let t and $t + 1$ be successive points in time, so $(t, t + 1)$ represents a time period; let n be an iteration number for when the equations have to be solved iteratively within a time period. The argument to be presented here partly follows that in Leonardi and Wilson (1989) and also relates to Wilson and Birkin (1985).

$$\Delta W_j^{n+1}(t) = \varepsilon_1\left[D_j^n(t) - C_j^n(t)\right]W_j^n(t) \tag{16}$$

$$\Delta p_j^{n+1}(t) = \varepsilon_2\left[C_j^n(t) - D_j^n(t)\right]p_j^n(t) \tag{17}$$

$$\Delta r_j^{n+1}(t) = \varepsilon_3\left[D_j^n(t) - C_j^n(t)\right]r_j^n(t) \tag{18}$$

$$W_j^{n+1}(t) = W_j^n(t) + \sigma\Delta W_j^{n+1}(t) \tag{19}$$

$$p_j^{n+1}(t) = p_j^n(t) + \sigma\Delta p_j^{n+1}(t) \tag{20}$$

$$r_j^{n+1}(t) = r_j^n(t) + \sigma\Delta r_j^{n+1}(t) \tag{21}$$

σ is a 'step length' - a constant between 0 and 1.

$$W_j(t + 1) = W_j^{n_1^{max}}(t) \tag{22}$$

$$p_j(t+1) = p_j^{n_2^{max}}(t) \tag{23}$$

$$r_j(t+1) = r_j^{n_3^{max}}(t) \tag{24}$$

n_1^{max}, n_2^{max}, and n_3^{max} are the maximum number of iterations to be carried out (at time t, or within a $(t, t+1)$-period) for W, p and r-variables respectively. Equations (16)-(21) are solved iteratively (if necessary), starting with $n = 0$ and

$$W_j^o(t) = W_j(t-1) \tag{25}$$

$$p_j^o(t) = p_j(t-1) \tag{26}$$

$$r_j^o(t) = r_j(t-1) \tag{27}$$

(or, for $t = 1$, whatever initial values are appropriate). At the end of the iteration, Equations (22)-(24) give the $t + 1$ values of the W, p and r-variables.

Different combinations of ε-parameters and n^{max}-values generate a great variety of dynamic models, many of which can be interpreted as representing interesting real processes for one kind of system or another. This, therefore, is the heart of the paper: we want to show what this variety is and to explore the properties of some of the more interesting models which can be generated.

For convenience, let us adopt the convention that $n_k^{max} = 0$ means that there is *no* adjustment of the k-th variable type. (This is equivalent to setting the associated ε_k parameter to zero, which is the more conventional way of expressing this). Then the following table shows how a variety of interesting cases can be generated.

	W	p	r
n^{max}-values	0	0	0
	1	1	1
	high	high	high

A model is any combination of three possible n^{max} values. Interesting combinations are

$$(n_1^{max}, n_2^{max}, n_3^{max}) = \begin{cases} (\text{high}, 0, 0) & (28) \\ (\text{high}, \text{high}, \text{high}) & (29) \\ (1, \text{high}, \text{high}) & (30) \\ (1, 1, 1) & (31) \end{cases}$$

The case (28) is essentially the Harris-Wilson model: W_j's only are adjusted until equilibrium is achieved (if it is possible, or by choosing sufficiently small σ) at each time. Indeed, if $\varepsilon_1 = 1$, the iterative process is precisely the usual iterative scheme for the Harris-Wilson model (see Wilson, 1985). The process is repeated, following any adjustment of exogenous variables or parameters, such as the E_i's or α or β, at $t + 1$.

Case (29) is comparative statics with W's, p's and r's adjusted simultaneously. The relative influence of stock, price or rent adjustment on the final outcome is determined by the relative magnitudes of ε_1, ε_2 and ε_3.

Case (30) represents a structure often used by economists (cf. Anas, 1989). From an initial disequilibrium state, there is an adjustment in stock values for one step in the direction of equilibrium, and the price and rent variables are iterated until equilibrium is forced at that time.

Case (31) represents constant disequilibrium with adjustments of one step being made to all variables and the possibility of changing exogenous variables at each time period. Of course, if there are no exogenous changes, then after a long period, a state will be reached which is the same as that represented by case (29).

We present results of numerical experiments for these cases in Section 6 and reserve further interpretation until then.

5. FURTHER EXTENSIONS

We begin by noting four points of (possibly important) detail which have to be tackled in any particular case. Then we consider one more substantial extension.

(1) At present, in Equations like (16)-(18), we have factors W_j, p_j and r_j respectively on the right hand side. These are particularly important in determining the 'pick up' of ΔW_j (say) for small W_j. For a more detailed exploration of this point, see Wilson (1981b).

(2) The ε-parameters are assumed constant. In some cases, it would be more enterprising to discover how different agents would 'learn' in different circumstances and thus how ε's would vary over time. (For instance, after a big 'overshoot', ε may be reduced).

(3) Demand should be made elastic in relation to price. One way to do this (cf. Wilson, 1985) is to take

$$\hat{p}_i = \frac{\sum_j p_j e^{-\beta c_{ij}}}{\sum_j e^{-\beta c_{ij}}} \tag{32}$$

as an exponentially-weighted average of p_j's to give a perceived price at i, and

$$E_i = E_i^o \left[\frac{\hat{p}_i(t)}{\hat{p}_i(t-1)} \right]^{\hat{\gamma}} \tag{33}$$

say.

(4) At present, the increments, say (16)-(18), are calculated simultaneously. It would be possible to generate more variants of the model by recalculating the flow, revenue and cost variables between calculations of increments. For example, in model (30) above, it may be appropriate to calculate $W_j(t+1)$ from its single step increment, and then to recalculate $\{S_{ij}\}$, $\{D_j\}$, and $\{C_j\}$ before solving for $\{p_j\}$ and $\{r_j\}$ iteratively.

The more substantial extension is to modify the mechanism for calculating the W_j's. At present, it can be argued (as in Wilson, 1985) that they are 'optimal' in the sense that the T_{ij} terms in the transportation problem of linear programming are optimal (cf. Evans, 1973, Wilson and Senior, 1974). We can find a more general, and typically more dispersed, $\{W_j\}$ solution by incorporating an entropy term, $-\sum_j W_j(\log W_j - 1)$. We show how to do this by taking the (high, 0, 0) model - case (28) above: that is, we neglect price and rent adjustments. The mathematical programming version of the Harris and Wilson (1978) model is (cf. Wilson, 1981b):

$$\begin{aligned} \underset{\{S_{ij}, W_j\}}{\text{Max}} \quad Z = & - \sum_{ij} S_{ij}(\log S_{ij} - 1) + \\ & + \sum_i a_i (E_i - \sum_j S_{ij}) + \\ & + \sum_j \lambda_j (W_j - \sum_i S_{ij}) + \\ & + \alpha (\sum_{ij} S_{ij} \log W_j - H) + \\ & + \beta (C - \sum_{ij} S_{ij} c_{ij}) \end{aligned} \tag{34}$$

We need to add the term $-\sum_j W_j(\log W_j - 1)$ into the objective function and relax the constraint

$$\sum_i S_{ij} = W_j = C_j \tag{35}$$

(where we have taken k_j as 1 for simplicity) which appears as $\sum_j \lambda_j(W_j - \sum_i S_{ij})$ in the Lagrangian. Given the kind of market imperfections which generate dispersion, we would expect $W_j > \sum_i S_{ij}$ and one way to express this is by taking

$$\sum_j (C_j - D_j)^2 = B \tag{36}$$

for some $B > 0$.

A term $\lambda[\sum_j (C_j - D_j)^2 - B]$ then replaces $\sum_j \lambda_j(W_j - \sum_i S_{ij})$ the mathematical programme (34), with $C_j = W_j$ and $D_j = \sum_i S_{ij}$. After some manipulation, it is possible in this case to obtain an explicit equation for W_j (rather as for T_{ij} in the doubly-constrained spatial interaction model relative to the transportation problem of linear programming) - and also a modified flow model:

$$S_{ij} = A_i E_i e^{-\lambda(D_j - W_j)} W_j^{\alpha} e^{-\beta c_{ij}} \tag{37}$$

$$A_i = 1/\sum_k e^{-\lambda(D_k - W_k)} W_k^{\alpha} e^{-\beta c_{ik}} \tag{38}$$

$$W_j = e^{\lambda(D_j - W_j)} e^{\alpha W_j/F_j} \tag{39}$$

Some preliminary results using this model are presented in Wilson (1985).

6. NUMERICAL EXPERIMENTS

The arguments presented in Sections 2-5 have been theoretical and somewhat abstract. As usual with this class of models, it is useful to undertake numerical experiments which help to fix ideas, and also point the way to new kinds of theoretical extensions (see Clarke and Wilson, 1983; also Campbell *et al*, 1985).

We proceed in the following way. Initially a set of comparative static cases are presented which are a special case of the framework of Section 4 corresponding to either n^{max} = (high, 0, 0), or the (1, 0, 0) case extrapolated over several time periods without exogenous change. By extracting a single example from the comparative static analysis, it is possible to show how the introduction of prices and rents greatly enriches the variety of spatial structures which arise. These models are basically of the (high, high, high) type in which the various actors may be seen as competing for a locational surplus, with each individual's economic strength represented by the relative magnitude of the associated ε parameter.

This is the kind of speed of change which has been discussed in the past, in relation to equations like (5) (e.g. Wilson, 1981a; Clarke and Wilson, 1983). However the introduction of different actors in the system allows us to consider a second kind of speed of change, which relates essentially to cross-sectional adjustment as a response to disequilibrium. Thus if an actor is associated with a high n^{max}, he is assumed to respond instantaneously to seek a new equilibrium. In a third set of numerical experiments we therefore present a scenario in which a fixed kind of exogenous change is taking place, and explore how the mode of response to disequilibrium affects the spatial structure of the state variables - prices, rents and floorspace.

6.1. *Comparative statics with the stock variable only*

Comparative static cases, in which structural equilibria are contrasted over a variety of parameter values, are special cases of the extended approach offered here: and first, prices and rents are excluded. This type of analysis provides a convenient starting point, not least in the sense that it directs us towards the areas of parameter space which are likely to prove worthy of further investigation.

Figure 1 shows the structure of our basic network, which takes the form of a 129-zone grid. It is assumed that each grid point generates an identical demand $\{E_i\}$, and initially that the friction of distance is uniform, although at a later stage this assumption is relaxed to allow preferential access to the city centre. Under these circumstances, and with an even distribution of initial floorspace levels $\{W_j^o\}$, Figure 2 demonstrates the equilibrium floorspace configuration for a plausible array of parameter values. Where the attractiveness of centres is strong relative to their size (α = 1.5) and travel is easy (β = 0.25), it is reasonable that a single centre should emerge (Figure 2.1.1). However, as differential attractiveness is reduced, or the cost of travel increases, more dispersed patterns begin to arise, until a rather more balanced pattern is attained for α = 1.05, β = 1.50 (Figure 2.3.3).

6.2. *Comparative statics with more than one variable*

Once non-zero values are introduced for the parameters ε_2 and ε_3, a competitive process is introduced in which patterns of spatial advantage may be exploited not only through the expansion of activity levels (via ε_1), but also through price reductions (ε_2), or the extraction of rents by landowners (ε_3). The objective of this section is to show how the relative strengths of the respective competitors influences the outcome of the spatial development process. In achieving this, we also take another trip through "parameter space", and hence further insights are attained into the operation of the models.

In the numerical simulations presented here, we assume the functional forms:

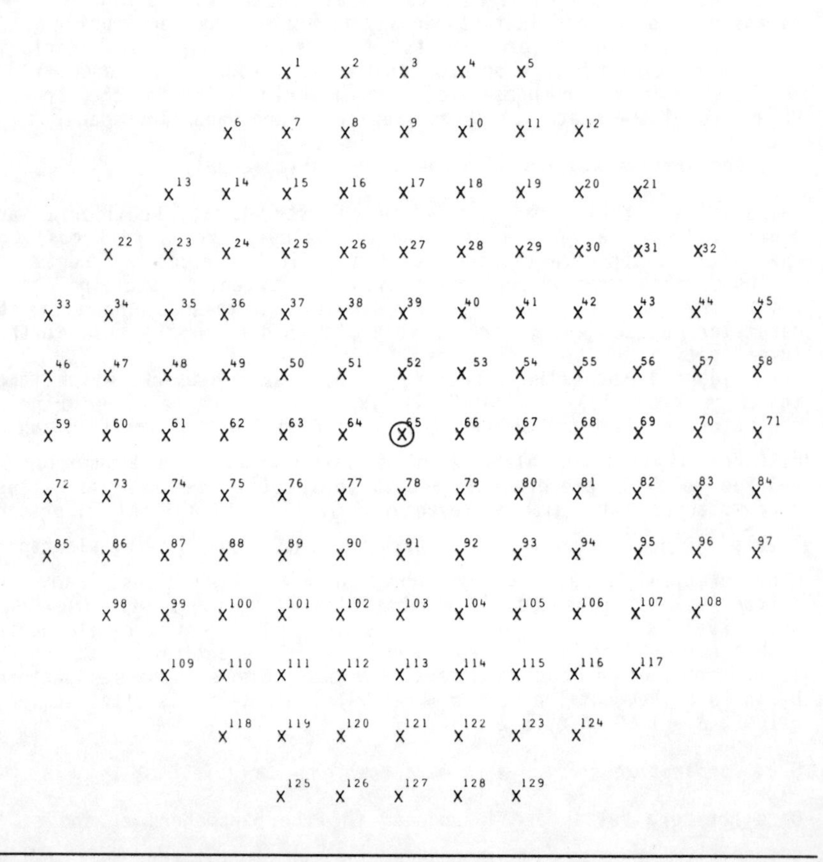

Fig. 1: The hypothetical grid system.

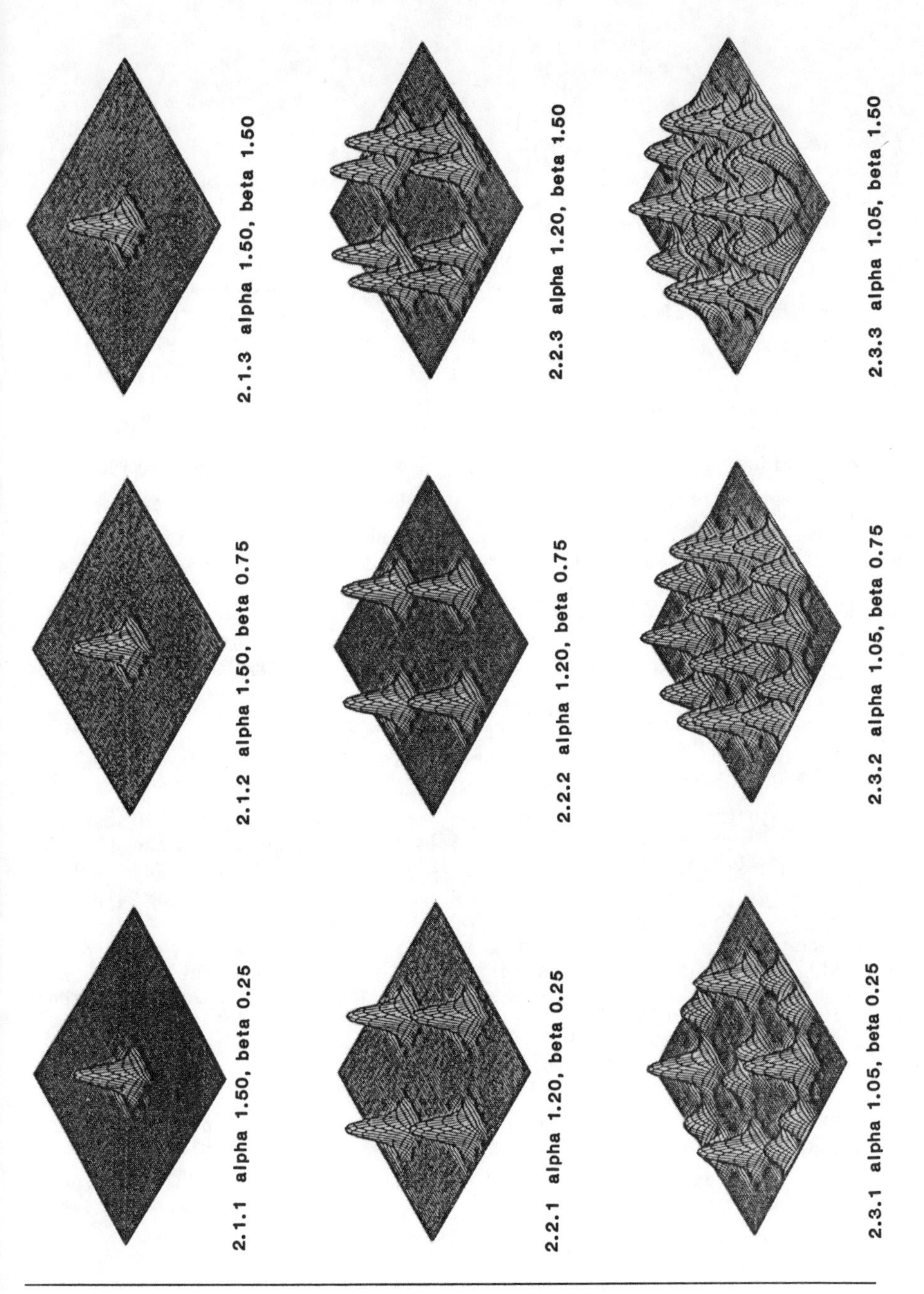

Fig. 2: Comparative statics with the stock variable only.

$$\dot{W}_j = \varepsilon_1(D_j - C_j) \tag{40}$$

$$\dot{p}_j = -\varepsilon_2(D_j - C_j)p_j \tag{41}$$

$$\dot{r}_j = \varepsilon_3(D_j - C_j) \tag{42}$$

While these forms could be varied, the important point to note here is that the ε's also function as scaling parameters: thus the values of ε_1 we deal with below are usually at least two orders of magnitude greater than ε_2 and ε_3 because it scales on the dimensions of floorspace, rather than price or rent indices.

Initially, we focus on the variation of the ε_2 and ε_3 parameters individually with respect to ε_1. First of all, let us consider the rent parameter, ε_3. Since this parameter diverts surplus revenue from reinvestment in spatial structure (through (40)) to an inactive rent stockpile (through (42)), we would expect increases in ε_3 to generate a dispersion of activities. As a pivotal case, we have therefore selected a fairly concentrated case from Figure 2, with α = 1.2 and β = 0.75 (Figure 2.2.2). Figure 3 then shows the effect on spatial structure of increasing ε_3 form 0.0001 through three orders of magnitude to 0.1.

For a low enough value of ε_3, we would expect its effect to be negligible, so that the structure of Figure 2.2.2 would be reproduced. ε_3 = 0.0001 is evidently a sufficiently small value (Figure 3.1.1). As the adjustment parameter increases, we can see the expected dispersion as the four initial centres multiply to 8 (Figure 3.1.2) and then 20 (Figure 3.1.3). Finally, for a rather large ε_3 of 0.1, the spatial order of activities loses this clear definition and becomes extremely fuzzy (Figure 3.1.4). In effect, all the locational advantage generated in the system is transferred to the landowners, so there is still some differentiation in the rent surface. The rents are shown, in a new graphical form, in Figure 3.2. Notice that although the rent surface of Figure 3.2.4 is more differentiated than the activity surface (Figure 3.1.4), it is still much less peaked than the corresponding distribution of Figure 3.2.3.

Thus one might argue that ε_3 = 0.1 represents an extreme, but unrealistic, representation of the rent accumulation process. Values of ε_3 = 0.01 and 0.001, on the other hand, show plausible outcomes to such a process, with corresponding implications for spatial structure, with ε_3 = 0.0001 as a limiting case in which rent is effectively eliminated.

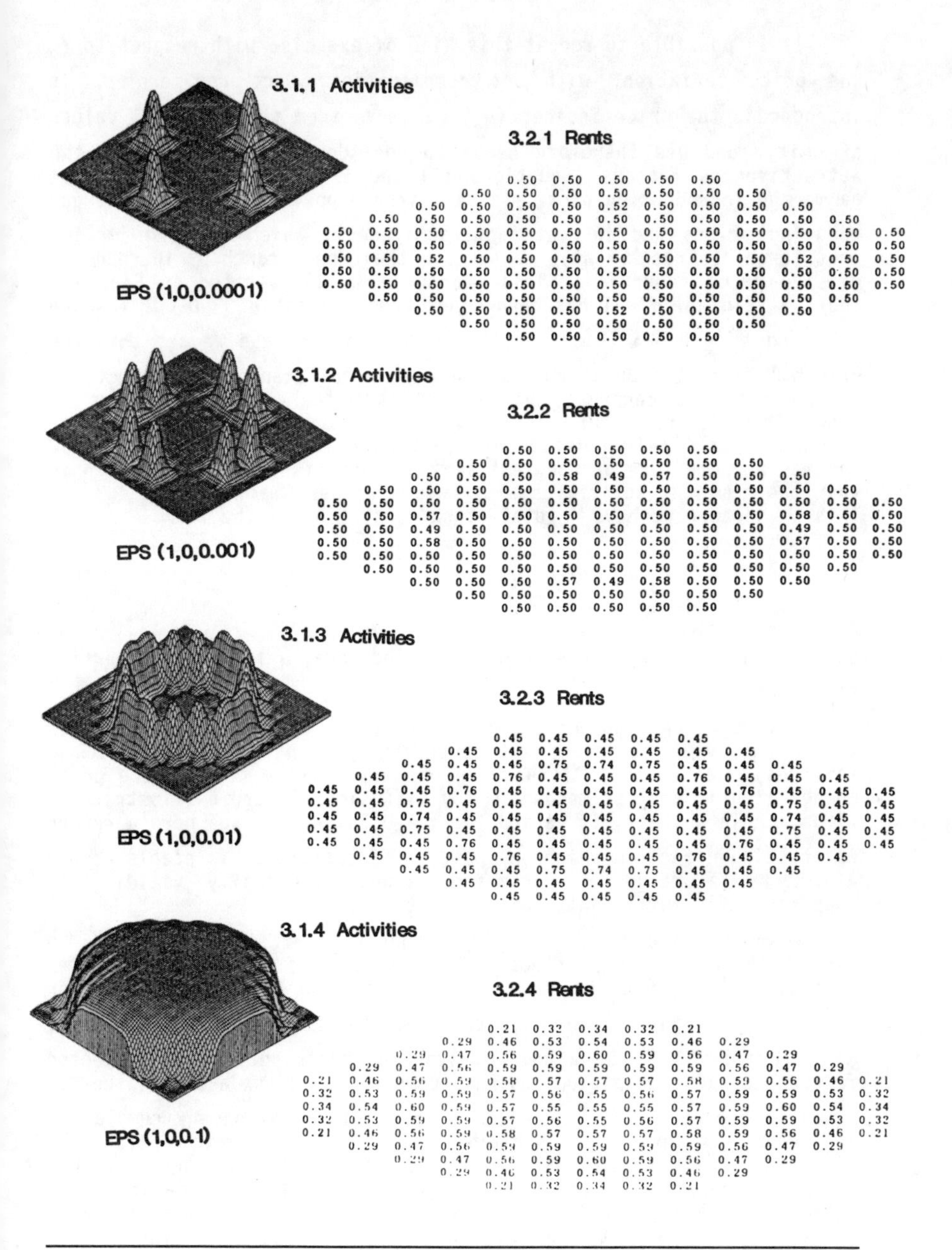

3.2.1 Rents

				0.50	0.50	0.50	0.50	0.50				
			0.50	0.50	0.50	0.50	0.50	0.50	0.50			
		0.50	0.50	0.50	0.50	0.52	0.50	0.50	0.50	0.50		
	0.50	0.50	0.50	0.50	0.50	0.50	0.50	0.50	0.50	0.50	0.50	
0.50	0.50	0.50	0.50	0.50	0.50	0.50	0.50	0.50	0.50	0.50	0.50	0.50
0.50	0.50	0.50	0.50	0.50	0.50	0.50	0.50	0.50	0.50	0.50	0.50	0.50
0.50	0.50	0.52	0.50	0.50	0.50	0.50	0.50	0.50	0.50	0.52	0.50	0.50
0.50	0.50	0.50	0.50	0.50	0.50	0.50	0.50	0.50	0.50	0.50	0.50	0.50
0.50	0.50	0.50	0.50	0.50	0.50	0.50	0.50	0.50	0.50	0.50	0.50	0.50
	0.50	0.50	0.50	0.50	0.50	0.50	0.50	0.50	0.50	0.50	0.50	
		0.50	0.50	0.50	0.50	0.52	0.50	0.50	0.50	0.50		
			0.50	0.50	0.50	0.50	0.50	0.50	0.50			
				0.50	0.50	0.50	0.50	0.50				

3.2.2 Rents

				0.50	0.50	0.50	0.50	0.50				
			0.50	0.50	0.50	0.50	0.50	0.50	0.50			
		0.50	0.50	0.50	0.58	0.49	0.57	0.50	0.50	0.50		
	0.50	0.50	0.50	0.50	0.50	0.50	0.50	0.50	0.50	0.50	0.50	
0.50	0.50	0.50	0.50	0.50	0.50	0.50	0.50	0.50	0.50	0.50	0.50	0.50
0.50	0.50	0.57	0.50	0.50	0.50	0.50	0.50	0.50	0.50	0.58	0.50	0.50
0.50	0.50	0.49	0.50	0.50	0.50	0.50	0.50	0.50	0.50	0.49	0.50	0.50
0.50	0.50	0.58	0.50	0.50	0.50	0.50	0.50	0.50	0.50	0.57	0.50	0.50
0.50	0.50	0.50	0.50	0.50	0.50	0.50	0.50	0.50	0.50	0.50	0.50	0.50
	0.50	0.50	0.50	0.50	0.50	0.50	0.50	0.50	0.50	0.50	0.50	
		0.50	0.50	0.50	0.57	0.49	0.58	0.50	0.50	0.50		
			0.50	0.50	0.50	0.50	0.50	0.50	0.50			
				0.50	0.50	0.50	0.50	0.50				

3.2.3 Rents

				0.45	0.45	0.45	0.45	0.45				
			0.45	0.45	0.45	0.45	0.45	0.45	0.45			
		0.45	0.45	0.45	0.75	0.74	0.75	0.45	0.45	0.45		
	0.45	0.45	0.45	0.76	0.45	0.45	0.45	0.76	0.45	0.45	0.45	
0.45	0.45	0.45	0.76	0.45	0.45	0.45	0.45	0.45	0.76	0.45	0.45	0.45
0.45	0.45	0.75	0.45	0.45	0.45	0.45	0.45	0.45	0.45	0.75	0.45	0.45
0.45	0.45	0.74	0.45	0.45	0.45	0.45	0.45	0.45	0.45	0.74	0.45	0.45
0.45	0.45	0.75	0.45	0.45	0.45	0.45	0.45	0.45	0.45	0.75	0.45	0.45
0.45	0.45	0.45	0.76	0.45	0.45	0.45	0.45	0.45	0.76	0.45	0.45	0.45
	0.45	0.45	0.45	0.76	0.45	0.45	0.45	0.76	0.45	0.45	0.45	
		0.45	0.45	0.45	0.75	0.74	0.75	0.45	0.45	0.45		
			0.45	0.45	0.45	0.45	0.45	0.45	0.45			
				0.45	0.45	0.45	0.45	0.45				

3.2.4 Rents

				0.21	0.32	0.34	0.32	0.21				
			0.29	0.46	0.53	0.54	0.53	0.46	0.29			
		0.29	0.47	0.56	0.59	0.60	0.59	0.56	0.47	0.29		
	0.29	0.47	0.56	0.59	0.59	0.59	0.59	0.59	0.56	0.47	0.29	
0.21	0.46	0.56	0.59	0.58	0.57	0.57	0.57	0.58	0.59	0.56	0.46	0.21
0.32	0.53	0.59	0.59	0.57	0.56	0.55	0.56	0.57	0.59	0.59	0.53	0.32
0.34	0.54	0.60	0.59	0.57	0.55	0.55	0.55	0.57	0.59	0.60	0.54	0.34
0.32	0.53	0.59	0.59	0.57	0.56	0.55	0.56	0.57	0.59	0.59	0.53	0.32
0.21	0.46	0.56	0.59	0.58	0.57	0.57	0.57	0.58	0.59	0.56	0.46	0.21
	0.29	0.47	0.56	0.59	0.59	0.59	0.59	0.59	0.56	0.47	0.29	
		0.29	0.47	0.56	0.59	0.60	0.59	0.56	0.47	0.29		
			0.29	0.46	0.53	0.54	0.53	0.46	0.29			
				0.21	0.32	0.34	0.32	0.21				

Fig. 3: Comparative statics, activity levels versus rent.

It is possible to repeat this kind of exercise with respect to ε_2 and price variation, with one complication. Once non-zero ε_2 is introduced, the price indices $\{p_j\}$ can move from their initial values of unity, and one therefore needs to consider the role of γ in the attractiveness factors (Equations (9) and (10)). For high values of gamma the introduction of ε_2 yields a strong positive feedback between reducing prices and increased attractiveness, which in turn implies that large centres are able to sustain their own growth by introducing massive price reductions. This mechanism is illustrated in Figure 4 where a value of $\gamma = 1$ has been considered against $\varepsilon_2 = 0.001$ (Figure 4.1) and $\varepsilon_2 = 0.025$ (Figure 4.2). In the former case we see that the original distribution is maintained with significant price reductions at the non-zero centres. Notice here that the prices which are in effect for centres of no activity operate as some kind of dual or shadow prices. Their relationship to the active prices in Figure 4.2 becomes extreme. One of the less sensible aspects of this situation is that the dominant activities also become rather small (Figure 4.2.1). This is because we have a total stock constraint of the form:

$$\bar{W} = \sum_j k_j W_j = \sum_j p_j D_j \tag{43}$$

Since both D_j and k_j are externally scaled here, $\bar{W}$ falls in proportion to reductions in p_j.

Although one might conceive of situations for which Figure 4.2 is not a wholly unrealistic representation (for high order goods and services), we choose to focus here on less extreme cases where price competition is another mechanism for diffusion. Figure 5 illustrates a situation in which ε_2 is varied for $\gamma = 0.25$. We can see here a rather interesting effect in which one pattern of activities is stable across a very wide range of parameter values, eventually yielding to dispersion for sufficiently high values of ε_2.

Having varied rents and prices separately, it is now possible to assess their combined effects. The results of Figure 3 suggested sensible values of ε_3 to range from 0.01 through to 0.0001, while for $\gamma = 0.25$, ε_2 might go from 0.01 to 0.001 (see Figure 5). For $\varepsilon_1 = 1.0$, $\alpha = 1.2$ and $\beta = 0.75$, a uniform starting distribution was projected forward for 30 time periods using a combination of ε_2 and ε_3 values as shown in Figure 6. Of course the two parameters now have a cumulative influence, so even low parameter values can generate relatively dispersed patterns (compare Figure 6.3.3 to Figure 3.1 and 5.5). One of the encouraging things here is that the pattern is sensitive to changes in both parameters, but as in Figure 5 there appear to be large areas of parameter space, e.g. for $\varepsilon_2 = 0.001$ as ε_3 varies from 0.01 to 0.0001 (Figures 6.3.1 - 6.3.3); $\varepsilon_2 = 0.005$ for $\varepsilon_3 = 0.01$ to 0.001 (Figures 6.2.1 and 6.2.2).

				0.00	0.00	0.00	0.00	0.00				
			0.00	0.00	0.00	0.00	0.00	0.00	0.00			
		0.00	0.00	0.00	0.00	252.77	0.00	0.00	0.00	0.00		
	0.00	0.00	0.00	0.00	0.00	0.00	0.00	0.00	0.00	0.00	0.00	
0.00	0.00	0.00	0.00	0.00	0.00	0.00	0.00	0.00	0.00	0.00	0.00	0.00
0.00	0.00	0.00	0.00	0.00	0.00	0.00	0.00	0.00	0.00	0.00	0.00	0.00
0.00	0.00	252.77	0.00	0.00	0.00	0.00	0.00	0.00	0.00	252.77	0.00	0.00
0.00	0.00	0.00	0.00	0.00	0.00	0.00	0.00	0.00	0.00	0.00	0.00	0.00
0.00	0.00	0.00	0.00	0.00	0.00	0.00	0.00	0.00	0.00	0.00	0.00	0.00
	0.00	0.00	0.00	0.00	0.00	0.00	0.00	0.00	0.00	0.00	0.00	
		0.00	0.00	0.00	0.00	252.77	0.00	0.00	0.00	0.00		
			0.00	0.00	0.00	0.00	0.00	0.00	0.00			
				0.00	0.00	0.00	0.00	0.00				

4.1.1 Eps2 0.001 Activity surface

				1.01	1.01	1.01	1.01	1.01				
			1.01	1.01	1.01	1.01	1.01	1.01	1.01			
		1.01	1.01	1.01	1.01	0.78	1.01	1.01	1.01	1.01		
	1.01	1.01	1.01	1.01	1.01	1.01	1.01	1.01	1.01	1.01	1.01	
1.01	1.01	1.01	1.01	1.01	1.01	1.01	1.01	1.01	1.01	1.01	1.01	1.01
1.01	1.01	1.01	1.01	1.01	1.01	1.01	1.01	1.01	1.01	1.01	1.01	1.01
1.01	1.01	0.78	1.01	1.01	1.01	1.01	1.01	1.01	1.01	0.78	1.01	1.01
1.01	1.01	1.01	1.01	1.01	1.01	1.01	1.01	1.01	1.01	1.01	1.01	1.01
1.01	1.01	1.01	1.01	1.01	1.01	1.01	1.01	1.01	1.01	1.01	1.01	1.01
	1.01	1.01	1.01	1.01	1.01	1.01	1.01	1.01	1.01	1.01	1.01	
		1.01	1.01	1.01	1.01	0.78	1.01	1.01	1.01	1.01		
			1.01	1.01	1.01	1.01	1.01	1.01	1.01			
				1.01	1.01	1.01	1.01	1.01				

4.1.2 Eps 0.001 Price surface

				0.50	0.50	0.50	0.50	0.50				
			0.50	0.50	0.50	0.50	0.50	0.50	0.50			
		0.50	0.50	0.50	0.50	0.50	0.50	0.50	0.50	0.50		
	0.50	0.50	0.50	0.50	0.50	0.50	0.50	0.50	0.50	0.50	0.50	
0.50	0.50	0.50	0.50	0.50	0.50	0.50	0.50	0.50	0.50	0.50	0.50	0.50
0.50	0.50	0.50	0.50	0.50	0.50	0.50	0.50	0.50	0.50	0.50	0.50	0.50
0.50	0.50	0.50	0.50	0.50	0.50	0.50	0.50	0.50	0.50	0.50	0.50	0.50
0.50	0.50	0.50	0.50	0.50	0.50	0.50	0.50	0.50	0.50	0.50	0.50	0.50
0.50	0.50	0.50	0.50	0.50	0.50	0.50	0.50	0.50	0.50	0.50	0.50	0.50
	0.50	0.50	0.50	0.50	0.50	0.50	0.50	0.50	0.50	0.50	0.50	
		0.50	0.50	0.50	0.50	0.50	0.50	0.50	0.50	0.50		
			0.50	0.50	0.50	0.50	0.50	0.50	0.50			
				0.50	0.50	0.50	0.50	0.50				

4.2.1 Eps 0.025 Activity surface

				0.00	0.00	0.00	0.00	0.00				
			0.00	0.00	0.00	0.00	0.00	0.00	0.00			
		0.00	0.00	0.00	27.93	0.00	27.93	0.00	0.00	0.00		
	0.00	0.00	0.00	0.00	0.00	0.00	0.00	0.00	0.00	0.00	0.00	
0.00	0.00	0.00	0.00	0.00	0.00	0.00	0.00	0.00	0.00	0.00	0.00	0.00
0.00	0.00	27.93	0.00	0.00	0.00	0.00	0.00	0.00	0.00	27.93	0.00	0.00
0.00	0.00	0.00	0.00	0.00	0.00	0.00	0.00	0.00	0.00	0.00	0.00	0.00
0.00	0.00	27.93	0.00	0.00	0.00	0.00	0.00	0.00	0.00	27.93	0.00	0.00
0.00	0.00	0.00	0.00	0.00	0.00	0.00	0.00	0.00	0.00	0.00	0.00	0.00
	0.00	0.00	0.00	0.00	0.00	0.00	0.00	0.00	0.00	0.00	0.00	
		0.00	0.00	0.00	27.93	0.00	27.93	0.00	0.00	0.00		
			0.00	0.00	0.00	0.00	0.00	0.00	0.00			
				0.00	0.00	0.00	0.00	0.00				

4.2.2 Eps 0.025 Price surface

Fig. 4: Comparative statics of activity size versus price, $\gamma = 1.0$.

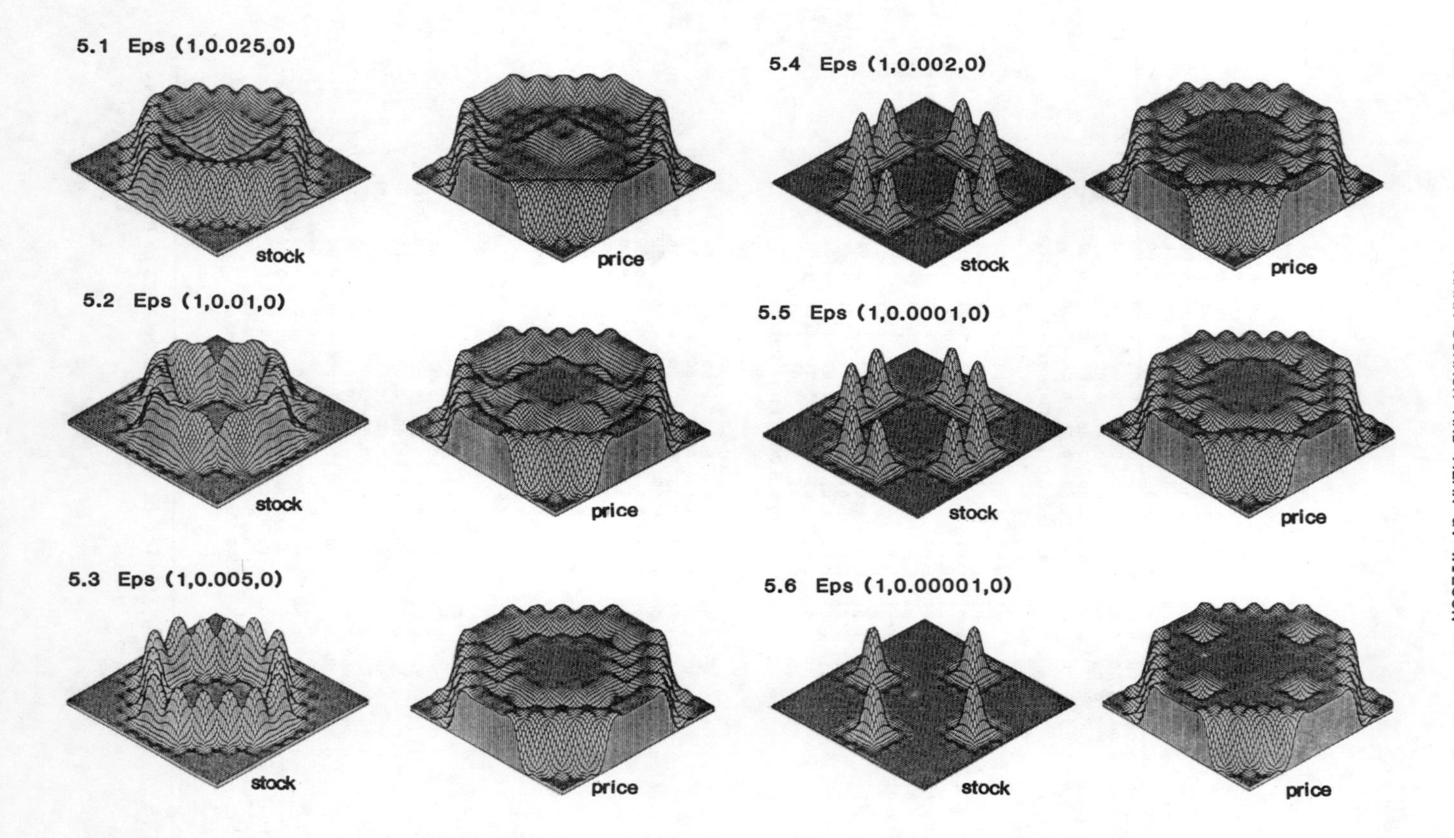

Fig. 5: Comparative statics, activity levels versus price for $\gamma = 0.25$.

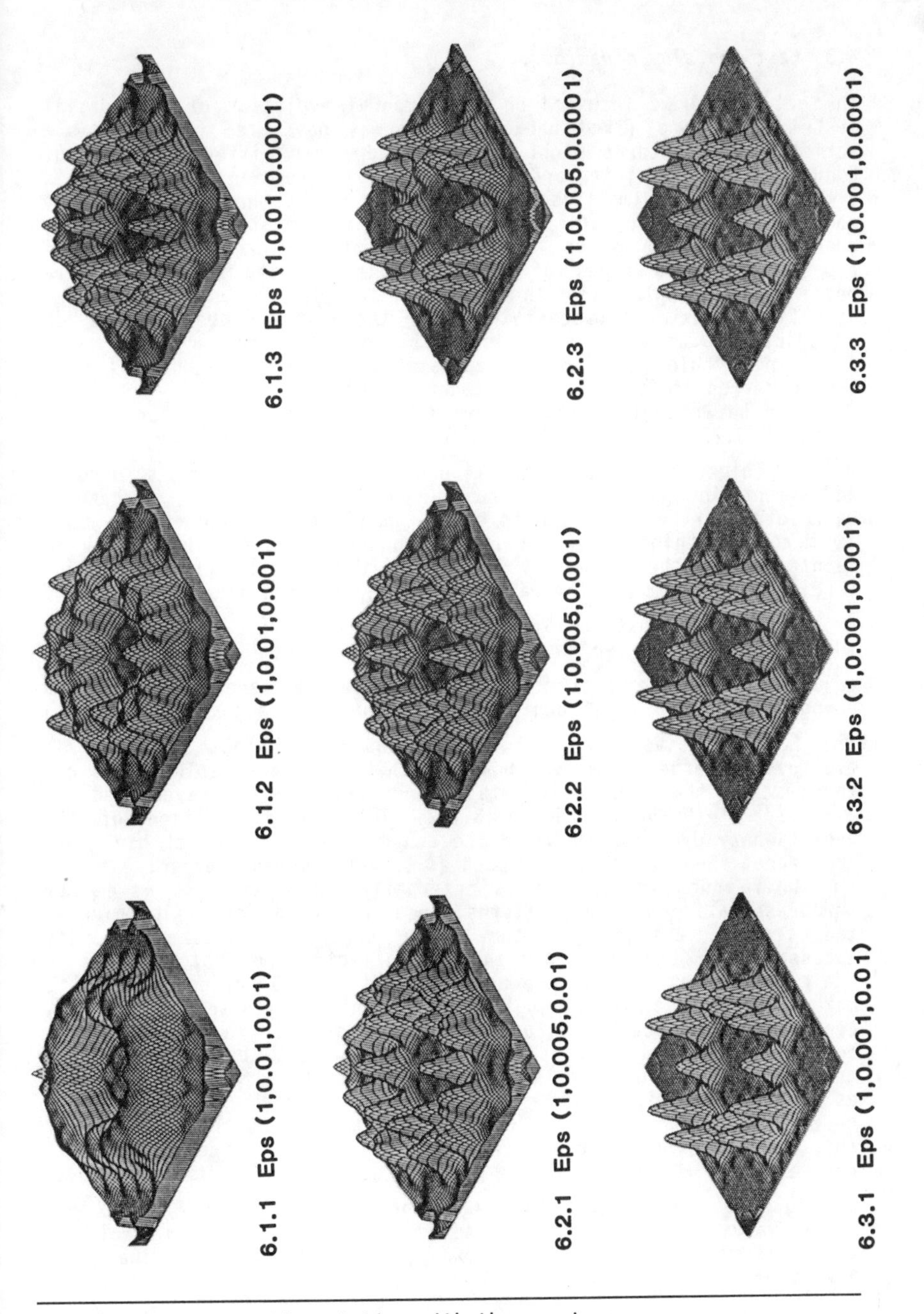

Fig. 6: Comparative statics with three actors.

6.3. *Fast and slow dynamics*

In Section 6.2 we focussed on the temporal evolution of our spatial system against a fixed backcloth. It was possible to show how a variety of structures might be generated under different speeds of adjustment of the state variables. As an approximation to reality this situation is oversimplified, as one would expect the backcloth to be constantly changing. In terms of dynamical systems theory, this changing backcloth is the "slow manifold". It was argued in Section 4 above that three types of adjustment in the state variables are possible: a single step in the direction of the equilibrium; an iterative adjustment process to return the system to equilibrium; or no adjustment at all. In Sections 6.1 and 6.2 we have covered cases involving single step or zero adjustments. We now extend this to a consideration of iterative adjustment procedures.

For the changing backcloth we focus on beta, which decreases in steps of 0.25 from its "high" of 1.5 down to 0.25 with $\alpha = 1.2$. The $\mathbf{n}^{max}$ = (high, 0, 0) case (Equation (28)) now represents structural adjustment in the style of Harris and Wilson, but to a changing backcloth. Further interest is generated if we consider exactly what is meant by "high", as a certain amount of disequilibrium may be maintained in the system by keeping this parameter finite and not forcing the system all the way to stability at each period.

Figure 7 illustrates our basic scenario. for $\mathbf{n}^{max}$ = (10, 0, 0) and $\varepsilon_1 = 1$. We can see two kinds of process at work here: first, the adjustment of the system to its equilibrium; and secondly, the changing nature of that equilibrium. Because the adjustments in $\{W_j\}$ are fast relative to the backcloth change, a stable solution is eventually reached. Observe, however, that the stable solution is *not* the same as the comparative static equilibrium which we derived for $\alpha = 1.2$, $\beta = 0.25$ in Figure 3.1.2. This is a function of the evolutionary dynamic, in which the centre is eliminated at high beta and cannot then force its way back as travel becomes cheaper.

While such an outcome is internally consistent, it is rather unrepresentative of real patterns, because we are failing to pick up the true advantages of the "city centre" in terms of its accessibility. One way round this is to factor the cost of trips to the city centre. Thus if one assumes that it is 5 percent cheaper to make a trip of a given length if that trip terminates at the city centre, we can attain a situation, as in Figure 8, in which the city centre becomes the focus of activity at the high beta extreme, and gradually forces out even the centres which compete from the periphery as beta is reduced.

A city centre factor of 0.95 is used throughout the remainder of this section. The principal challenge to the dominance of the city centre is now through the price and rent mechanisms as in Section 6.2. In Figure 9, we consider the case where $\mathbf{n}^{max}$ = (10, 10, 1) so that rents vary as a single step adjustment to equilibrium, while a trade-off between floorspace and price cutting determines the iterative shift towards equilibrium. Even for a low value of

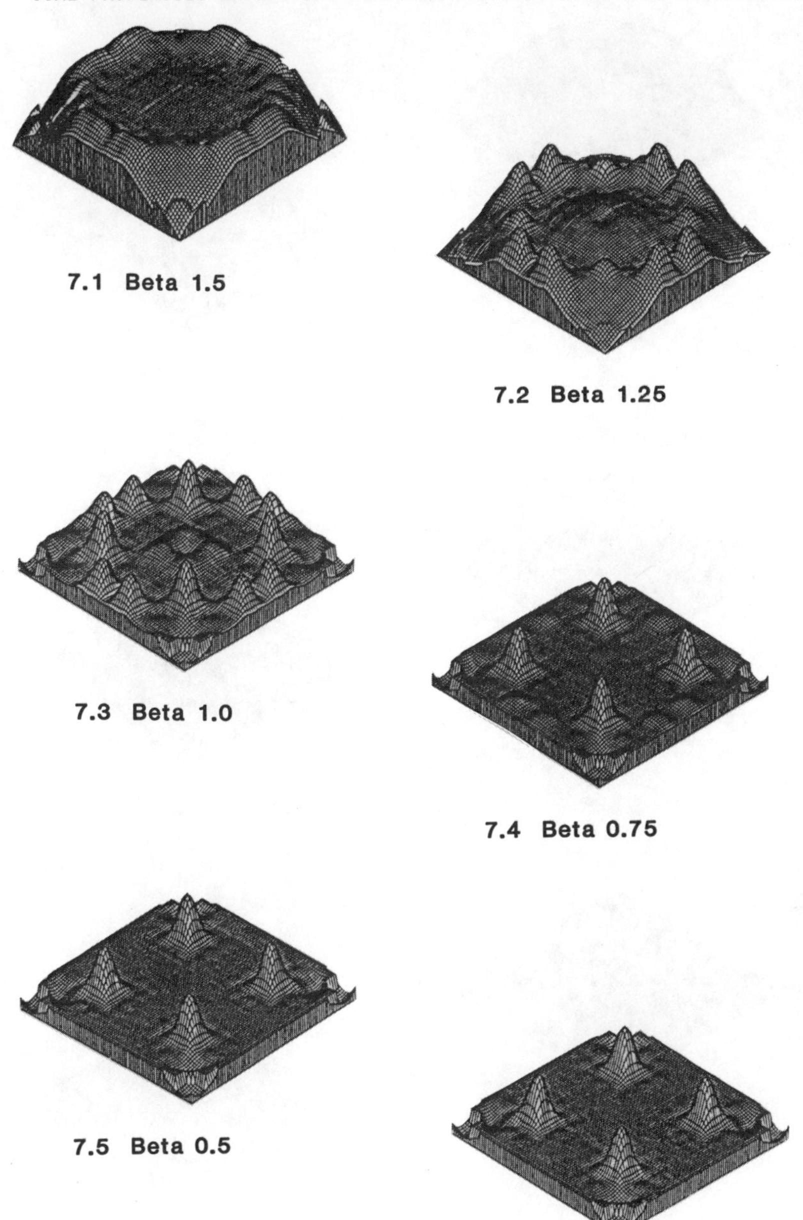

Fig. 7: Stock dynamics with backcloth change.

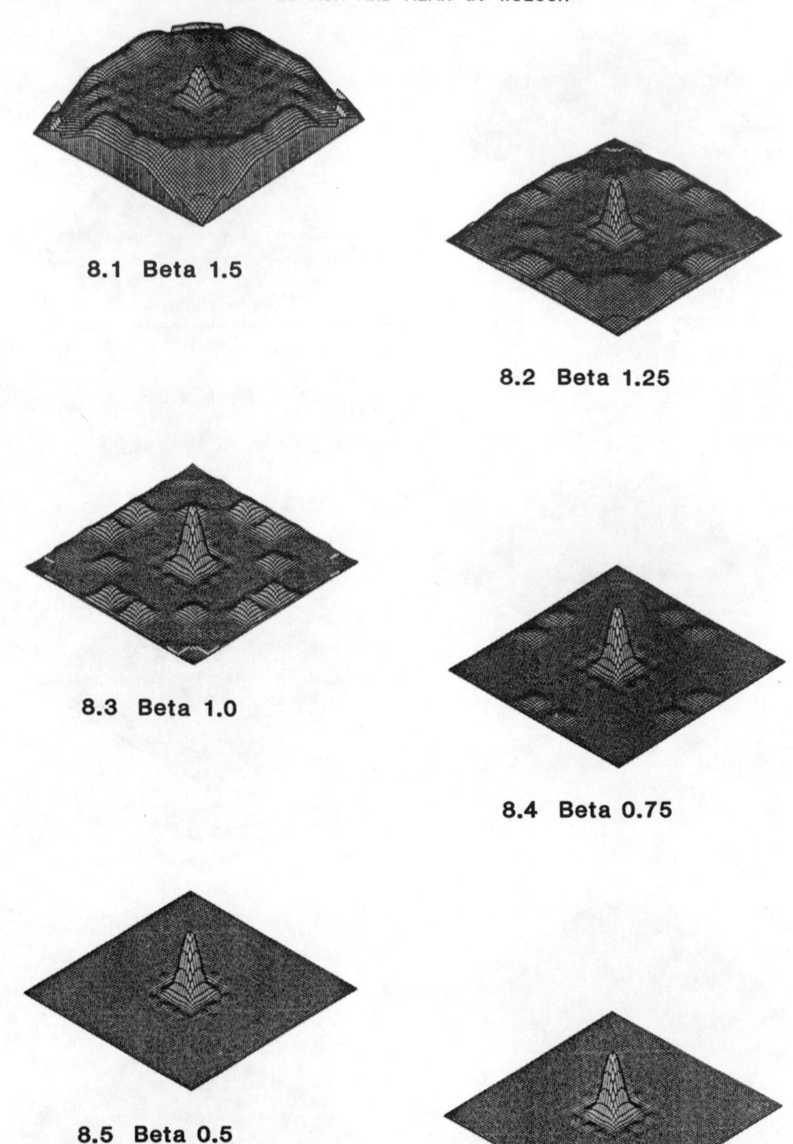

Fig. 8: Stock dynamics with backcloth change and city centre factoring.

$\varepsilon_2 = 0.001$, large price differentials are seen to arise, where typically prices are forced up as declining centres endeavour to maintain competitiveness, while more advantageously placed producers reinforce their dominance through price cutting.

Figure 10 takes the case of Figure 9.2 with $\boldsymbol{\varepsilon} = (1, 0.01, 0.001)$ and increases the number of cross-sectional rent iterations to 10 (hence $\mathbf{n}^{max} = (10, 10, 10)$). Peripheral development is now slightly more well-defined throughout and we obtain the beginnings of a peaked rent surface. However in terms of real world pattern, again one would perhaps like to see still greater differentiation in the rent surface, and a little more uniformity of price.

In Figure 11 we have assumed rent adjustment to be the dominant equilibriating mechanism with $\varepsilon = (1, 0.001, 0.01)$, while floorspace adjustment has been restricted to a single step towards equilibrium ($\mathbf{n}^{max} = (1, 10, 10)$). For the first time period only, an iteration vector of (10, 0, 0) was applied to obtain a non-uniform, disequilibrium initial allocation of $\{W_j\}$. An intuitively pleasing pattern of development results, with central growth which is tempered not only by the slow reaction of activity levels to disequilibrium, but also by the formation of an increasingly peaked rent surface. The rents fall most rapidly in the inner city ring, rising to minor peaks in the periphery. The rent surface is now much more differentiated than the price surface, which nonetheless reflects the same pattern of comparative advantage.

7. CONCLUSIONS

In this paper we have demonstrated, both theoretically and numerically, how the variety of spatial-structural-dynamic urban models can be extended through the introduction of new kinds of economic agent. Further refinement is still desirable, however, in the ways discussed in Section 5. Thus we might concern ourselves in future with alternative surplus allocation mechanisms, different kinds of dynamic adjustment, the methods in which the various procedures are nested, and especially the introduction of entropic dispersion. In the new, and more explicitly dynamic models we also need to consider carefully the possiblity of instability and multiple equilibria. This paper has been only illustrative, therefore, of the potential of economic-dynamic extensions to models of urban spatial structure.

School of Geography
University of Leeds
Leeds
United Kingdom

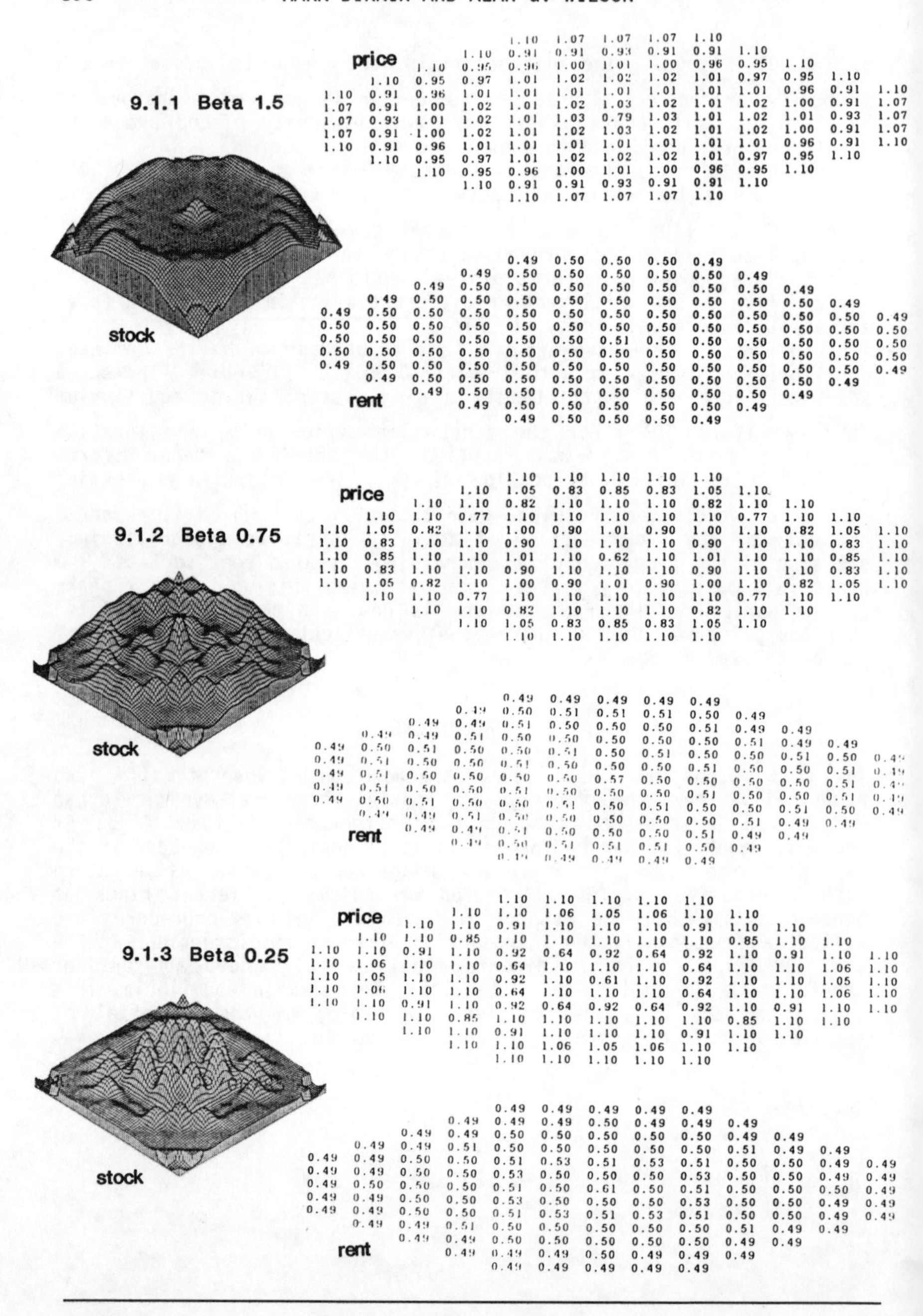

Fig. 9.1: Dynamic adjustment, $\mathbf{n}^{max} = (10,10,1)$, $\boldsymbol{\varepsilon} = (1,0.01,0.01)$.

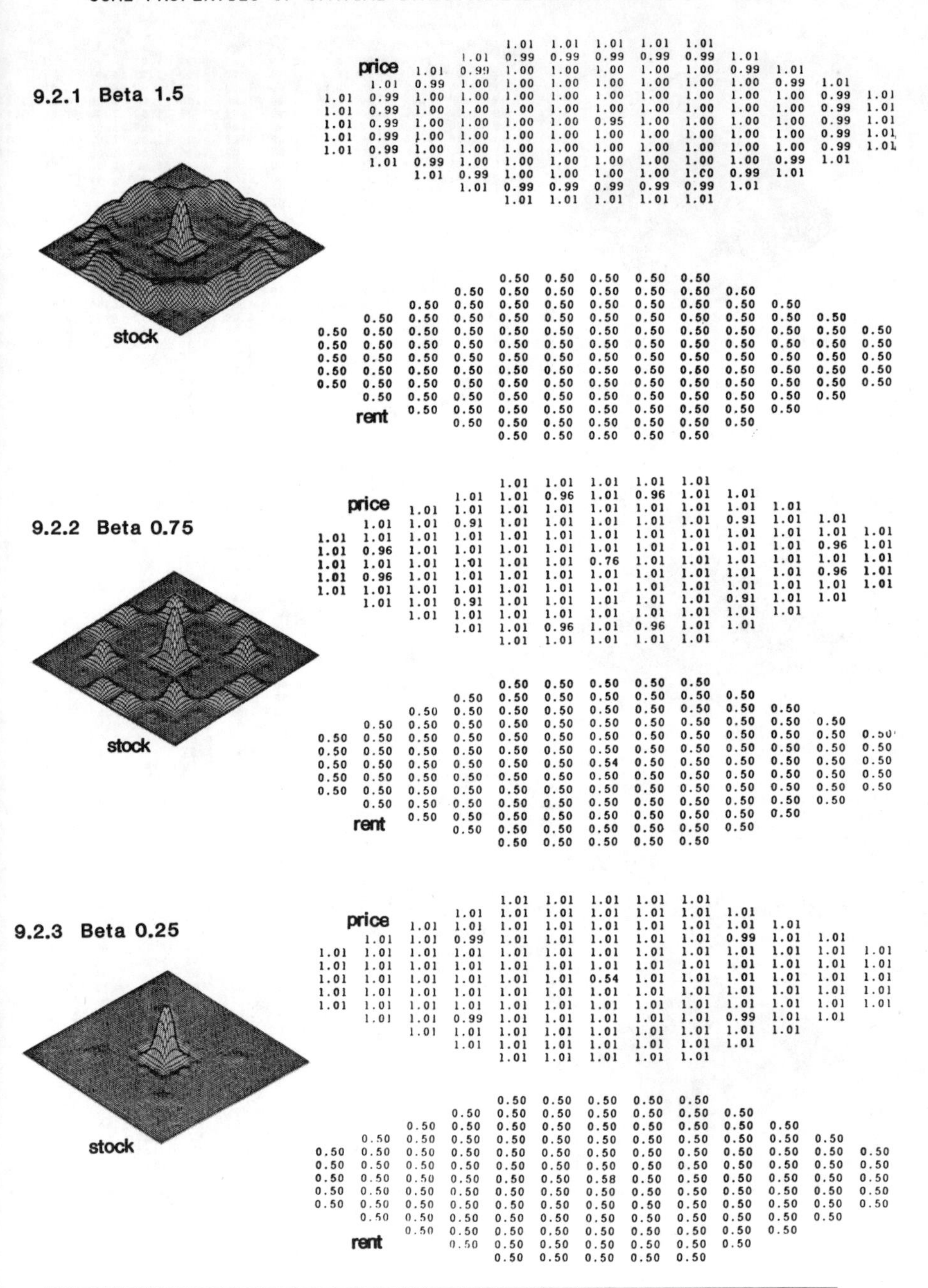

Fig. 9.2: Dynamic adjustment, $\mathbf{n}^{max} = (10,10,1)$, $\boldsymbol{\varepsilon} = (1,0.001,0.01)$.

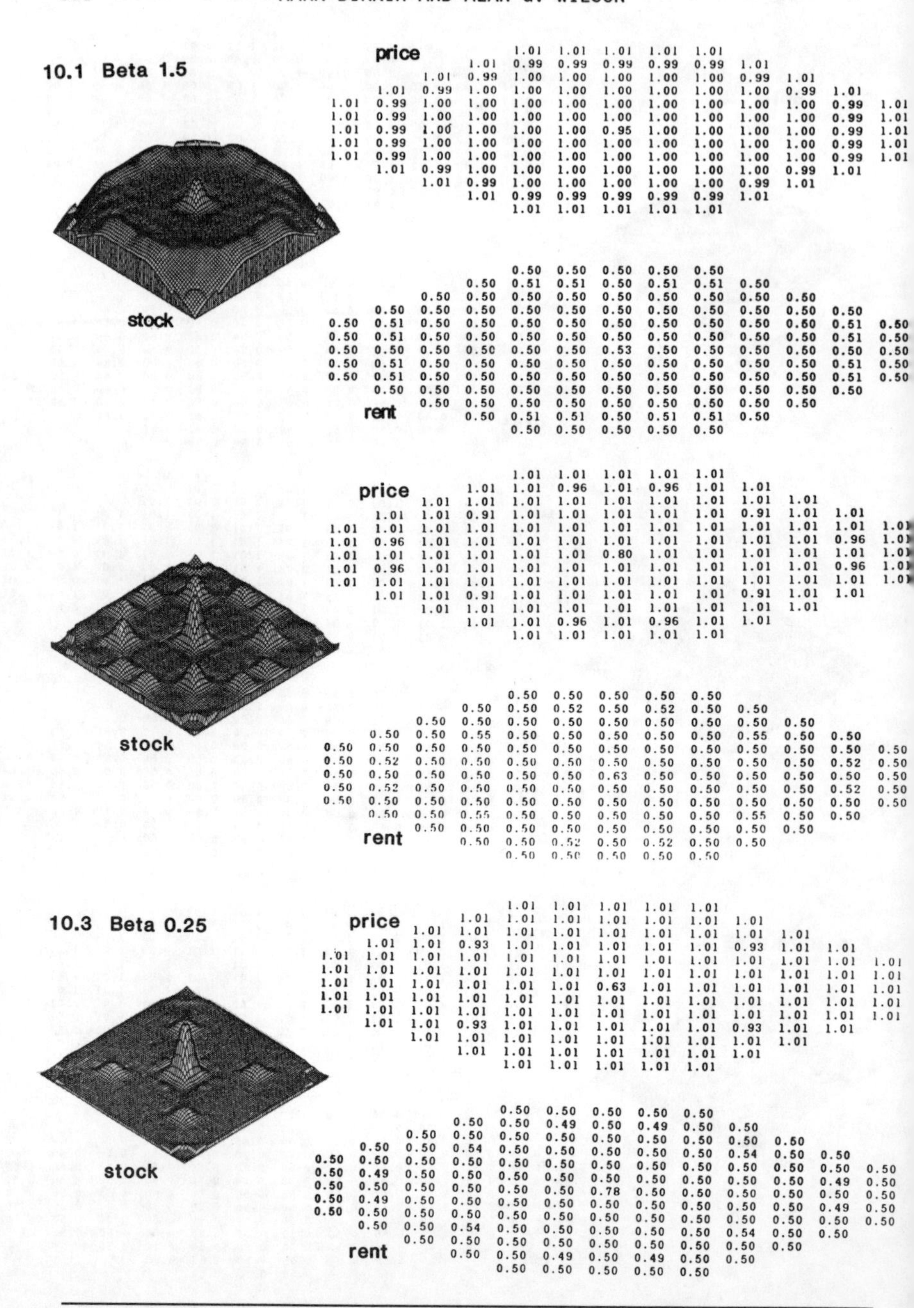

Fig. 10: Dynamic adjustment, $\mathbf{n}^{max} = (10,10,10)$, $\boldsymbol{\varepsilon} = (1,0.001,0.01)$.

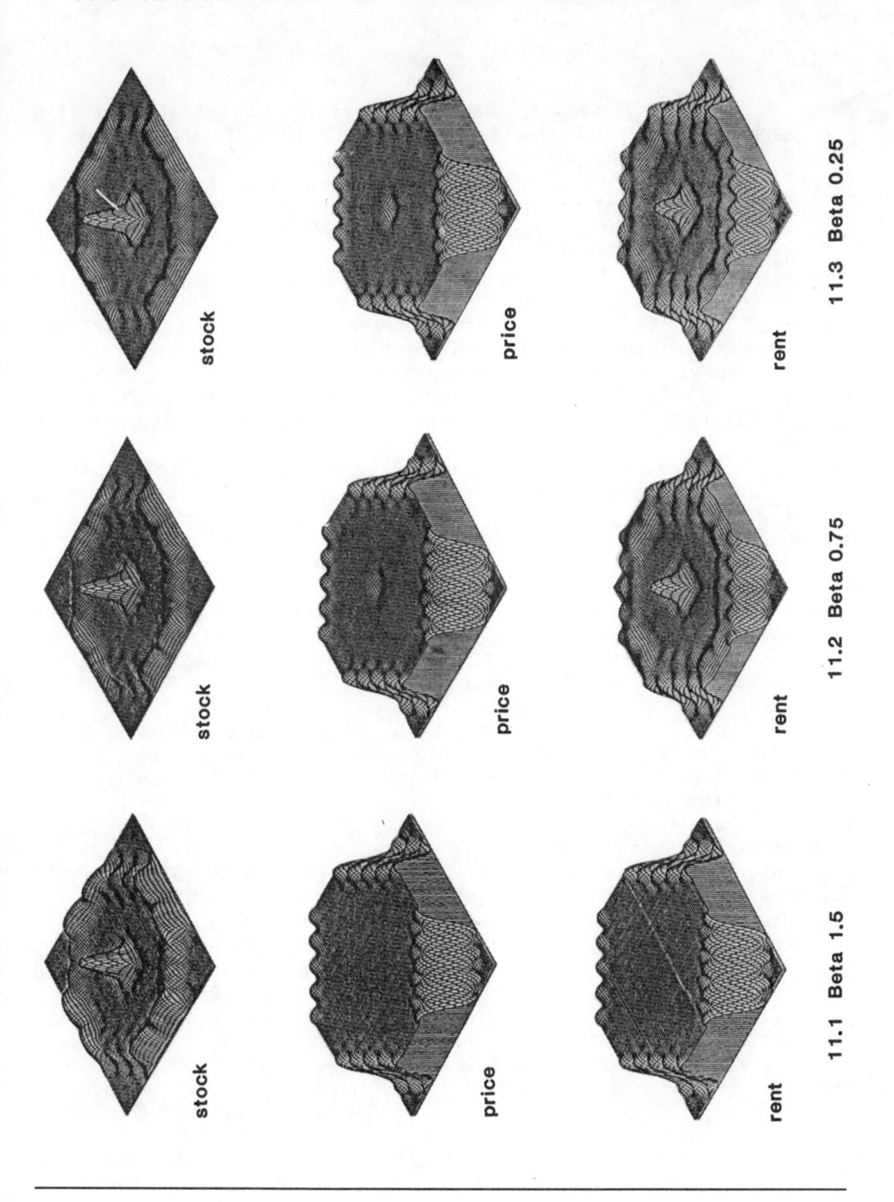

Fig. 11: Dynamic adjustment: $n^{max} = (1,10,10)$, $\varepsilon = (1,\ 0.001,\ 0.01)$.

BIBLIOGRAPHY

Anas, A.: 1989, 'A dynamic economic model of a regulated housing market', in C. S. Bertuglia, G. Leonardi and A. G. Wilson (eds.), *Modelling of Urban Dynamics*, forthcoming.

Campbell, D., J. Crutchfield, D. Former, and E. Jen: 1985, 'Experimental mathematics: the role of computation in non-linear science', *Communications of the Association for Computing Machinery* **28**, 374-384.

Chudzynska, I. and Z. Slodkowski: 1984, 'Equilibrium of a gravity demand model', *Environment and Planning* **A 16**, 185-200.

Clarke, G. P.: 1986, 'Retail centre useage and structure: empirical, theoretical and dynamic explorations', PhD thesis, School of Geography, University of Leeds.

Clarke, M.: 1984, 'Integrating dynamic models of urban structure and activities', PhD thesis, School of Geography, University of Leeds.

Clarke, M. and A. G. Wilson: 1983, 'The dynamics of urban spatial structure: progress and problems', *Journal of Regional Science* **13**, 1-18.

Evans, S.: 1973, 'A relationship between the gravity model for trip distribution and the transportation problem in linear programming', *Transportation Research* **7**, 39-61.

Fotheringham, A. S.: 1985, 'Equilibrium and the competing destination model', paper presented to the IBG Annual Conference, Leeds.

Haag, G.: 1989, 'Services 2: a master equations approach', in C. S. Bertuglia, G. Leonardi, and A. G. Wilson (eds.), *Modelling of Urban Dynamics*, forthcoming.

Harris, B., J-M. Choukroun, and A. G. Wilson: 1982, 'Economics of scale and the existence of supply-side equilibria in a production-constrained spatial-interaction model', *Environment and Planning* **A 14**, 813-827.

Harris, B. and A. G. Wilson: 1978, 'Equilibrium values and dynamics of attractiveness terms in production-constrained spatial-interaction models', *Environment and Planning* **A 10**, 371-388.

Kaashoek, J. F. and A. C. F. Vorst: 1984, 'The cusp catastrophe in the urban retail model', *Environment and Planning* **A 16**, 851-862.

Leonardi, G.: 1981a, 'A unifying framework for public facility location problems - Part I: a critical review and some unsolved problems', *Environment and Planning* **A 13**, 1085-1108.

Leonardi, G.: 1981b, 'A unifying framework for public facility location problems - Part 2: some new models and extensions', *Environment and Planning* **A 13**, 1085-1108.

Leonardi, G. and A. G. Wilson: 1989, 'Approaches to integrated models', in C. S. Bertuglia, G. Leonardi, and A. G. Wilson (eds.), *Modelling of Urban Dynamics*, forthcoming.

Lombardo, S. R. and G. A. Rabino: 1983, 'Non-linear dynamic models for spatial interaction: the results of some numerical experiments', paper presented to the 23rd European Congress, Regional Science Association, Poitiers, France.

May, R. M.: 1976, 'Simple mathematical models with very complicated dynamics', *Nature* **261**, 459-467.

Phiri, P.: 1980, 'Calculation of the equilibrium configuration of shopping facility sizes', *Environment and Planning* **A 12**, 983-1000.

Pumain, D., T. Saint-Julien, and L. Sanders: 1984, 'Vers une modélisation de la dynamique intra-urbaine', *L'Espace Geographique* **2**, 125-135.

Rijk, F. J. A. and A. C. F. Vorst: 1983a, 'Equilibrium points in an urban retail model and their connection with dynamical systems', *Regional Science and Urban Economics* **13**, 383-399.

Rijk, F. J. A. and A. C. F. Vorst: 1983b, 'On the uniqueness and existence of equilibrium points in an urban retail model', *Environment and Planning* **A 15**, 475-482.

Roy, J. R. and J. F. Brotchie: 1984, 'Some supply and demand considerations in urban spatial interaction models', *Environment and Planning* **A 16**, 1137-1147.

Roy, J. R. and B. Johansson: 1985, 'On planning and forecasting the location of retail and service activity', *Regional Science and Urban Economics* **14**, 433-452.

Wilson, A. G.: 1981a, 'Some new sources of instability and oscillation in dynamic models of shopping centres and other urban structures', *Sistemi Urbani* **3**, 391-401.

Wilson, A. G.: 1981b, *Catastrophe Theory and Bifurcation: Applications to Urban and Regional Systems*, Croom Helm, London; University of California Press, Berkeley.

Wilson, A. G.: 1985, 'Structural dynamics and spatial analysis: from equilibrium balancing models to extended economic models for both perfect and imperfect markets', Working Paper 431, School of Geography, University of Leeds.

Wilson, A. G. and M. Birkin: 1985, 'Dynamic models of agricultural location in a spatial interaction framework', Working Paper, 399, School of Geography, University of Leeds.

Wilson, A. G. and M. Clarke: 1979, 'Some illustrations of catastrophe theory applied to urban retailing structures', in M. Breheny (ed.), *Developments in Urban and Regional Analysis*, Pion, London, 5-27.

Wilson, A. G. and M. L. Senior: 1974, 'Some relationships between entropy maximising models, mathematical programming models and their duals', *Journal of Regional Science* **14**, 207-215.

SILVANA T. LOMBARDO AND GIOVANNI A. RABINO

URBAN STRUCTURES, DYNAMIC MODELLING AND CLUSTERING

1. INTRODUCTORY REMARKS ON DYNAMIC SPATIAL INTERACTING ACTIVITY MODELS

In this paper we continue our studies (Lombardo and Rabino 1983, 1984, 1986) of the dynamic spatial interacting activity (S.I.A.) model of the Harris and Wilson (1978) type. This model, usually related to the location of shopping centres, is a general paradigm for the analysis of locational problems of several other urban activities: e.g. industries (Wilson and Birkin, 1983), residences (Clarke and Wilson, 1983), public facilities (Wilson and Crouchley, 1983).

In its most usual form, the model is expressed by the following differential equation:

$$\dot{W}_j = \varepsilon\left(\sum_i e_i P_i \frac{W_j^\alpha \exp(-\beta c_{ij})}{\sum_j W_j^\alpha \exp(-\beta c_{ij})} - k_j W_j\right) \tag{1}$$

where W_j is the size of the supply centre at j;

$e_i P_i$ is the demand for services expressed by zone i (with e_i as per capita demand);

c_{ij} is the cost of travel from i to j;

α is a measure of the economies of scale of consumers;

β is a measure of the deterrence of the costs of travel;

k_j is the cost per unit of supplying the centre at j;

ε is a measure of the speed of response of the system to the exogenous variations.

The first term (within the brackets) of the right hand side of Equation (1) is the demand (D_j) for services in zone j, computed by means of a spatial interaction model from the demand expressed by the consumers in their zone of residence. The second term (within the brackets) of the right hand side of Equation (1) is the supply (S_j) of services in zone j and is assumed, as a first approximation, proportional to the size of the supply centre.

From an economic point of view, the model represents a process of adjustment to a general equilibrium in a competitive market, where the sizes of the supply centres grow or decline until the profits in each centre are equal to zero. It can be seen easily that the stationary state of Equation (1):

$$\dot{W}_j = 0$$

implies that

J. Hauer et al. (eds.), Urban Dynamics and Spatial Choice Behaviour, 203–217.

$$D_j - S_j = 0$$

which corresponds to the annullment of the Walrasian function of excess of demand:

$$E_j = D_j - S_j$$

A more accurate economic treatment should take into account also the dependence of demand and supply on the prices of the services in the various zones, that is:

$$E_j(\mathbf{p}) = D_j(\mathbf{p}) - S_j(\mathbf{p})$$

where **p** is the vector of the prices.

In an urban context, this dependence appears quite unimportant, because the prices can be considered - at least to a first approximation - as an exogenous constant, defined in a market area which is much larger than an urban area. Therefore, this aspect is usually neglected in the modelling. However, some versions of the model taking prices into account were developed by Wilson (1985).

From the viewpoint of dynamic system analysis, the model represents an adaptive process in which many of the causes of changes in service centres location and size are made endogenous. This is obtained by assuming both that supply is a function of demand:

$$\dot{W}_j = f(\mathbf{D}, \mathbf{W})$$

and that demand is a function of supply:

$$D_j = g(\mathbf{W})$$

Economic theory allows us to say that the existence of externalities in the dependence of demand on supply (namely: economies of scale and saturation effects) does not guarantee the existence, uniqueness, stability and optimality - in a Paretian sense - of a general equilibrium toward which the adaptive process would have to converge. From the aspects mentioned above the interest in analysing the behaviour of model (1) is clear.

A number of results have already been obtained by means of both numerical simulations and analytical studies. The most important finding, concerning the equilibrium solutions, is the demonstration that, when the scale economies factor α is smaller than or equal to 1, the model presents an equilibrium solution which is:

- positive, that is $W_j \neq 0,\ \forall\ j$ (Harris and Wilson, 1978);
- unique and globally stable (Rijk and Vorst, 1983a);
- a continuous function of the values of the parameters (Kaashoek and Vorst, 1984).

On the contrary, when α is greater than 1, it was demonstrated that the system presents one or more positive equilibria, of which at least one must be unstable (Rijk and Vorst, 1983b). Numerical simulations (Lombardo and Rabino, 1986; Clarke and Wilson, 1981) showed however that usually there is only one equilibrium, which is therefore unstable, except for very particular cases of the spatial system's symmetry and for certain values of the parameters.

In other words, for $\alpha > 1$, the unique stable equilibrium is non-positive (i.e. with one or more $W_j = 0$), so that the service centres are located in few zones. Another interesting result, concerning the spatial structures simulated with the model, is that when α is smaller than 1, an increase of mobility (that is a decrease of β) leads to an increase in the size of the small centres instead of an increase in the large centres, as expected (Lombardo and Rabino, 1984).

In this paper, we continue along this line of research, analysing in more detail the spatial structures resulting from the equilibrium solutions of the model. We will focus on the range of stability of these structures, on the speed of convergence to these solutions, on the shape of the market areas of the service centres and on the related patterns of journeys to services.

This analysis will show the ability of the model to highlight the latent structures of urban systems. On this basis, in the final part of this paper, we show that the model itself can be considered as a clustering technique and profitably used first of all for the regionalization of geographical systems, but also for the partitioning of any kind of data.

2. URBAN SPATIAL STRUCTURES

In this section, an analysis of the urban spatial structures simulated by the dynamic S.I.A. model is reported. The results discussed are based on numerical simulations carried out for various urban systems, both real and fictitious, but the data presented in this paper refer to the metropolitan area of Rome as the real system and to a hexagonal grid as the fictitious system. The main characteristics of the Rome area are described in Lombardo and Rabino (1986). The fictitious system (Fig. 1) is made up of 61 hexagonal zones, whose centres are connected by a uniform network with the size of the links assumed equal to 1.

A total population of 1,000,000 inhabitants is distributed among the zones, with a density exponentially decreasing from the core zone to the peripheral zones.

In Figure 2 service centre size (S_j) trajectories from the initial state to the equilibrium state are shown for three zones of the hexagonal system (case $\alpha = 1.3$, $\beta = 2.5$). In the same figure the trajectories of the demand (D_j) are also represented by a dashed line. These curves clearly illustrate the two types of trajectories obtained in all simulations when the system evolves from an initial disequilibrium to an equilibrium, all values of the parameters being constant. Some zones (dominant) move from the initial to the final state following quasi-exponential paths; the others follow a humped path.

In case (A) of Figure 2, demand exceeds supply from the outset, so that the service centre of the zone grows until the supply meets the demand. On the contrary, in case (B), supply always exceeds demand, so that the zone declines monotonically to zero. In case (C), initially demand exceeds supply, so that supply increases towards an equilibrium. In the meanwhile, however, the changes of supply in the other zones reduces demand in the zone considered by attracting it elsewhere. As a consequence, the supply in the zone exceeds the demand and starts to decline towards zero.

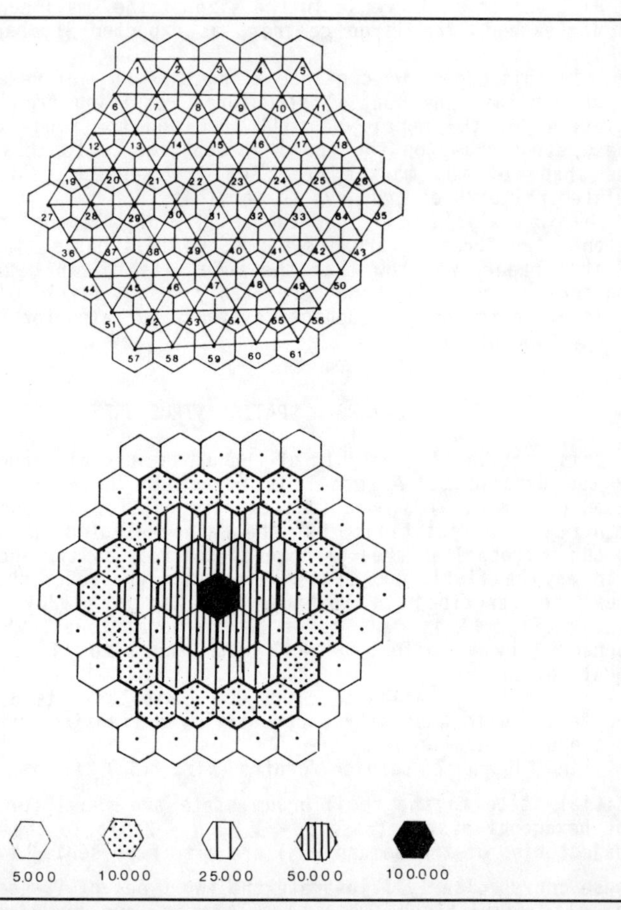

Fig. 1: Hexagonal spatial system: zone numbering and population by place of residence.

It is worthwhile observing that, in this case, an interaction process is established between demand and supply which recalls the well known predator-prey interaction. This is not surprising because of the strong similarity between the Volterra model and the present one (see for instance Wilson, 1981). But, as in the present model the interactions are nonlinear, the oscillatory trajectories of the Volterra model, which are caused by the linearity of the interactions, do not appear.

An analysis of the speed of convergence to the equilibrium solution was carried out for different values of the parameters. In particular, we were interested in the effect of the parameters α and

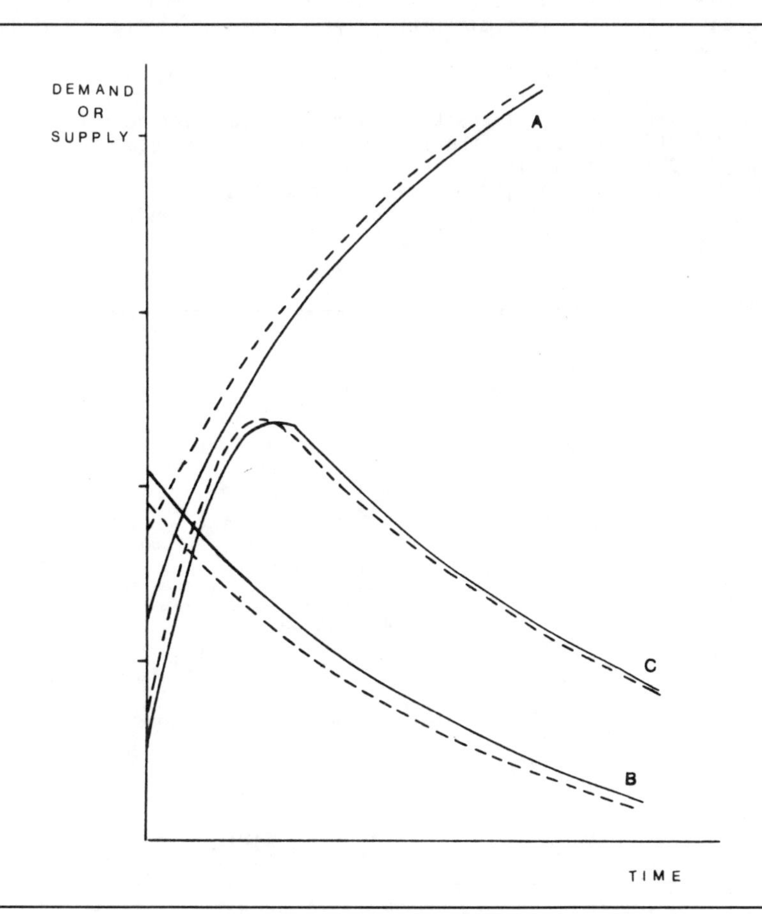

Fig. 2: Trajectories to equilibrium state (hexagonal system, zones 1, 10, 31).

β. It is possible to recognize again two completely different situations corresponding to $\alpha > 1$ and $\alpha \leq 1$. In the latter case, the simulations showed that the time needed to reach the equilibrium, evaluated in terms of number of steps, does not vary significantly even for large variations of β. Obviously the time depends on the initial conditions and is proportional to the adopted speed of response of the system: ε.

The case of $\alpha > 1$ is illustrated in Figure 3, with reference to the hexagonal system. First of all it is possible to see that the speed of convergence is highly dependent on the value of β and generates a complex profile with peaks and valleys.

Each valley corresponds to one spatial structure, that is to one of those non-positive equilibrium solutions of the model occurring for α greater than 1 (such structures, presented in Figure 4, will be discussed below).

Each peak occurs when the value of β is critical, that is when we are in those points in the space of the parameters in which sudden jumps appear in the transition from one equilibrium solution to another. This means that the speed of convergence becomes lower and lower as β approaches its critical values.

In Figure 3 it is possible to see also that the ranges of values of β within which each structure is stable can be very different. So, for instance, structure A is stable in the range $0 < \beta < 1.76$, while

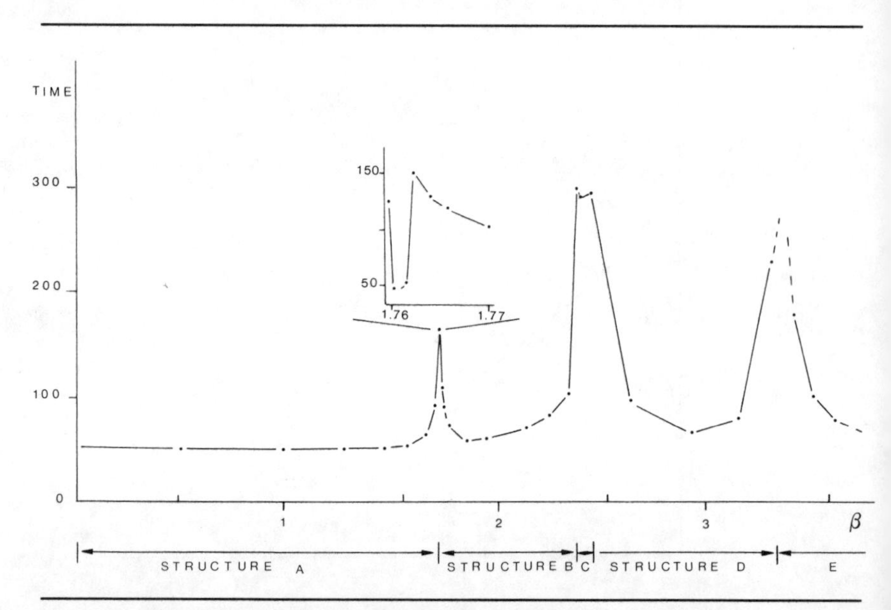

Fig. 3: Speed of convergence to equilibrium state (hexagonal system, $\alpha > 1$).

structure B is stable only in the range $1.76 < \beta < 2.4$ and a third structure, similar to the previous one, is stable only in the very narrow range $1.7605 < \beta < 1.7620$. We see also that the different structures cannot be reached with the same speed. So, for instance, to reach the equilibrium solution C, we need a number of iterations at least six times larger than for other structures. From the above observations it appears that some structures could easily be overlooked; therefore a systematic classification of all obtainable structures would be a very hard task. Keeping this in mind, Figure 4 presents the location of the service centres for the six most

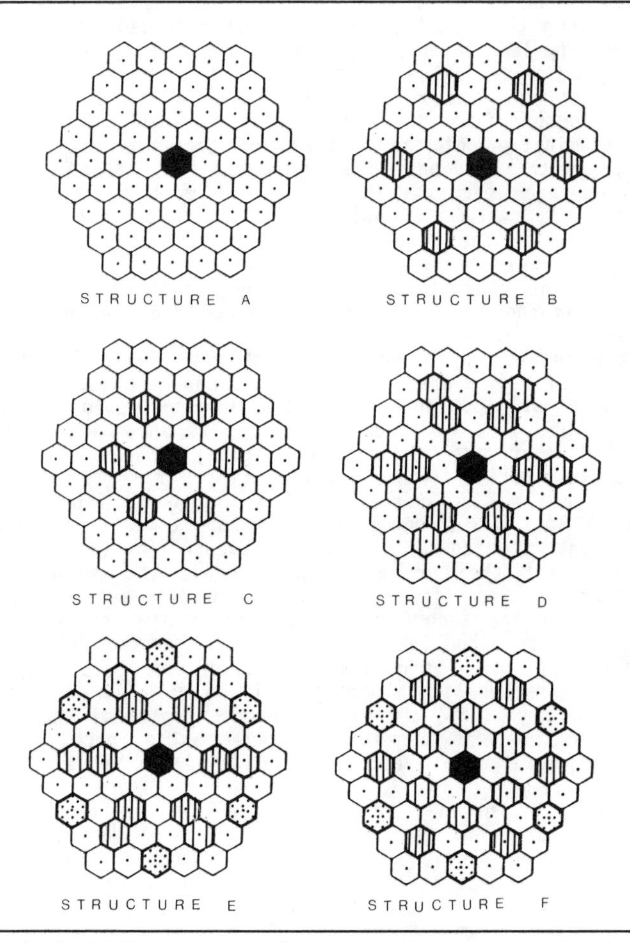

Fig. 4: Spatial structures (location of service centres).

meaningful equilibrium solutions of the hexagonal system, obtained for β ranging from 0.0 to 5.0 (with $\alpha = 1.3$). As expected, an increase in β (equivalent to reducing the mobility of population) leads to an increasingly diffuse location of the service centres. This spreading of the service centres, although obtained for a fictitious system, presents some features which can be found also in real systems. Case A corresponds to a unique large centre which supplies all the system. In case B, it can be seen that, when accessibility to the core zone is reduced, relatively small service centres emerge in the peripheral areas to supply the less-favoured zones. However the core zone still supplies about 75% of the population. Another increase of β leads to a noticeable change of the pattern (case C). The core zone of case A is now substituted by a set of seven zones (one is the core zone and the others belong to a semiperipheral ring). The core zone now supplies approximately 40% of the population, while the remaining 60% is shared uniformly among the other six zones.

In Figure 5, the main journeys to service are represented for a selected number of the structures analysed; namely for cases B, C and D. An examination of the patterns of the flows confirms the main findings of the previous locational analysis, adding details on the areas served by each service centre. It must be stressed, however, that, in this case, the analysis of the market areas is relatively easy because of the very clear pattern of flows generated by each zone. In general the flows are oriented to one single service centre (generally the nearest one) when they originate in a zone without services, or they remain within the zone itself when it is supplied with services.

This result is counter-intuitive, for, while in the case of a high value of β (low mobility) the flows can be expected to end in the same zone of origin, for a small value of β we would expect a pattern of flows that is more spread out. But, as shown above, for a small value of β, there are few zones with supply centres, so that the flows are clearly directed towards these centres.

For case A, not represented in Figure 5, the pattern is the simplest. Flows are directed from each zone towards the only existing service centre, which is located in the core zone (see Figure 4a).

In case B the core zone supplies all its own population, the population of the first ring and, on the average, 90% of the population of the second ring. The six supply centres of the third ring supply more than 75% of their population and all the population of the fourth ring. The remaining zones of the third ring (e.g. zones 17 and 25) are supplied almost equally by the nearest service centre of the same ring (55%) and by the service centre of the core zone. By considering then the largest flow coming out of each zone and headed for a service centre, we obtain the market areas represented in Figure 6 (case B).

In case C, the core zone supplies all its own population plus 85% of the population of the first ring. The service centres located in the second ring supply the majority of their population (85%) and almost all the population of the zones belonging to the third and fourth rings. Each residential zone of the second, third and fourth rings is served primarily by a single service centre. Exceptions are, for instance, zones 24 and 18, which, because of the symmetry of the system, gravitate equally towards two supply centres (namely centres 16 and 33). A small quantity of the population of the second, third and fourth rings then reaches the core zone.

The simulated pattern of case C has a great deal of similarity with real patterns, given that in many real urban systems we can recognize a central service area, which supplies its surroundings and is also used by the whole urban system, being of a large size and having rare services available. In this case, a set of semiperipheral service areas usually supply semiperipheral and peripheral zones. The

Fig. 5: Main journeys to service. Fig. 6: Market areas.

market areas resulting from the journeys to services in case C are represented in Figure 6c.

By comparing Figure 6c with Figure 6b, the smaller size of the market area of the core zone and the partial overlapping of the other market areas, due to the symmetry of the system, are noticeable. We will not discuss case D in the same detailed way as the previous cases. It is sufficient to say that case D is very similar to case C. An important difference though is that the flows produce a hierarchic structure of services. Indeed, it can be seen (Figure 5d) that the zones of the fourth ring gravitate mostly towards the six service centres of the third ring and these, in their turn, gravitate towards the service centres of the second ring. This hierarchic structure leads to market areas which are less clearly defined than the previous ones. Overall, the ability of this simple model to reproduce hierarchic patterns of flows, which can usually be recognized in real urban systems, is very marked.

3. CLUSTERING

To introduce this topic it is useful to recall that, in general terms, a clustering procedure is an allocation process in which the entities to be classified are assigned to clusters on the basis of the optimization of some function of the distances between the entities in the multidimensional space of the characteristics of the entities.

Typically, the objective function assumed in the classification processes (when homogeneous partitions of the set of the entities must be obtained) is the minimization of the sum of the squares of the internal distances of the elements of the clusters: in other words, the minimization of the variance of the clusters.

It is possible to see that the definition of the clustering procedure described above is different from the allocation obtained by the customary single constrained spatial interaction model in two aspects. Firstly, in an allocation performed by a S.I.A. model, the destination poles are predetermined whereas the clusters are unknown. Secondly, a S.I.A. model generates a probabilistic allocation to the different destinations, while a clustering algorithm assigns each entity to one and only one cluster.

This latter difference can be overcome by means of the concept of a probabilistic cluster where the association of each element to the clusters is given in terms of probability. Alternatively, we can proceed from a probabilistic allocation to an "all or nothing" allocation, by means of some appropriate device (for instance assigning the element to the pole which presents the highest probability).

In the case of the former difference, we note that the equilibrium solutions of the dynamic S.I.A. model are sets of service centres endogenously defined on the basis of the population which use the centres (see Sections 1 and 2). Indeed the size of the service centres is given by:

$$S_j = \sum_i T_{ij}$$

where T_{ij} is the amount of the demand of i which is allocated to j. By interpreting it in terms of clustering techniques, this can be seen as an assignment of entities to poles not predetermined, but defined as to their existence and size by the relations of distance and by the entities which are assigned to them. The endogenous mechanism specific to the dynamic S.I.A. model turns out to be the key factor which changes a model of allocation to fixed poles into a clustering technique.

Further evidence of the equivalence of the dynamic S.I.A. model and a clustering technique can be found in the context of dynamic analysis. As suggested by Bellacicco (1984), we can think of a clustering technique as a reallocation process in a compartmental system, where the compartments are the different feasible clusters. The differential equations for the dynamics of this system are:

$$\dot{p}_i^k = \sum_j p_j^k r_{ji}^k - p_i^k \sum_j r_{ij}^k \tag{2}$$

where p_i^k is the probability that the entity k is assigned to the compartment i, and r_{ij}^k is the probability that the entity k shifts (in the unit of time) from an allocation in compartment i to an allocation in compartment j.

A suitable expression for the r_{ij}^k's is:

$$r_{ij}^k = \lambda \exp(\beta[U_j^k - U_i^k])$$

In other words the probability that an entity shifts from one compartment to another is assumed to be a function of the difference in the utilities, U_j^k and U_i^k, resulting from the assignment of the entity to those compartments.

Let us then assume that U_j^k is:

$$U_j^k = \frac{\alpha}{\beta} W_j - \ln c_{jk} \tag{3}$$

with $W_j = \sum_k p_j^k$, which implies that utility is a decreasing function of the distance between entity and compartment (cluster) and is an increasing function of the size of the compartment (the size being defined in terms of the number of entities assigned - in a probabilistic sense - to the compartment itself). It is now possible to show (Lombardo, 1986) that, with the above assumptions, this system (or the dynamic clustering process) is identical in conceptual terms to the dynamic S.I.A. model and to demonstrate that the two models have the same equilibrium solutions.

It should be noted that the assumption that the utilities are increasing functions of the size of the clusters stems from the consideration that, when the distances are equal, a bigger cluster implies that it has a large "radius" and is therefore closer to the entity to aggregate.

All the above remarks allow an interpretation of the dynamic S.I.A. model's fundamental parameters, α and β, in terms of clustering technique in the following way.

For parameter α, we can easily see that the use of $\alpha > 1$, instead of $\alpha \leq 1$, favours the assignment of the entities to bigger clusters. Since, on average, the marginal increase of variance decreases for increasing sizes of cluster (in other words, the addition of a new entity increases the variance less for a big cluster than for a smaller one), the above assumption corresponds to the objective of minimizing the variances of the clusters. In turn, this demonstrates why the use of $\alpha > 1$ (or, more generally, the assumption of a positive return-to-scale in the attractiveness factors of the dynamic S.I.A. model) is the key determinant of the appearance of the clusters. This also explains why the dynamic S.I.A. model produces all the spatial structures presented in Section 2, when α is greater than 1.

For parameter β, we recall that in all the dynamic or static S.I.A. models this parameter can be interpreted as the Lagrangian multiplier associated to the constraint:

$$\sum_{i,j} T_{ij} c_{ij} / \sum_{i,j} T_{ij} = \bar{c} \tag{4}$$

i.e. a constraint on the mean (or total) cost of transport. However, the equivalence already demonstrated between clustering techniques and the dynamic S.I.A. model implies that constraint (4) may also be interpreted as a mean size of the clusters. This explains why, for the same set of entities, different values of β produce a different number of clusters. Parameter β is, therefore, the factor defining the number of clusters whilst parameter α is the factor determining the minimization of the variance of clusters.

An example of use of the dynamic S.I.A. model as a clustering technique is represented by the market areas illustrated in Figure 6. The areas show how the hexagonal spatial systems can be divided into a different number of clusters by varying the parameter β. The analysis of speed of convergence to equilibrium states and the range of stability of these states, carried out in Sections 1 and 2, can be interpreted here as a methodology for evaluating the stability of the different way of partitioning a set of spatial data.

Another example of the use of the dynamic S.I.A. model as a clustering technique is an application carried out on the metropolitan area of Rome. In this application, for the sake of simplicity, we considered the interzonal travel time matrix as a proxy for every kind of matrix of distances between characteristics of the zones.

The main results of the application are shown in Figure 7 (a, b, c, d) for a varying number of clusters (or, in other words, for different values of β). We can see that, for $\beta = 0.08$, the metropolitan area is divided into 2 clusters: the zone 38 and the aggregate of all the other zones. This result fits very well with the

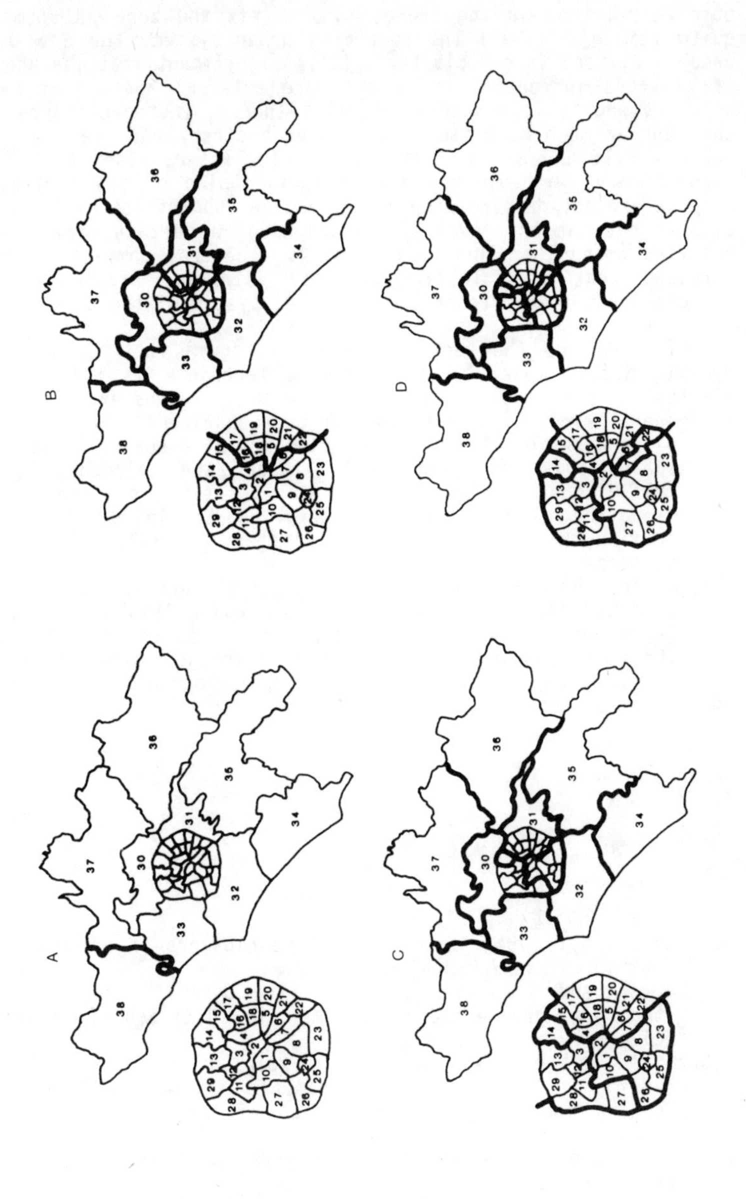

Fig. 7: Clusters (Rome urban system).

observation that in the travel time matrix the zone 38 seems to be quite remote from all the remaining urban system. For $\beta = 0.12$ the area is divided into 9 clusters. It is significant that, as the number of clusters increases, it is the more external zones that begin to separate and to form single clusters. Indeed, in the distance matrix, they appear much more isolated from each other than the inner zones. For $\beta = 0.15$ and $\beta = 0.16$ the number of clusters rises to 11 and 13, respectively, with an increase of partitioning of the central area. Here we are not dealing with a progressive subdivision of the previous cluster but, at any level of partitioning, a certain reallocation of entities among the clusters occurs. As a general comment, it must be stressed that the clusters obtained with our procedure fit the expectations of experts on the transport system of the Rome area quite well.

To conclude the section, we stress again that the use of the dynamic S.I.A. model as a clustering technique is not necessarily limited to spatial data (such as in the two previous examples).
Two general kinds of application can be recognized:

i. given a set of elements, characterized by a series of attributes, the matrix of distances between the elements (in the multidimensional space of the attributes) is computed and then, using the model, clusters are detected. Varying β, it is possible to find the required number of clusters or explore the whole set of cases.

ii. if the above distances are generators of interactions between the elements (for instance, flows) then using this interaction data (by means of constraint Equation 4), it is possible to estimate the corresponding value of β and to choose, among the different partitioning schemes, the one most appropriate to the existing level of interaction.

IRES
Turin
Italy

BIBLIOGRAPHY

Bellacicco, A.: 1984, 'Exponential type clusters in networks in time domain', *Sistemi Urbani* **6**, 95-102.

Clarke, M. and A.G. Wilson: 1981, 'The dynamics of urban spatial structure: progress and problems', W.P.313, School of Geography, University of Leeds, Leeds.

Clarke, M. and A.G. Wilson: 1983, 'Exploring the dynamics of urban housing structure: a 56 parameter residential location and housing model', W.P. 363, School of Geography, University of Leeds, Leeds.

Harris, B. and A.G. Wilson: 1978, 'Equilibrium values and dynamics of attractiveness terms in production-constrained spatial interaction models', *Environment and Planning* **A 10**, 371-388.

Kaashoek, J.F. and A.C.F. Vorst: 1984, 'The cusp catastrophe in the urban retail model', *Environment and Planning* **A 16**, 851-862.

Lombardo, S.T.: 1986, 'New developments of a dynamic urban retail model with reference to consumers mobility and costs for developers', in R. Haining and D.A. Griffith (eds.), *Transformations through Space and Time*, NATO ASI series, Nijhoff, Dordrecht.

Lombardo, S.T. and G.A. Rabino: 1983, 'Some simulations of a central place theory model', *Sistemi Urbani* **5**, 315 -332.

Lombardo, S.T. and G.A. Rabino: 1984, 'Nonlinear dynamic models for spatial interaction: the results of some empirical experiments', *Papers of the Regional Science Association* **55**, 83-101.

Lombardo, S.T. and G.A. Rabino: 1986, 'Calibration procedures and problems of stability in nonlinear dynamic spatial interaction modeling', *Environment and Planning* **A 18**, 341-350.

Rijk, R.J.A. and A.C.F. Vorst: 1983a, 'On the uniqueness and existence of equilibrium points in an urban retail model', *Environment and Planning* **A 15**, 475-482.

Rijk, R.J.A. and A.C.F. Vorst: 1983b, 'Equilibrium points in an urban retail model and their connection with dynamic systems', *Regional Science and Urban Economics* **13**, 383-399.

Wilson, A.G.: 1981, *Catastrophe Theory and Bifurcation: Application to Urban and Regional Systems*, Croom Helm, London.

Wilson, A.G.: 1985, 'Structural dynamics and spatial analysis: from equilibrium balancing models to extended economic models for both perfect and imperfect market', W.P. 431, School of Geography, University of Leeds, Leeds.

Wilson, A.G. and M. Birkin: 1983, 'Industrial location theory: explorations of a new approach', W.P. 361, School of Geography, University of Leeds, Leeds.

Wilson, A.G. and R. Crouchley: 1983, 'The optimum sizes and locations of schools', W.P. 369, School of Geography, University of Leeds, Leeds.

GUNTER HAAG AND PIERRE FRANKHAUSER

A STOCHASTIC MODEL OF INTRAURBAN SUPPLY AND DEMAND STRUCTURES

1. INTRODUCTION

The dynamic modelling of socio-economic processes has recently attracted wide attention. The description of such processes is usually based on plausible assumptions about the macro-state of the system and the interaction of the aggregate variables (cf. for example Birkin and Wilson, 1985; Wilson, 1971, 1974; Wilson and Clarke, 1979; Clarke and Wilson, 1983; Harris and Wilson, 1978; Lombardo and Rabino, 1983; Pumain, Saint-Julien, and Sanders, 1984; Johansson and Nijkamp, 1984). In particular, this principle of model building has been applied successfully to systems in equilibrium. But it is questionable whether this concept can also be generalized to socio-economic processes far from equilibrium. On the other hand there exists considerable literature concerning the behaviour of individuals with respect to their decisions between alternative attitudes on the micro-scale.

The present paper aims at a combination of these two different points of view on a more general level. The application of the master equation approach provides this general framework for the dynamic modelling of socio-economic processes (Weidlich and Haag, 1983). We assume that individuals will adopt a given socio-economic attitude (to be specified later) with a certain probability. Individual decision processes may then lead to changes in this attitude so that the probability distribution for the various possible attitudes will change, too. The stochastic description of the dynamics of a set of such individual decision-making processes within a group of individuals can be formulated in terms of a master equation; this is an equation of motion for the time-dependent probability distribution of the attitudes over members of the society.

This stochastic description leads to a macro-configuration characterizing the socio-economic system on a global macroeconomic level, which, however, is determined by the individual decision processes. Thus the micro- and the macro-level are coupled in such a way that the individual processes depend on the socio-economic "field" manifest in the society. Recently such concepts have been applied to urban and regional systems (Weidlich and Haag, 1978, 1983; Dendrinos and Haag, 1984; Haag and Dendrinos, 1983; Haag, 1983, 1989a,b; Haag and Weidlich, 1983, 1984; Leonardi, 1989; Sonis, 1985; Haag and Wilson, 1986).

In order to provide insight into the complicated dynamics of intraurban models we first consider an urban system consisting of L zones with a given transportation network. The transportation costs between zone i and zone j are denoted by c_{ij}. We then study in detail the dynamic behaviour of the expenditure flows $T_{ij}(t)$, and their consequences for the change of demand $D_j(t)$ in zone j, as well as the time evolution of the facility stocks $Z_j(t)$ in the different areas j.

J. Hauer et al. (eds.), Urban Dynamics and Spatial Choice Behaviour, 219–246.

This means that we focus our attention on both the "supply side" and the "demand side" of the urban system. For example, in retailing it is important to understand and model how the demand for different types of goods changes over time and how consumers' activity patterns (expenditure flow distribution) respond to changes in the structure of the supply side. Thus, the demand side and the supply side must be seen as highly interdependent.

The macroeconomic description of the urban system (expenditure flows T_{ij}, facility stock Z_j) is the result of a certain aggregation of microeconomic variables. At the microeconomic scale the decision processes of consumers, who decide about their choices (control of expenditure flows), and investors, who decide about the provision of facilities (control of facility stock), are related to the concept of "utility". Since the expected utility gain influences the decision behaviour of the "agents", the functional dependence of "utility" on certain socio-economic variables is of crucial importance and must be investigated (Haag and Weidlich, 1983, 1984; Haag, 1989a,b).

The decision process of consumers is mainly dependent on their residential location and the supply of services and goods within the different zones (Wilson, 1971). In the present model we assume the residential location to be fixed, though this assumption can easily be relaxed (Haag, 1989a,b). The question how prices of goods or services and land rents influence the dynamics will not be treated here. For the treatment of this problem, see Haag and Wilson (1986).

Of course, we are not able to describe the individual decisions on a fully deterministic level, but a probabilistic treatment is possible and adequate. As a consequence, the resulting theory is stochastic. This means, we expect that the theory yields as a main result the evolution of a probability distribution over the possible configurations of consumers and investors. It will then be possible to derive macroeconomic equations of motion for the expenditure flow and facility stock from such a time-dependent distribution function. This will be done in Section 2. A new exact stationary solution leading to fundamental economic consequences in urban modelling is presented in Section 3. In Section 4 we consider the influence of the transportation costs, as expressed by the spatial deterrence parameter β. It will be shown that for different values of β there arise completely different patterns of the stationary expenditure flow. Even more important is the time dependence of the flows and the facility stock if the urban system is disturbed from outside. Using the same initial conditions $Z_j(0)$, $T_{ij}(0)$ we simulate numerically:

1. The effect of a sudden increase of the scale of provision of facilities in one particular zone for different values of the deterrence parameter β.
2. The effect of an improvement of the infrastructure between two zones by a reduction in the transportation costs between these zones, assuming
    - in a first run that the initial values of the scale of provision in the different zones will not be changed;
    - in a second run a sudden simultaneous increase of the scale of provision in one zone.

It is worthwhile at this stage to emphasize that the short-time behaviour of the expenditure flow $T_{ij}(t)$ and the facility stock $Z_j(t)$ as well as the long-time dynamics bear some unexpected results. This is mainly due to the fact that the underlying dynamic equations of motion exhibit a continuum of critical points. Therefore, each set of initial conditions and time-scaling parameters describing the speed of adjustment of the subsystems will lead to a different stationary urban flow and stock pattern.

2. THE STOCHASTIC SERVICE SECTOR MODEL

In this section we review briefly the stochastic service sector model of Haag (1989a,b), with emphasis on the stochastic approach to applications in urban and regional analysis. For details on the stochastic approach and further applications see Weidlich and Haag (1983).

2.1. *The master equation*

The urban system is decomposed into L zones. The $T_{ij}(t)$ describe expenditure flows from a zone i (zone of consumers' residence) to zone j (the consumer is using facilities in j) at time t. The expenditure configuration is described by the array $\{T_{ij}\} = \mathbf{T}$. Let Z_j be the scale of provision of facilities in zone j at time t. Then the facility stock configuration is described by the vector $\{Z_j\} = \mathbf{Z}$, and the probability distribution that a configuration $\mathbf{c} = \{\mathbf{T},\mathbf{Z}\}$ will occur at time t will be denoted by $P(\mathbf{T},\mathbf{Z};t)$.

The configuration $\mathbf{c}$ can be represented in our case as a point in the (L^2+L)-dimensional socio-configuration space ζ. The contributions to the transition probability $w(\mathbf{c}+\mathbf{k},\mathbf{c}) \geq 0$ from a socio-configuration $\mathbf{c} \in \zeta$ to a configuration $(\mathbf{c}+\mathbf{k}) \in \zeta$, where $\mathbf{k} = (\mathbf{k}_1,\mathbf{k}_2)$, are due to transfer of expenditures and birth/death of facilities. If the transition probabilities are specified - and this is the crucial problem of model-building - the equation of motion for the evolution of the probability distribution can be derived

$$\frac{dP(\mathbf{T},\mathbf{Z};t)}{dt} = \sum_{\mathbf{k}} w(\mathbf{T},\mathbf{Z};\mathbf{T}+\mathbf{k}_1,\mathbf{Z}+\mathbf{k}_2)\; P(\mathbf{T}+\mathbf{k}_1,\mathbf{Z}+\mathbf{k}_2;t) - \sum_{\mathbf{k}} w(\mathbf{T}+\mathbf{k}_1,\mathbf{Z}+\mathbf{k}_2;\mathbf{T},\mathbf{Z})\; P(\mathbf{T},\mathbf{Z};t) \tag{1}$$

where the sum on the right-hand side extends over all $\mathbf{k}$ with nonvanishing transition probabilities (cf. Weidlich and Haag, 1983).

The master equation (1) can be interpreted as a probability balance equation. The change of the probability of socio-configuration $\mathbf{c} = (\mathbf{T},\mathbf{Z}) \in \zeta$ is due to two counteractive effects, namely to the probability flux from all neighbouring configurations $(\mathbf{c}+\mathbf{k}) =$

$(\mathbf{T}+\mathbf{k}_1,\mathbf{Z}+\mathbf{k}_2) \in \zeta$ into $\mathbf{c}$ and to the probability flux from $\mathbf{c}$ to all $(\mathbf{c}+\mathbf{k})$. The probability distribution is normalized

$$\sum_{\mathbf{c}} P(\mathbf{c};t) = 1 \tag{2}$$

It is possible to derive directly from the master equation (1) separate sets of equations of motion for the mean values of the expenditure flows $\bar{T}_{ij}(t)$ and for the mean values of the facility stock $\bar{Z}_j(t)$. This is justified as long as it can be assumed that the probability distribution $P(\mathbf{T},\mathbf{Z};t)$ is unimodal, which means, it has only one peaked maximum. The mean values are defined by

$$\bar{T}_{ij}(t) = \sum_{\mathbf{c}} T_{ij}\, P(\mathbf{c};t) \tag{3}$$

$$\bar{Z}_j(t) = \sum_{\mathbf{c}} Z_j\, P(\mathbf{c};t) \tag{4}$$

where the bar indicates that the mean over this variable has been taken.

Equations of motion for these mean values can be obtained by multiplying Equation (1) with T_{ij}, Z_j, respectively and summing up over all configurations $\mathbf{c} = (\mathbf{T},\mathbf{Z})$ (cf. Weidlich and Haag, 1983).

2.2. *The total transition probability*

The total transition probability $w(\mathbf{c}+\mathbf{k};\mathbf{c})$ is in our case the sum over contributions of different socio-economic processes.

$$w(\mathbf{T}+\mathbf{k}_1,\mathbf{Z}+\mathbf{k}_2) = \sum_{i,j,k=1}^{L} w_{ik,ij}(\mathbf{T}+\mathbf{k}_1,\mathbf{Z};\mathbf{T},\mathbf{Z}) + \sum_{j=1}^{L} [w_j^+(\mathbf{T},\mathbf{Z}+\mathbf{k}_2;\mathbf{T},\mathbf{Z}) + w_j^-(\mathbf{T},\mathbf{Z}+\mathbf{k}_2;\mathbf{T},\mathbf{Z})] \tag{5}$$

The $w_{ik,ij}(\mathbf{T}+\mathbf{k}_1,\mathbf{Z};\mathbf{T},\mathbf{Z})$ refers to changes in the expenditure flow configuration $\mathbf{T}$ (demand side), due to individual decisions of consumers to change from a state having residence in i and using facilities in j to a state of still having residence in i but now using facilities in k. Thus changes of residential location are not taken into account, since it is reasonable to assume that the housing mobility will be considerably slower than the shopping mobility of the population. Indeed decisions to buy in another zone can be taken much easier than changing an apartment.

On the supply side, the w_j^+, w_j^- describe decisions of entrepreneurs to add or to remove one unit of the facility stock, respectively.

The crucial task is now to specify the transition rate. Here the experience of the model-builder will influence the manner in which socio-economic processes are introduced. Let us consider for this aim separately and in more detail the demand side and the supply side.

2.3. *The demand side*

The actual expenditure flow array

$$\mathbf{T}(t) = \{T_{11}(t), T_{12}(t), \ldots, T_{ij}(t), \ldots T_{LL}(t)\} \tag{6}$$

will change in the course of time, due to decisions of consumers to use facilities of another zone.

The total revenue attracted to each zone is obtained by

$$D_j(t) = \sum_{i=1}^{L} T_{ij}(t) \tag{7}$$

and the total expenditure $E_i(t)$ of consumers living in zone i is

$$E_i(t) = \sum_{j=1}^{L} T_{ij}(t) \tag{8}$$

We shall assume that such decision processes are determined by utility functions $v_{ij}(t)$, where $v_{ij}(t)$ is a measure of the attractivity of state ij at time t. The utility functions $v_{ij}(t)$ have two subscripts since they relate to both housing and service sectors. The first subscript refers to the attractivity of housing in zone i, the second subscript to the attractivity for shopping in zone j. Thus the utility functions $v_{ij}(t)$ will be influenced by the supply side variables. As discussed above, it is plausible to assume that changes of residential location need not to be taken into account in a first approach for the description of shopping dynamics. The individual transition rate $p_{ik,ij}(\mathbf{T},\mathbf{Z})$ for changing the shopping area is now assumed to depend on the expected utility gain $(v_{ik}(t) - v_{ij}(t))$ as follows

$$p_{ik,ij}(\mathbf{T},\mathbf{Z}) = \varepsilon_1^*(t)\, e^{v_{ik}(t)-v_{ij}(t)} > 0 \tag{9}$$

Therefore the probability transition rate is a function of the corresponding utilities and a frequency parameter $\varepsilon_1^*(t)$, which scales the basic flows.

Let $n_{ij}(t)$ be the number of individuals having residence in zone i, $i = 1,2,\ldots,L$ and using facilities in j. Then

$$n_i(t) = \sum_{j=1}^{L} n_{ij}(t) \tag{10}$$

is the number of consumers living in i, and the total number of all consumers in the city area is

$$N(t) = \sum_{i=1}^{L} n_i(t) \tag{11}$$

Then for the total transition rate per unit of time we obtain

$$w_{ik,ij}(\mathbf{T}+\mathbf{k}_1,\mathbf{Z};\mathbf{T},\mathbf{Z}) = n_{ij}\, p_{ik,ij}(\mathbf{T},\mathbf{Z}) \tag{12}$$

In many cases it is reasonable to assume that the expenditure flows $T_{ij}(t)$ are related to the consumer activities $n_{ij}(t)$ by

$$T_{ij}(t) = g(t)\, n_{ij}(t) \tag{13}$$

where the factor $g(t)$ can be determined by a comparison of the total stocks

$$\sum_{j=1}^{L} D_j(t) = \sum_{i,j=1}^{L} T_{ij}(t) = g(t)\, N(t) \tag{14}$$

Thus, $g(t)$ can be interpreted as an average of individual needs. Using the relation (13) the total transition rate per unit of time finally reads

$$w_{ik,ij}(\mathbf{T}+\mathbf{k}_1,\mathbf{Z};\mathbf{T},\mathbf{Z}) = \begin{cases} T_{ij}\, \varepsilon_1(t)\, e^{v_{ik}(t)-v_{ij}(t)} \\ \text{for } \mathbf{k}_1=(0,\ldots,1_{ik},\ldots,0,\ldots,(-1)_{ij},\ldots,0) \\ 0 \text{ for all other } \mathbf{k}_1 \end{cases} \tag{15}$$

2.4. *The supply side*

In the next step we consider the dynamics of $Z_j(t)$, i.e. the facility stock. The facility stock dynamics at each possible location j depends on the decisions of investors to open or to close facilities at j. That is the rate at which $Z_j(t)$ is increased (birth rate) or decreased (death rate). Since there is no "flow" of facilities between the zones, the transition rates must be considered as birth/death rates.

Let **Z** be the facility configuration

$$\mathbf{Z}(t) = \{Z_1(t), Z_2(t), \ldots, Z_L(t)\} \tag{16}$$

which describes one possible realization of a facility pattern at time t. By decisions of shop owners this configuration will change in the course of time. As done above, we shall now again introduce the corresponding rates to these processes. A rather general formulation for the birth/death rates may be the following. Let $w_j^+(T,Z+k_2;T,Z)$ be the birth rate for creating an additional unit of facility stock by a shop owner in zone j and $w_j^-(T,Z+k_2;T,Z)$ be the death rate for closing such a unit

$$w_j^+(T,\mathbf{Z}+\mathbf{k}_2;T,\mathbf{Z}) = \begin{cases} 0.5\ f_j(\mathbf{Z}) \exp \Phi_j(T,\mathbf{Z}) \text{ for } \mathbf{k}_2=(0,..,1_j,..,0) \\ 0 \text{ for all other } \mathbf{k}_2 \end{cases}$$

$$w_j^-(T,\mathbf{Z}+\mathbf{k}_2;T,\mathbf{Z}) = \begin{cases} 0.5\ f_j(\mathbf{Z}) \exp -\Phi_j(T,\mathbf{Z}) \text{ for } \mathbf{k}_2=(0,..,-1_j,..,0) \\ 0 \text{ for all other } \mathbf{k}_2 \end{cases} \tag{17}$$

with

$$f_j(\mathbf{Z}) = \varepsilon_2 Z_j > 0 \tag{18}$$

and

$$\Phi_j(\mathbf{Z}) = \lambda(D_j - C_j) \tag{19}$$

We assume that the propensity of an investor to add or remove one unit of the facility stock in a zone j depends on the imbalance between revenue attracted $D_j(t)$ and the providing cost of supply $C_j(t)$. We assume that the costs of supply are proportional to the number of units of the facility stock

$$C_j(t) = k_j Z_j(t) \tag{20}$$

Then k_j is a measure for the cost of providing one unit of facilities in area j. The factor $f_j(\mathbf{Z})$ describes the speed of adjustment. Since the birth/death rates may not be negative, the condition $f_j > 0$ has to be fulfilled. Here it is assumed that the speed of adjustment depends on the scale of provision in a linear way with a time-scaling parameter ε_2. The function $\Phi_j(\mathbf{Z})$ takes into account the imbalance between cost supply and the revenue attracted to facilities in the zone under consideration. If there is an economic surplus $D_j > C_j$ it is more likely that the facility stock is expanded; and vice versa. The parameter λ describes the response to economic surplus for the facility stock. Other functional relationships are discussed in Haag (1989a,b).

2.5. *The mean value equations*

By insertion of the transition rates (15) and birth/death rates (17) into the master equation (1), we obtain the evolution in time of the probability distribution $P(\mathbf{T},\mathbf{Z};t)$ for $t > 0$ from an initial distribution $P(\mathbf{T},\mathbf{Z};0)$.

Since it is difficult to handle this solution in practice, we derive the corresponding mean value equations for the expenditure flow $\bar{T}_{ij}(t)$ and $\bar{Z}_j(t)$ as discussed above. Then the equation of motion of the mean values reads for the expenditure flows $\bar{T}_{ij}(t)$

$$\dot{\bar{T}}_{ij}(t) = \varepsilon_1(t) \left\{ \sum_k e^{v_{ij}(t)-v_{ik}(t)} \bar{T}_{ik}(t) - \sum_k e^{v_{ik}(t)-v_{ij}(t)} \bar{T}_{ij}(t) \right\}, \text{ for } i,j = 1,2,\ldots,L \tag{21}$$

and for the facility stock $\bar{Z}_j(t)$

$$\dot{\bar{Z}}_j(t) = f_j(\bar{\mathbf{Z}}) \sinh \Phi_j(\bar{\mathbf{Z}}), \qquad \text{for } j = 1,2,\ldots,L \tag{22}$$

or after insertion of (18) and (19), in (22)

$$\dot{\bar{Z}}_j(t) = \varepsilon_2(t)\, \bar{Z}_j(t) \sinh [\lambda(\bar{D}_j - k_j \bar{Z}_j(t))] \tag{23}$$

where $\bar{D}_j$ is the mean value of the revenue attracted. The hyperbolic sin functions lead to an amplification of the reactions on economic disequilibrium $D_j \neq C_j$, as compared to the linear relation. Near equilibrium $|\Phi_j| \ll 1$ (23) yields the Harris and Wilson hypothesis (Harris and Wilson, 1978).

By summing up Equation (21) over the index j it can easily be seen that the mean value of the total expenditure stock $\bar{E}_i$ of zone i does not change with time

$$\dot{\bar{E}}_i(t) = 0, \qquad \text{for } i = 1,2,\ldots,L \tag{24}$$

and therefore satisfies a conservation law. From an economic point of view this is due to the fact that changes of residential location of individuals are not considered. Thus the total expenditure stocks $\bar{E}_i$ can be considered as given constants.

Equations (21) and (23) together constitute the dynamic service sector model. The estimation of the model parameters from empirical socio-economic data can be done by a generalized least squares procedure (Haag, 1989a,b; Weidlich and Haag, 1988). In the next section we shall discuss in more detail the stationary pattern of the expenditure flows as well as the scale of provision.

3. THE STATIONARY PATTERN OF THE SERVICE SECTOR MODEL

By considering the stationary versions of (21) and (23), we obviously obtain a complicated nonlinear transcendental system of equations for the stationary expenditure flows $\hat{T}_{ij}$ and the stationary scale of provision of facilities $\hat{Z}_j$. Stationary values of the variables are denoted by a hat. It can be seen that the stationary version of (21) is satisfied by

$$\hat{T}_{ij} = \hat{A}_i \, e^{2\hat{v}_{ij}} \tag{25}$$

where $\hat{v}_{ij}$ is $v_{ij}(t)$ evaluated at the stationary state. We can find the normalization coefficient $\hat{A}_i$ by the requirement (8) yielding

$$\hat{A}_i = \hat{E}_i \Big/ \sum_{j=1}^{L} e^{2\hat{v}_{ij}} \tag{26}$$

so that

$$\hat{T}_{ij} = \hat{E}_i \, e^{2\hat{v}_{ij}} \Big/ \sum_{l=1}^{L} e^{2\hat{v}_{il}} \tag{27}$$

On the other hand the stationary solution of (23)

$$\hat{D}_j = k_j \hat{Z}_j \tag{28}$$

must be fulfilled simultaneously. Therefore there exists a coupling between the stationary demand $\hat{D}_j$ and the stationary values of the facility stock $\hat{Z}_j$ for each zone j. In other words, for each scale of provision $\hat{Z}_j$, there exists an appropriate demand.

By means of (7) the demand $\hat{D}_j$ can be obtained from Equation (27) by summing up over the residential area index i. Inserting then Equation (28), we get a relation for the stationary facility stock $\hat{Z}_j$

$$k_j \hat{Z}_j = \sum_{i=1}^{L} \hat{E}_i \, e^{2\hat{v}_{ij}} \Big/ \sum_{l=1}^{L} e^{2\hat{v}_{il}}, \qquad \text{for } j = 1,2,\ldots,L \tag{29}$$

3.1. *The utility functions*

The model and its stationary solution become explicit if the utilities are specified as functions of socio-economic variables. If the utilities $\hat{v}_{ij}$ are functions of the facility stock $\hat{Z}_j$, the algebraic system of equations (27) leads by means of (7), (8) and (28) in general to a coupled system for the expenditure flows, where each expenditure flow $\hat{T}_{ij}$ may depend on all other flows $\hat{T}_{kl}$. In other words, the expenditure flows of individuals living in zone i and using facilities in zone j are influenced by the expenditure flows of individuals having residence in another zone k and using facilities in any other zone l. This structure occurs for example when the following assumptions for the utility function are used (Haag and Wilson, 1986)

$$v_{ij}^{(W)} = 0.5\,\alpha\,\ln Z_j - 0.5\,\beta\,c_{ij} \tag{30}$$

$$v_{ij}^{(H)} = 0.5\,\alpha\,Z_j - 0.5\,\beta\,c_{ij} \tag{31}$$

From an economic point of view, it seems very plausible to assume that the shopping attitude of an individual living in zone i will primarily depend on the offer of facilities in a certain retailing area j and his travelling costs c_{ij} from i to j. In comparison to these effects the shopping attitude of individuals living in any other residential area should not have an important influence and thus could be neglected in a first approximation. Therefore Frankhauser (1986) proposed another plausible assumption for the utilities

$$v_{ij}^{(F)} = 0.5\,\ln (k_j Z_j) - 0.5\,\beta\,c_{ij} - 0.5\,\ln \tau_j \tag{32}$$

with

$$\tau_j = \sum_{i=1}^{L} e^{-\beta c_{ij}} \tag{33}$$

By insertion of (32) and (8) into (27) we obtain the stationary flows

$$\hat{T}^{(F)}_{ij} = k_j \hat{Z}_j \, (\tau_j)^{-1} \, e^{-\beta \, c_{ij}} \tag{34}$$

which do not depend on any other stationary flows $\hat{T}^{(F)}_{ik}$ $(k \neq j)$. Thus the stationary flows depend only on k_j and $\hat{Z}_j$, both being typically economic variables of the shopping area j, as well as on the transport-costs matrix c_{ij}.

It can easily be seen that equation (34) by means of (33) is not only a correct solution of (27) but also fulfills the requirement (28), so that (34) can be written alternatively as

$$\hat{T}^{(F)}_{ij} = \hat{D}_j \, (\tau_j)^{-1} \, e^{-\beta \, c_{ij}} \tag{35}$$

It must be stressed that the L equations (28) are satisfied by (34) in an identical way. Thus there would remain only the set of L^2 equations (34) to determine the stationary pattern $(\hat{T},\hat{Z})$, however, containing L^2+L variables. Therefore L values could be chosen arbitrarily. Using this set the L^2 remaining values then could be computed by (34). For example, an arbitrarily fixed set of $\hat{Z}_j$ $(j=1,2,\ldots,L)$ could be inserted in (34) to calculate the L^2 values $\hat{T}_{ij}$.

However, the dynamic conservation law (24) for the expenditure distribution leads by summing up (34) over the index j to a system of L equations, which have to be satisfied too

$$\sum_{j=1}^{L} k_j \, (\tau_j)^{-1} \, e^{-\beta \, c_{ij}} \, \hat{Z}_j = \hat{E}_i \tag{36}$$

Thus for given parameters β, c_{ij}, k_j, the closed set of equations (34), (36) leads to a unique set of stationary values $\hat{T}_{ij}$, $\hat{Z}_j$.

The utilities (32) depend on the transport cost matrix in the same way as in the above-mentioned utility functions $v^{(W)}_{ij}$ and $v^{(H)}_{ij}$. Since τ_j represents the sum over the transportation costs to reach this particular zone j, τ_j can be seen as a measure for the accessibility to reach the rest of the urban system from zone j. The precise form of how the facility stock Z_j is involved in the utility function (32) seems to be similar to the utility assumption of Wilson for $\alpha = 1$. However in (30) and (31) the offer in zone j is only introduced by the variable Z_j, usually interpreted as an arbitrary quantitative measure, e.g. the supply surface. On the other hand, in (32), instead of Z_j the product $k_j Z_j$ is used, representing the cost of providing C_j. This is a well defined economic variable, for which data usually will be available. Hence the utility function depends only on monetary variables. Taking into account the meaning of the factor k_j (the cost per unit of providing Z_j), another interpretation becomes obvious. In this interpretation k_j can be understood as a qualitative measure for the level of the quality and price standard in the area considered. Thus it can be expected that k_j will be higher in a city centre than in suburban centres. The product $k_j Z_j$ therefore includes qualitative as well as quantitative aspects.

3.2. *The number of global parameters*

To describe the importance of an economic variable, parameters are often used as weighting exponents. Thus a more general assumption could be

$$v_{ij}^* = 0.5\,\alpha\,\ln(k_j Z_j) - 0.5\,\beta\,c_{ij} - 0.5\,\ln\tau_j \tag{37}$$

yielding

$$\hat{T}_{ij} = (k_j \hat{Z}_j)^{\alpha}\,(\tau_j^*)^{-1}\,e^{-\beta\,c_{ij}} \tag{38}$$

Moreover, if the discussed coupling of the stationary expenditure flows among one another is avoided, the condition (28) has to be fulfilled again in the same way and we obtain

$$\tau_j^* = \tau_j\,(k_j \hat{Z}_j)^{\alpha-1} \tag{39}$$

That must be interpreted as a rescaling of the denominator τ_j, depending, however, now on the facility stock $\hat{Z}_j$. Therefore in this case there exists only one non-trivial deterrence parameter β, describing the influence of the travelling costs.

3.3. *The influence of the global parameter* β

Indeed the structure of the stationary flow pattern is highly influenced by the value of β. To illustrate this point the stationary expenditure flows for two different values of β have been computed for a simple urban system. We consider in the following a city centre (1) surrounded by three suburban zones (2, 3, 4) with the following properties (see Figure 1):

- the travelling costs c_{ij} between neighbouring zones are assumed to be twice as high as the costs c_{ii} within each zone;
- the facility stock $\hat{Z}_1$ as well as the factor k_1 is three times higher in the centre than in the suburban zones, $\hat{Z}_j$, k_j for $j = 2, 3, 4$.

Figure 2a shows the expenditure flow distribution $\hat{T}_{ij}$ for $\beta = 0.35$. In the front row the flows T_{i1} into the city centre (1) are depicted. Obviously the relatively small value of β suggests, in general, a high interaction between the zones. Therefore many customers are attracted to the city centre. A quite different pattern can be observed for $\beta = 70$ (Figure 2b). There are still considerable flows into zone 1, but an additional diagonal structure arises. For high values of β the flows within each zone dominate. This gives smaller centres a good chance of surviving, as we shall see later.

The observed structure can easily be proved analytically for the limits $\beta \to 0$ and $\beta \to \infty$. For $\beta \to 0$ we obtain

$$\lim_{\beta \to 0} \hat{T}_{ij} = 1/L \; k_j Z_j \tag{40}$$

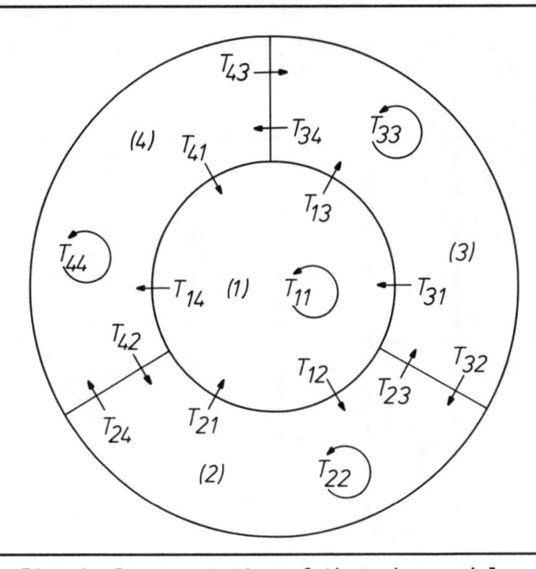

Fig. 1: Representation of the urban model.

L being the number of the zones. In this case the expenditure flow $\hat{T}_{ij}$ only depends on the shopping area j as can be observed in Figure 2a. To calculate the case $\beta \to \infty$ it is preferable to substitute the discrete zoning system by a continuous description

$$c_{ij} \to |x_i - x_j|a, \qquad a > 0 \tag{41}$$

x_i, x_j being different points in the urban system. After a few manipulations it can be seen that a diagonal structure appears

$$\lim_{\beta\to\infty} \hat{T}_{ij} = \lim \left\{ \frac{\beta\, a}{2}\, e^{-\beta |x_i - x_j| a} \right\} = \begin{cases} 1 & \text{for } i = j \\ 0 & \text{otherwise} \end{cases} \tag{42}$$

This result corresponds to the pattern arising in Figure 2b for increasing β.

3.4. *Normalization of parameters and variables*

To obtain comparable results it will be necessary to introduce dimensionless variables and parameters. Since all T_{ij} and all Z_j are only fixed to a common constant, respectively, the most convenient normalization will be

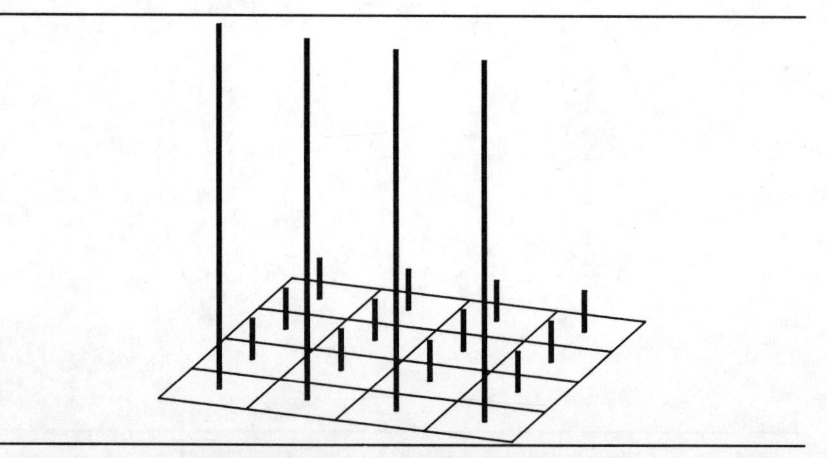

Fig. 2a: Stationary expenditure flow distribution $\hat{T}_{ij}$ for β=0.35.

$$\sum_{ij} \tilde{T}_{ij} = 1$$

$$\sum_{j} \tilde{Z}_{j} = 1 \tag{43}$$

Then each $\tilde{T}_{ij}$, $\tilde{Z}_j$ represents the relative share of the expenditure flows and facility stocks, respectively. The rescaling leads to the new variables and parameters

$$\tilde{T}_{ij} = \frac{T_{ij}}{\sum_{mn} T_{mn}}$$

$$\tilde{Z}_{j} = \frac{Z_{j}}{\sum_{m} Z_{m}} \tag{44}$$

$$\tilde{k}_{j} = \frac{k_{j}}{\sum_{m} k_{m} \tilde{Z}_{m}}$$

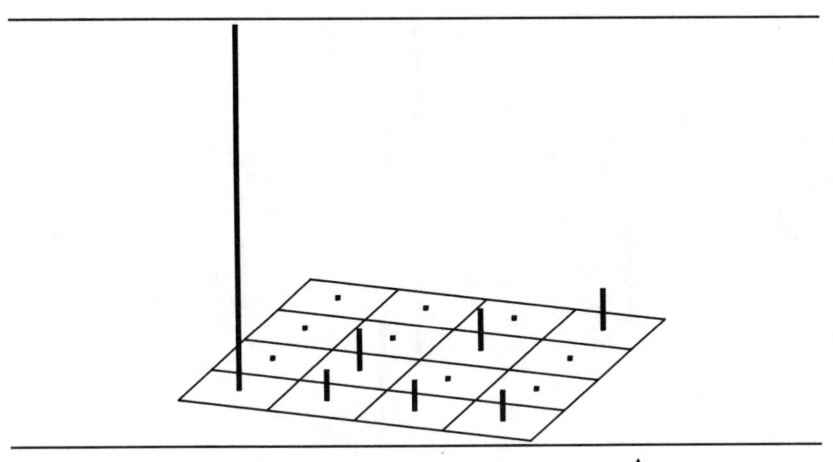

Fig. 2b: Stationary expenditure flow distribution $\hat{T}_{ij}$ for β=70.

The travelling cost matrix c_{ij} can be normalized in a similar way by

$$\tilde{c}_{ij} = \frac{c_{ij}}{\sum_{mn} c_{mn}} \tag{45}$$

These technical steps conclude the general presentation of the model. In the next section we discuss some aspects of the dynamic behaviour.

4. EXAMPLES FOR TIME EVOLUTION IN THE NEIGHBOURHOOD OF STATIONARY SOLUTIONS

In order to illustrate the time evolution of the economic variables $T_{ij}(t)$, $D_j(t)$, $Z_j(t)$ we assume that the equilibrium solution for the urban system already discussed in Section 3 (Figure 1) will be perturbed. Thus the stationary values for Z_j (Figure 3a), c_{ij} and k_j are chosen as in Section 3, whereas for the deterrence parameter β the value 14 was chosen, assuring an intermediate situation with repect to the influence of the travelling costs. The corresponding stationary expenditure flows are represented in Figure 3b, the rows representing the flows into the different facility zones starting with zone 1 in the front row. The flow distribution indicates a persisting interaction between the zones but a small preference of individuals for buying in their own residential area.

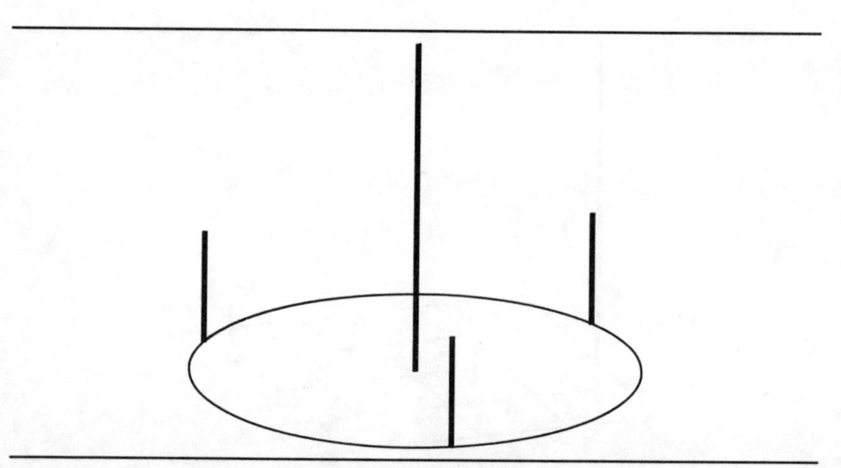

Fig. 3a: Stationary facility distribution $\hat{Z}_j$ for β=14.

4.1. *Variations in the facility stock*

In a first series of analyses we test the consequence of a sudden doubling of supply in the peripheral zone 2. This corresponds to the rise of new shopping facilities in zone 2. Three scenarios with different global transport cost situations are considered:

- β will not be changed (β = 14);
- β is enhanced to 70 at t = 0. The travelling costs suddenly become very important;
- β is diminished to 0.35 at t = 0. The influence of the travelling costs can virtually be neglected.

In particular, the characteristic expenditure flows T_{11}, T_{21} and T_{22} as well as the demands D_j and the supplies Z_j for the zones 1, 2, 3 are compared for these three cases. Since zone 4 is symmetric to zone 3, it need not be discussed separately. All variables are renormalized after each integration step.

For all curves in this series some common traits can be observed. It seems that all variables creep into their stationary solution over a rather long time period, which may, however, differ considerably for the different variables. This leads in many cases to additional interesting short-time dynamics. In this time period transient phenomena may occur. From equations (34) and (36), it can be seen that changing the value of β leads to new stationary patterns, which differ from the unperturbed initial stationary solution.

The time evolution of the system could be influenced by the value of the parameter λ, weighting the dynamic influence of the imbalance between demand and supply. Thus a high value of λ will assume a strong

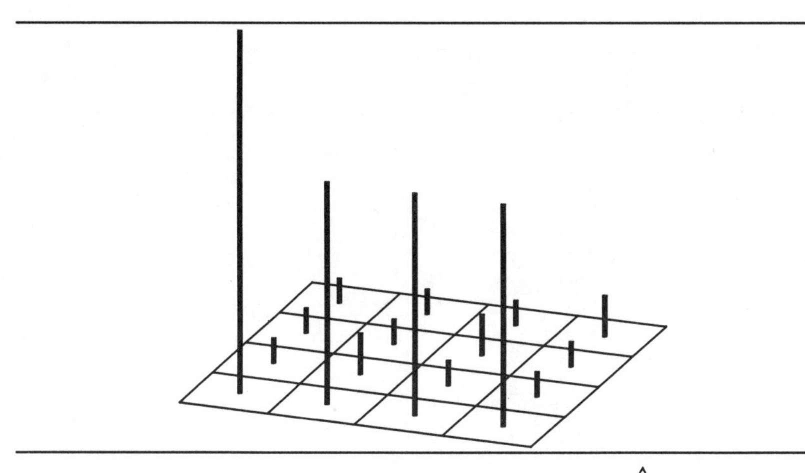

Fig. 3b: Stationary expenditure flow distribution $\hat{T}_{ij}$ for β=14.

reaction of the shop owners. We chose $\lambda = 12$ for the figures discussed, since this assures on the one hand an adjustment in a time period which still can be surveyed, whereas on the other hand short time effects will not be suppressed. The way in which the value $\lambda = 12$ influences the dynamics of the facility stock in the time interval under consideration is similar to the way a value $\beta = 14$ of the deterrence parameter influences the dynamics of the expenditure flows. For $\lambda = 1$ a comparative test has been made. The time evolution $T_{ij}(t)$, $Z_{ij}(t)$ here shows almost the same behaviour, except for the time scale being enlarged by a factor of about 50. The mobility has been taken as fixed ($\varepsilon_1 = \varepsilon_2 = 1$) in all cases.

In reality we expect lower values of λ than those assumed here. In this case the initial transient dynamics will become important. In many cases it could even happen that the stationary state is never reached, because externally driven changes of transport costs or other variables may occur in the meantime.

4.2. *The influence of low travelling costs*

Let us now discuss the details of Figure 4, 5 and 6. Obviously the patterns of demand and supply do not differ very much for $\beta = 14$ (straight line) and $\beta = 0.35$ (dashed line) In some cases both curves cannot be distinguished (in those cases only one straight line has

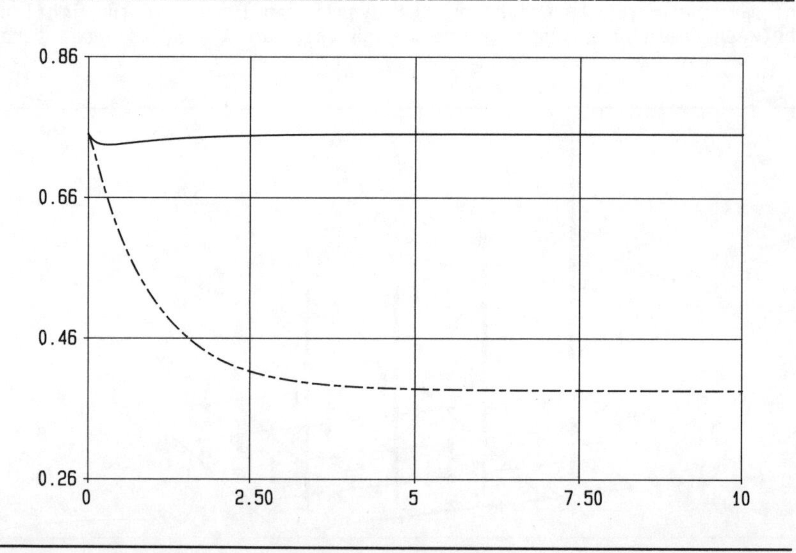

Fig. 4a: Demand $D_1(t)$ for β=0.35, β=14 (——) and β=70 (-··-).

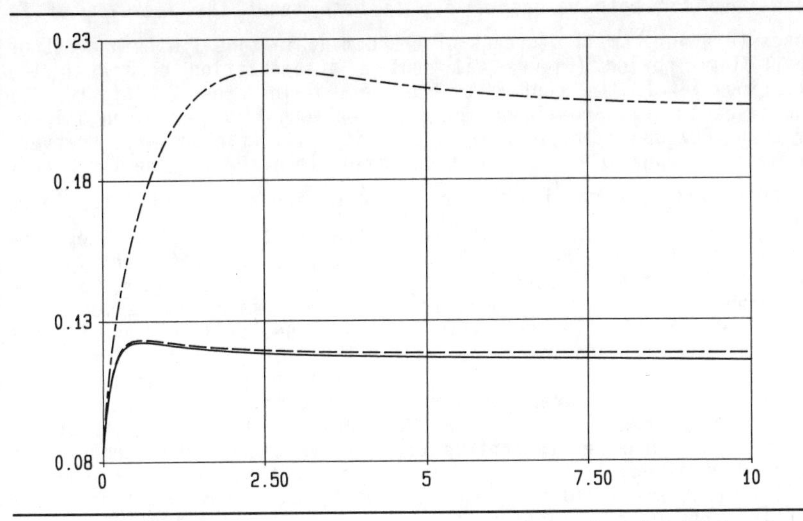

Fig. 4b: Demand $D_2(t)$ for β=0.35 (---), β=14 (——) and β=70 (-•-).

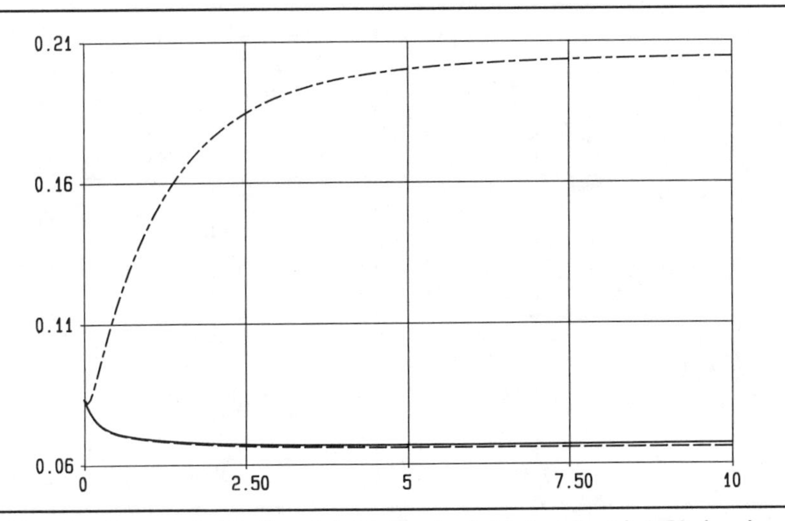

Fig. 4c: Demand $D_3(t)$ for β=0.35 (---), β=14 (——) and β=70 (-•-).

been used for both values of β). In both cases the increase of Z_2 leads to a very small decrease of the demand in zone 1 within a rather short time period (Figure 4a), but a stabilization occurs on the starting level. In contrast, the demand in zone 2 (Figure 4b) increases quickly, remaining for a rather long time period on a level of about 40% above its initial value, after a little decline. However, it must be emphasized that the high demand level D_1 in zone 1 can only be preserved by a considerable increase of the supply Z_1 (cf. Figure 5a). On the other hand, an increase of the starting level of supply by a factor of two in zone 2 cannot create a sufficiently high attractivity, thus leading to a reduction of supply (Figure 5b). The importance of zone 3 is now decreasing, and remains for a long time period on a low level, especially for low values of β (compare Figures 4c, 5c). This makes evident the influence of low transportation costs: customers will not hesitate to go shopping in another area if β is small. This effect can also be observed by comparison of the expenditure flows for β = 14 and β = 0.35 (see Figures 6a-6c). It can be seen that big retail centres can survive even if new competition arises. Newly developed retail centres have a good chance to grow up, too, whereas small old shopping areas will lose their importance. Thus for this choice of parameters the model describes a situation of hard competition calling for investors' activity.

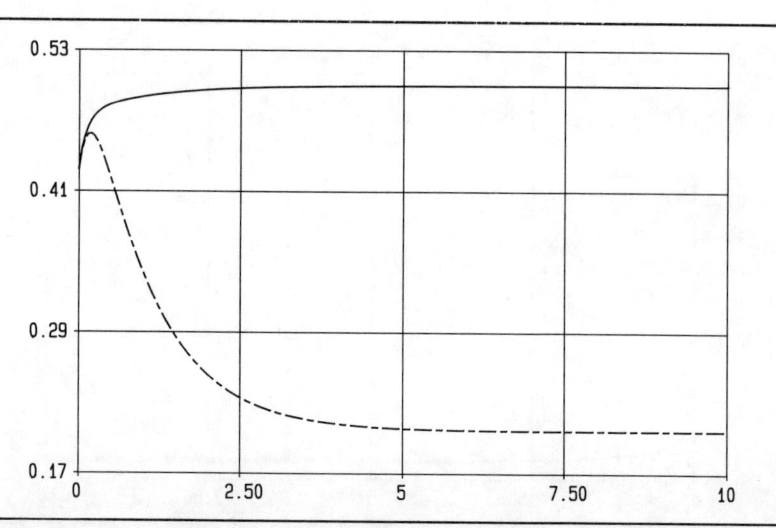

Fig. 5a: Scale of provision of facilities $Z_1(t)$ for β=0.35 (---), β=14 (——) and β=70 (-••-).

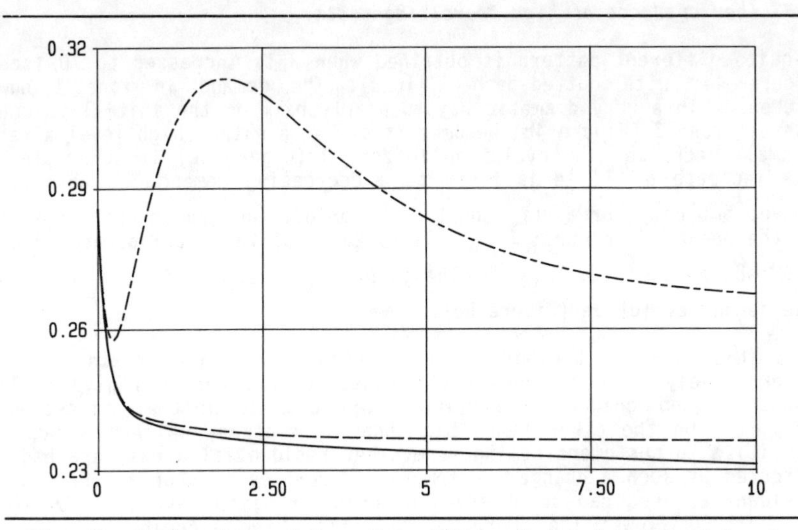

Fig. 5b: Scale of provision of facilities $Z_2(t)$ for β=0.35 (---), β=14 (——) and β=70 (-•-).

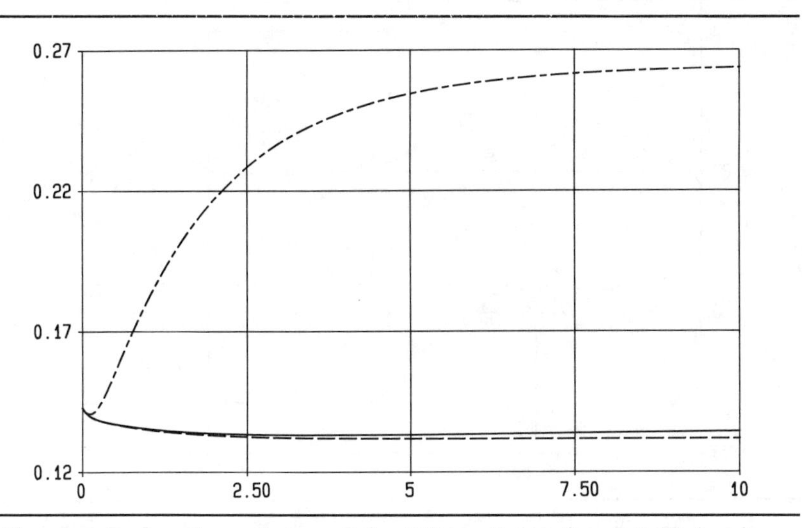

Fig. 5c: Scale of provision of facilities $Z_3(t)$ for β=0.35 (---), β=14 (——) and β=70 (-•-).

4.3. The influence of high travelling costs

A quite different pattern is obtained when β is increased to 70 (see Figure 4a; with dotted-dashed lines). The demand in zone 1 now decreases in a very dramatic way to nearly half of the initial value, whereas zone 2 (Figure 4b) becomes stable at a rather high level after a small decrease. The evolution in zone 3 (Figure 4c) is surprising. The increase of Z_2 leads first to a decreasing demand, as observed above, but only for a very short time period. Then an abrupt increase of the demand D_3 occurs: from 8.3% to 20.3% of the total demand, i.e. 145% above its initial value. The supply Z_3 in zone 3 follows roughly the demand evolution (Figure 5c).

This makes it evident that high travelling costs or - alternatively - poorly developed transportation networks give small zones a good chance to survive, even without initial investment activity. On the other hand, big shopping centres, which are rather sensitive to customers coming from other residential areas, are badly affected by such a change. Customers' tendencies to shop in their own residential area can be observed by considering the evolution of the expenditure flows. The improved possibility for shopping in zone 2 leads to a high level of T_{22} (Figure 6c), but also T_{11} (Figure 6a) increases. Therefore the flows from the suburban zones to the city

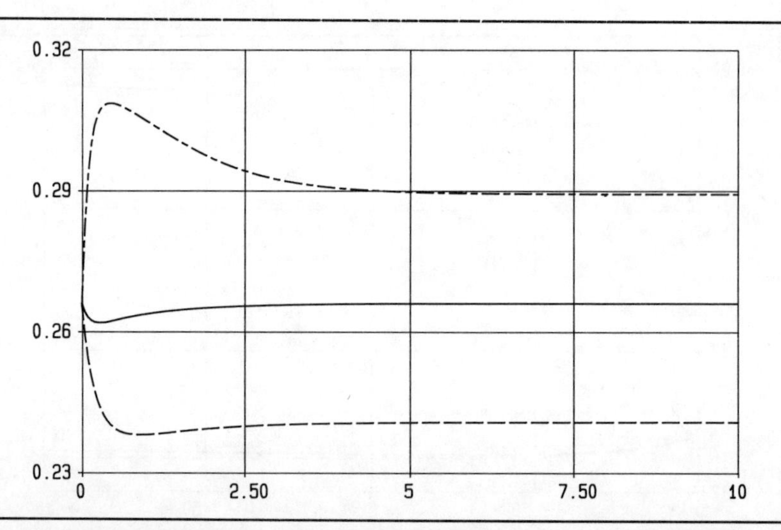

Fig. 6a: Expenditure flow $T_{11}(t)$ for β=0.35 (---), β=14 (——) and β=70 (-•-).

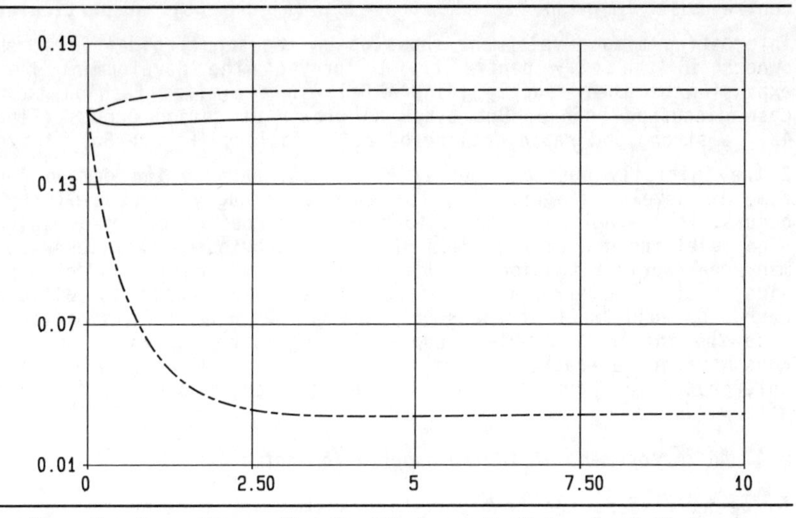

Fig. 6b: Expenditure flow $T_{21}(t)$ for β=0.35 (---), β=14 (——) and β=70 (-•-).

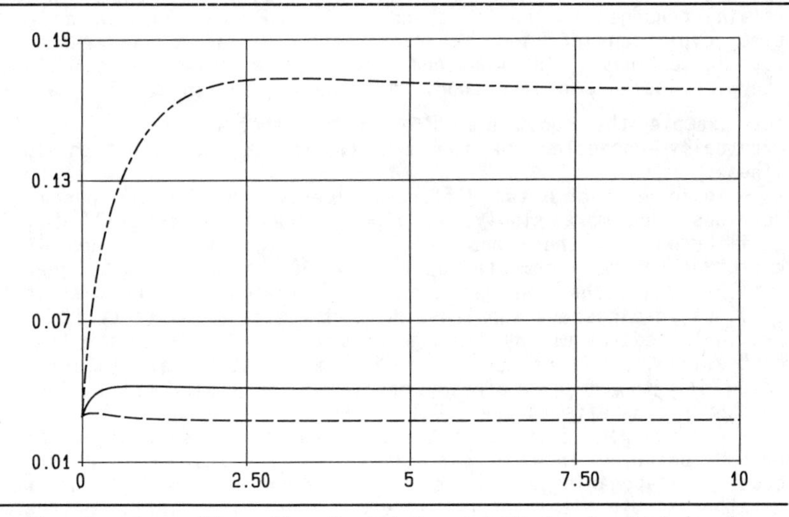

Fig. 6c: Expenditure flow $T_{22}(t)$ for β=0.35 (---), β=14 (——) and β=70 (-•-).

centre must diminish, as shown by T_{21} (Figure 6b). A particularly interesting time development shows up on the supply side. First shop owners in the city centre try to prevent the development by an expansion of their supply (Figure 5a), in a similar fashion to the case discussed before. But since no growth of demand occurs (Figure 4a), a strong and rapid decline of $Z_1(t)$ follows (Figure 5a). In zone 2 the initially doubled supply is reduced, because the demand needs time to develop (Figure 5b). But then a strongly increasing supply occurs, following the demand situation, overshooting to an excessively high level and decreasing again. This is an obvious effect observed in many real-world situations. Sudden changes of the demand and/or supply side lead to overreactions and oscillations, caused by different speeds of adjustment of the supply and the demand. In this particular case the initially doubled supply must again be augmented when more consumers are attracted. But this big growth of demand, which can be interpreted as "curiosity effect" of the customers, will diminish afterwards.

4.4. *An improvement of the transportation network*

As we have seen, the transportation system has a major influence on the time evolution and the trade pattern which occurs. Therefore a further situation has been analysed, starting from the same initial stationary scenario as above. In this case it is assumed that between the city centre and zone 2 the transportation system suddenly changes, for example by an improvement of the transportation network, like the opening of a new street or a suburban train. The other conditions remain unchanged in the first scenario. In Figures 7a, 7b, and 8 the time evolution of some characteristic variables is represented (straight lines). In a second scenario we assume a simultaneous stepwise increase of the supply in zone 2 (doubling of Z_2), assuming for example the construction of a new shopping centre, which is eventually connected to the new transportation situation (dashed lines).

Again we notice two different time scales. But the process now develops much more slowly. At $t = 50$ the state is still far from equilibrium, but the trend can be seen (Figures 7a, 7b and 8). The evolution has been computed up to $t = 250$, where it can be seen that both initial situations lead to an identical stationary distribution of flows, demands and supplies. Hence the stationary state ($t \to \infty$) is not influenced at all by a change of supply, but only by the transport cost matrix, as it can be expected from the stationary solution (34) and (36). However, as already mentioned, in real-world applications the short time effects are likely to be more important.

For the demand in zones 1 and 3, roughly the same time evolution can be observed in both cases. In zone 1, first a small increase occurs, disappearing after a certain time period ($t = 30$), followed by a rather stable situation. In zone 3 a nearly exponential decrease of the supply and demand can be observed. In the same - very long - time interval of decreasing D_3, the demand in zone 2 builds up. However, if

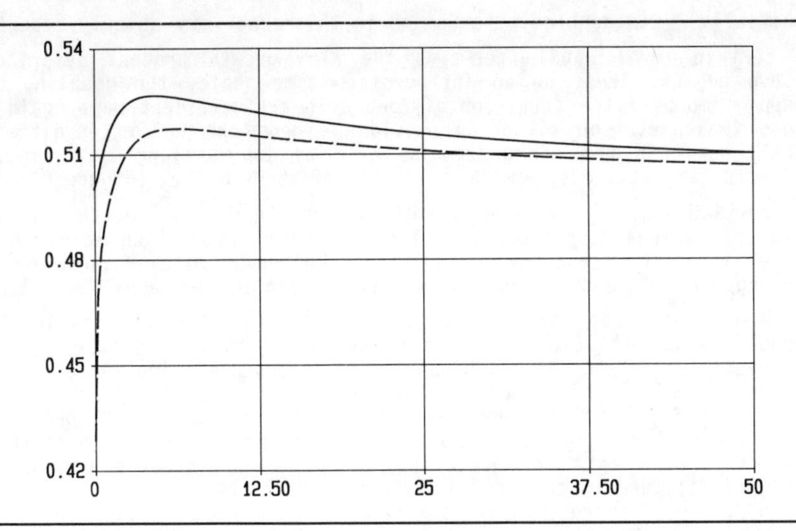

Fig. 7a: Scale of provision of facilities $Z_1(t)$ for different initial conditions $Z_2(0) = \hat{Z}_2$ (——) and $Z_2(0) = 2\,\hat{Z}_2$ (---).

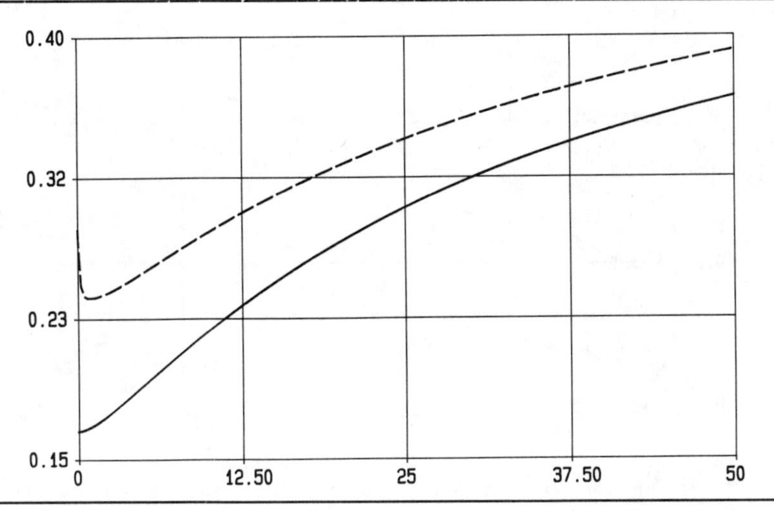

Fig. 7b: Scale of provision of facilities $Z_2(t)$ for different initial conditions $Z_2(0) = \hat{Z}_2$ (——) and $Z_2(0) = 2\,\hat{Z}_2$ (---).

initially Z_2 is doubled, the demand in this zone very quickly reaches a certain level, and afterwards the slow growth process described above begins. Thus, on an intermediate time scale, the doubling of supply improves the situation of zone 2 in a significant way. Again a good transportation situation provides a good chance for investment activities, in the short term as well as in the long term. These effects can be studied in the evolution of the flow T_{22} (Figure 8). Z_2 is doubled suddenly, and immediately a trend to stay in the consumers' own area arises. Improvement of the transport system then leads to a reversal of the trend, which can also be observed in the case without enlarged Z_2. In both cases T_{22} increases again and sustains for a long time a higher level compared to the situation without changes in the supply. This evolution seems to be connected to an increase in the attractivity of zone 2, which can indeed be observed simultaneously. A comparison of the time evolution patterns of the supplies in the zones 1 (Figure 7a) and 2 (Figure 7b) makes it clear that zone 2 benefits from the new transport situation in the long run. The reaction of shop owners in the city centre quickly leads to an increase of supply (Figure 7a), and to the previously described higher demand in zone 1. Then Z_1 slowly approaches its starting value, while Z_2 still continues to increase (Figure 7b).

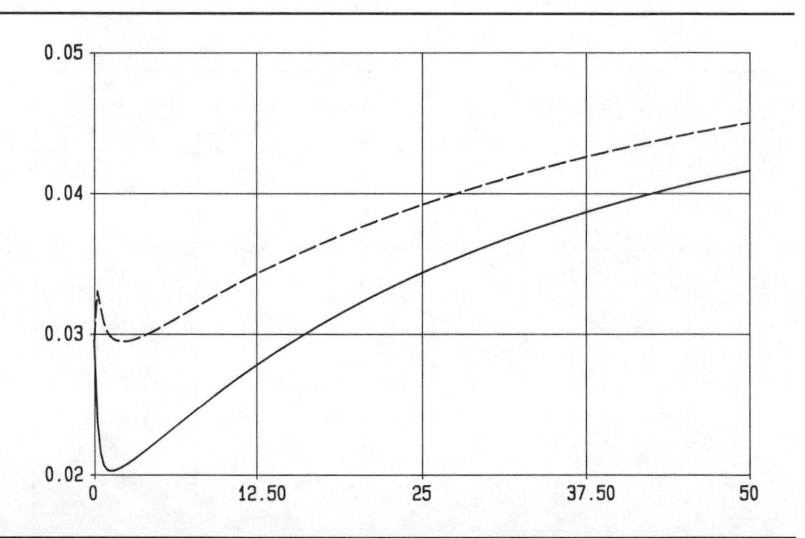

Fig. 8: Expenditure flow $T_{22}(t)$ for different initial conditions $Z_2(0) = \hat{Z}_2$ (——) and $Z_2(0) = 2\,\hat{Z}_2$ (---).

4.5. *Economic conclusions*

In summary our numerical simulation has confirmed, on one hand, that a decrease of transportation costs between two zones leads to pronounced competition between these zones. In the short run, the tendency of investors in the zone of smaller facility stock to increase their supply has a positive effect on the demand attracted. This investor behaviour is directed towards the equilibrium situation of the intraurban system and thus accelerates this adjustment process. In the long run, the stationary state is independent of initial changes in the scale of provision. The long relaxation time of the intraurban system, however, makes it clear that in general such an equilibrium situation will not be reached. Parameters involved in socio-economic models can only be assumed to be constant for moderately long time intervals. Changes of model parameters, e.g. a change of the transport cost matrix c_{ij}, may lead to different stationary supply and demand patterns. More important, however, is the fact that perturbations of parameters as described above or stepwise change in, for example, the scale of provision, lead to amplified or dampened oscillations on both the demand and supply side.

ACKNOWLEDGEMENT

The authors are very grateful to Professor W. Weidlich for many stimulating discussions and a critical reading of the manuscript.

Institute for Theoretical Physics
University of Stuttgart
Stuttgart
West Germany

BIBLIOGRAPHY

Birkin, M. and A.G. Wilson: 1985, 'Some properties of spatial-structural-economic-dynamic models', Working Paper 440, School of Geography, University of Leeds.

Clarke, M. and A.G. Wilson: 1983, 'The dynamics of urban spatial structure: progress and problems', *Journal of Regional Science* **13**, 1-18.

Dendrinos, D. and G. Haag: 1984, 'Toward a stochastic dynamical theory of location: empirical evidence', *Geographical Analysis* **16**, 287-300.

Frankhauser, P.: 1986,, 'Entkopplung der stationären Lösung des Haag-Wilson-Modells durch einen neuen Ansatz für die Nutzen-Funktion', Arbeitspapier, 2. Institut für Theoretische Physik, Universität Stuttgart.

Haag, G.: 1983, 'The macroeconomic potential', Proceedings of the IIASA-Meeting on "Long waves, depression and innovation", Siena-Florence.

Haag, G.: 1989a, 'Master equations', in C.S. Bertuglia, G. Leonardi and A.G. Wilson (eds.), *Modelling of Urban Dynamics*, forthcoming.

Haag, G.: 1989b, 'Services 2: a master equations approach', in C.S. Bertuglia, G. Leonardi and A.G. Wilson (eds.), *Modelling of Urban Dynamics*, forthcoming.

Haag, G. and D. Dendrinos: 1983, 'Toward a stochastic dynamical theory of location: a nonlinear migration process', *Geographical Analysis* **15**, 269-286.

Haag, G. and W. Weidlich: 1983, 'An evaluable theory for a class of migration problems', Collaborative Paper CP-83-58, IIASA, Laxenburg.

Haag, G. and W. Weidlich: 1984, 'A stochastic theory of interregional migration', *Geographical Analysis* **16**, 331-357.

Haag, G. and A.G. Wilson: 1986, 'A dynamic service sector model - a master equations' approach with prices and land rents', Working Paper 447, School of Geography, University of Leeds.

Harris, B. and A.G. Wilson: 1978, 'Equilibrium values and dynamics of attractiveness terms in production-constrained spatial-interaction models', *Environment and Planning* **A 10**, 371-388.

Johansson, B. and P. Nijkamp: 1984, 'Analysis of episodes in urban event histories', Contribution to the metropolitan study, 9, IIASA.

Leonardi, G.: 1989, 'Housing 3: stochastic dynamics', in C.S. Bertuglia, G. Leonardi and A.G. Wilson (eds.), *Modelling of Urban Dynamics*, forthcoming.

Lombardo, S.R. and G.A. Rabino: 1983, 'Nonlinear dynamic models for spatial interaction: the results of some numerical experiments', paper presented to the 23rd European Congress, Regional Science Association, Poitiers.

Pumain, D., T. Saint-Julien, and L. Sanders: 1984, 'Dynamics of spatial structure in French urban agglomerations', *Papers of the Regional Science Association* **55**, 71-82.

Sonis, M.: 1985, 'Quantitative and qualitative methods for relative spatial dynamics', CERUM lectures in Regional Science, Nordic workshop.

Weidlich, W. and G. Haag: 1978, 'Migration behaviour of mixed population in a town', *Collective Phenomena* **3**, 89.

Weidlich, W. and G. Haag: 1983, *Concepts and Models of a Quantitative Sociology: the Dynamics of Interacting Populations*, Springer-Verlag, Berlin.

Weidlich, W. and G. Haag (eds.): 1988, *Interregional Migration - Dynamic Theory and Comparative Evaluation*, Springer Series in Synergetics, Springer-Verlag, Berlin.

Wilson, A.G.: 1971, 'A family of spatial interaction models and associated developments', *Environment and Planning* **A 3**, 1-32.

Wilson, A.G.: 1974, *Urban and Regional Models in Geography and Planning*, John Wiley, Chichester.

Wilson, A.G.: 1985, 'Structural dynamics and spatial analysis: from equilibrium balancing models to extended economic models for both perfect and imperfect markets', Working Paper 431, School of Geography, University of Leeds

Wilson, A.G. and M. Clarke: 1979, 'Some illustrations of catastrophe theory applied to urban retailing structures', in M. Breheny (ed.), *Developments in Urban and Regional Analysis*, Pion, London.

MARIE-GENEVIEVE DURAND AND MARYVONNE LE BERRE

SPATIAL SYSTEMS MODELLING: FROM LAND USE PLANNING TO A GEOGRAPHICAL THEORY APPROACH

1. INTRODUCTION

The modelling of micro-regions via a system dynamics approach is such a recent phenomenon in geography that its utility can hardly be evaluated yet. However, the construction of the AMORAL model (Groupe AMORAL, 1984) and the progress made in building a model for the Aigueblanche- Valmorel basin (Groupe AMORAL, 1984) suggest that a systems approach is suitable for analyzing the interactions of complex systems, as it takes into account their dynamic nature and the increasing complexity of space undergoing transformation.

These experiments have proven to be enlightening for geographical research. The construction of the AMORAL model, which had been conceived as a tool to facilitate decision-making in land-use planning (applied geography), quickly led the research group to define concepts relating to space itself, to its organization and functioning, as well as to enlarge its arsenal of methodological approaches (theoretical geography). In doing so, a dialectical relation seemed to emerge between applied geography and theoretical geography, paving the way for a deepening of research in these two areas. The application of a methodology (system dynamics) to a specific region (the French Southern PreAlps, and one other example, the Aigueblanche-Valmorel basin in Savoie in Northern Alps) implies a number of in-depth theoretical studies, which in turn relate to both methodological aspects and to studies of land-use in the region: a fine feedback effect! In the first case study, the "pays" of the Southern PreAlps are regarded as homogeneous spatial units. In the second case of study, an attempt at generalization is made, whereby each micro-region is considered to be spatially heterogeneous. Thirdly, the theoretical functioning of spatial systems - however heterogeneous in character or whatever the scale chosen - is considered. This chapter presents an overview of these various studies.

2. THE AMORAL MODEL: A TOOL FOR DECISION-MAKING IN LAND-USE PLANNING

Modelling the functioning of a micro-region should lead to the construction of an image. This image should closely reflect the major dynamic elements; it should have a simple pattern to facilitate the use of the model; it should work efficiently to simulate the effects of those events which are likely to change the course of socio-economic developments. At present, system dynamics is one of the very few methods that meets these conditions. Thus, models simulating the working of complex systems can be built, and the method may be applied as a tool for decision-making.

The modelling system used here was developed in response to a request from the D.A.T.A.R. (Délégation à l'Aménagement du Territoire et à l'Action Régionale). The task consists of measuring the impact of

J. Hauer et al. (eds.), Urban Dynamics and Spatial Choice Behaviour, 247–268.

possible land-use planning operations on the general functioning and development of the Southern PreAlps region.

2.1. *Methodological Aspects*

The process of building the AMORAL model finds its origin in the remarks and conclusions of the 'Schéma d'Orientation et d'Aménagement des Alpes du Sud' (D.A.T.A.R., 1979), which reveals a clearly degraded situation in the southern PreAlps as regards demography, economic and social problems. It contrasts with the more active sectors in the higher mountain areas and the Durance valley. In the PreAlps, the situation is characterized by a very low population growth rate, a predominantly elderly population due to a long tradition of permanent out-migration, and a socio-economic pattern with no inherent capacity for renewal.

In an attempt to reverse this process of land abandonment, the "Schéma" underlines the necessity of implementing a few remedial measures such as: helping young adults to find housing and to settle down, creating jobs and improving general conditions of life through a constant fight against social isolation. Such measures, which are considered indispensable, have been integrated into the model, along with more specific policies. The "Schéma" also encourages the imple-

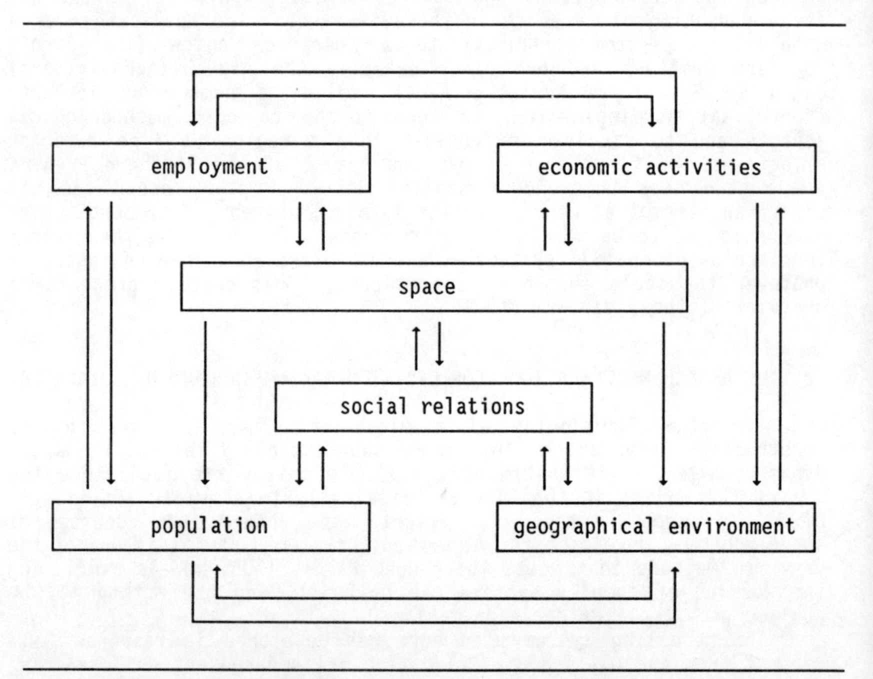

Fig. 1: The AMORAL model: global conceptualization.

mentation of comprehensive planning operations in the framework of micro-regions whose inhabitants are faced with similar problems.

In this model, the "pays" has been used as the essential areal unit suitable for this type of planning. It consists of a group of several sparsely populated communities situated in an isolated mountain valley and surrounding a village or, more rarely, a small market town, acting as a central place.

The modelling process consisted of clarifying the functioning of the "pays" as a spatial system through a study designed to reveal its inner structures. A global approach was used which moved from the general to the particular: the main interacting phenomena were first selected to understand the general functioning of the "pays". Then, as the need arose, finer interactions were considered, down to the most refined level of specification which appeared suitable for our objectives. Figures 1, 2, and 3 illustrate the various steps in that process.

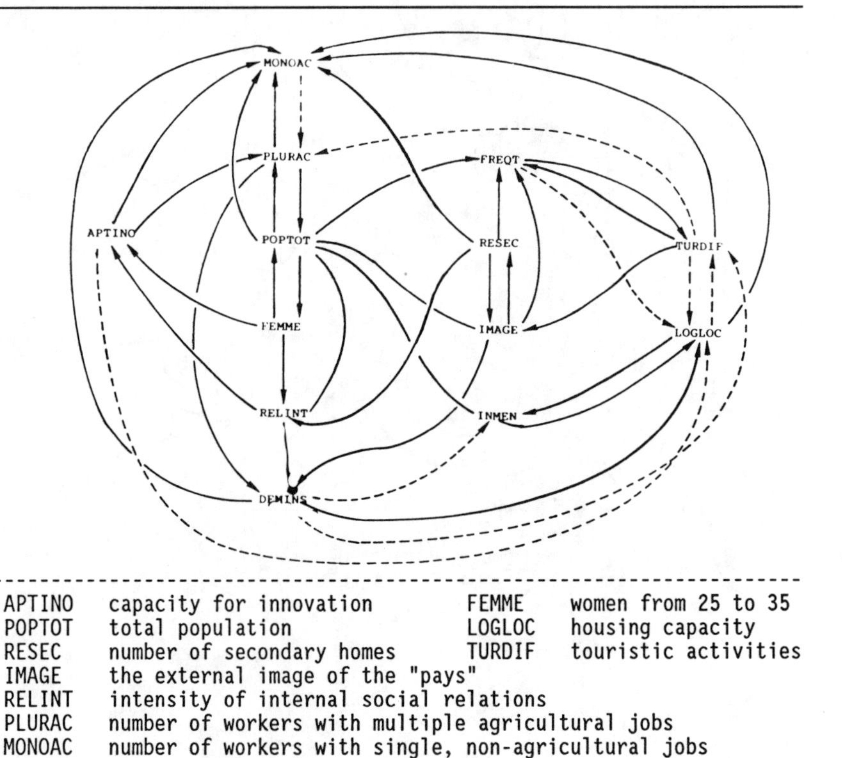

Fig. 2: Simplified graph of the relations in the AMORAL model.

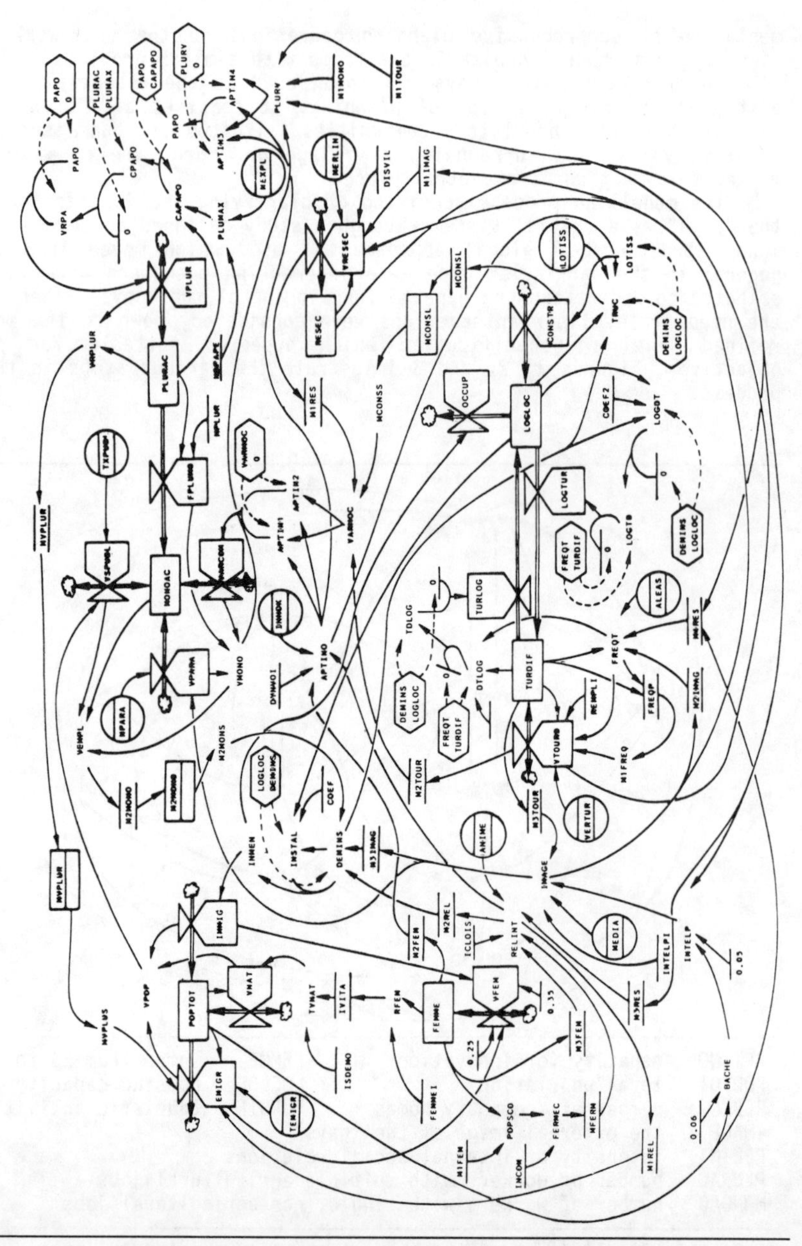

Fig. 3: The complete graph of the AMORAL model.

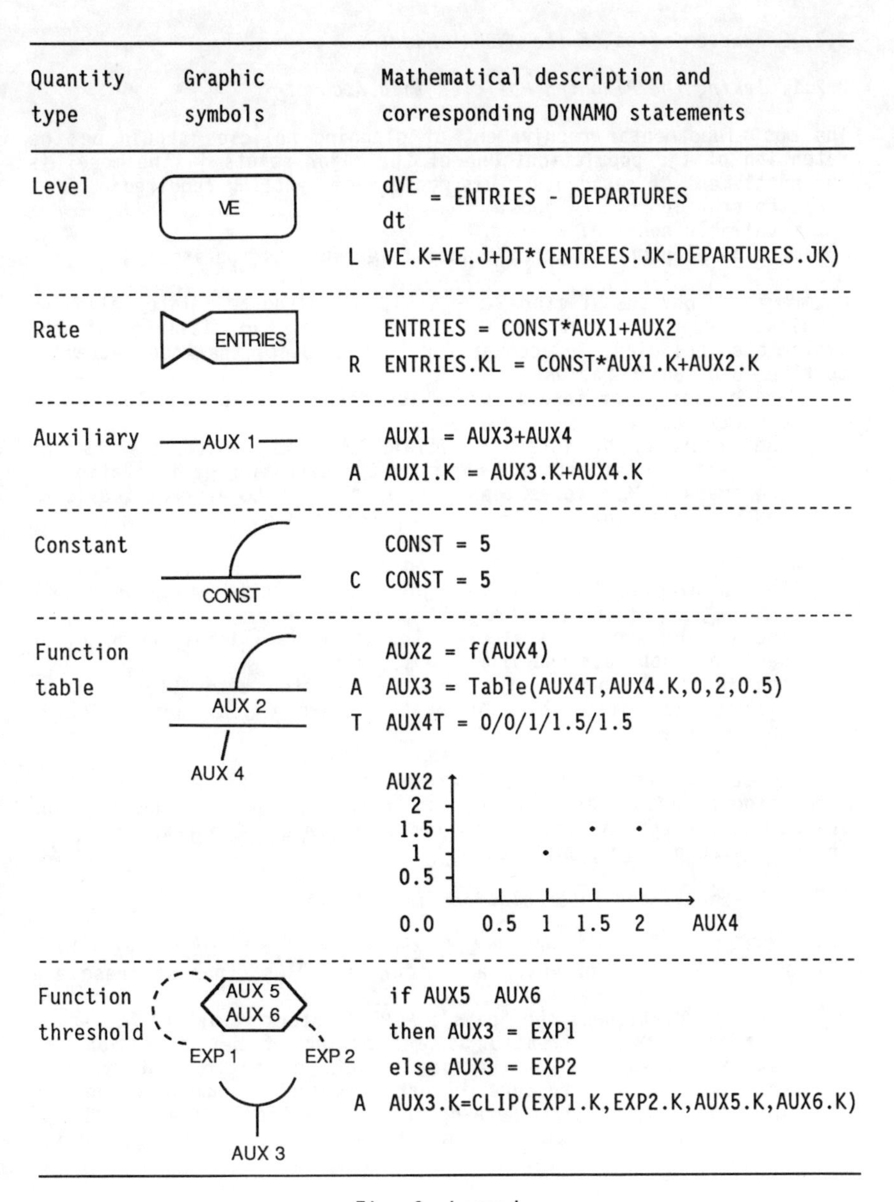
Quantity type
Graphic symbols
Mathematical description and corresponding DYNAMO statements
Level
VE
dVE/dt = ENTRIES - DEPARTURES
L VE.K=VE.J+DT*(ENTREES.JK-DEPARTURES.JK)
Rate
ENTRIES
ENTRIES = CONST*AUX1+AUX2
R ENTRIES.KL = CONST*AUX1.K+AUX2.K
Auxiliary
AUX 1
AUX1 = AUX3+AUX4
A AUX1.K = AUX3.K+AUX4.K
Constant
CONST
CONST = 5
C CONST = 5
Function table
AUX 2
AUX 4
AUX2 = f(AUX4)
A AUX3 = Table(AUX4T,AUX4.K,0,2,0.5)
T AUX4T = 0/0/1/1.5/1.5
AUX2
2
1.5
1
0.5
0.0
0.5
1
1.5
2
AUX4
Function threshold
AUX 5
AUX 6
EXP 1
EXP 2
AUX 3
if AUX5 AUX6
then AUX3 = EXP1
else AUX3 = EXP2
A AUX3.K=CLIP(EXP1.K,EXP2.K,AUX5.K,AUX6.K)

Fig. 3: Legend.

2.2. *Characteristics of the AMORAL Model*

2.2.1. *Taking the Planning Policies into Account*

The most fundamental requirement of planning policies should be the retention of the population. One of the major points in the model is the settlement of residents. This process of settling required:
- the creation of new jobs;
- available housing;
- the development of certain activities such as tourism.

Other less obvious factors are seldom introduced into planning studies. Yet they contribute to the success or failure of the implemented policies. To account for their impact, the model attempts to take into consideration:
- the capacity for innovation, which determines the development of economic activities in the "pays";
- the intensity of internal social relations, which assists the settling of residents and channels the diffusion of innovation;
- the image of the "pays", as far as it helps to attract tourists, new categories of residents, and possibly industrial firms.

The planning policies to be considered by the model mostly concern:
- the development of multiple job opportunities which may help retain part of the population;
- the development of scattered touristic facilities which offer secondary jobs derived from new activities;
- the introduction into the area of selected activities such as light industries, which generate numerous jobs and may have spread effects.

The AMORAL model is organized around the six following entities: population, social relations, jobs, economic factors, housing and space. Its conception aims to confirm or to refute the feasibility of the suggested planning policies.

2.2.2. *Variables and Interrelations in the Model*

The AMORAL model is constructed around thirteen major variables (Figure 2, 3), seven of which are stock variables. Four of these are presented below in more detail:
- Total population of the "pays": *POPTOT* (stock variable)
  This variable is treated as a function of net migration and natural increase. Given the time lapse considered (twenty years), the demographic structure is not supposed to change enough to interfere with the development of *POPTOT*. The level of out-migration is modified by the quality of internal social relations, *RELINT*, and by the varying numbers of agricultural workers with multiple sources of employment, *PLURAC*. In-migration results from the settling of new residents. If the demand is higher than the housing capacity, *LOGLOC*, no more people can settle. Then, in this case, the natural increase is in proportion

to the total population, *POPTOT*, taking into account the structural fertility index evaluated from the changes in the number of women aged 25 to 35, *FEMME*.

- Women from 25 to 35: *FEMME* (stock variable)
  Women aged 25 to 35 are a major factor in the dynamic pattern of the "pays", not only from a purely demographic point of view, but also because they strongly contribute to the creation and retention of internal social relations, *RELINT*, and consequently to the dynamics of the "pays". Their number varies according to migration flows. They also influence another dynamic element: the opening or closing of school classes in the "pays", which in turn is related to the quality of internal social relations, *RELINT*. Class closing depends on a minimum number of school-going children, which is in proportion to the number of women, *FEMME*.
- Secondary homes: *RESEC* (stock variable)
  The development in the number of secondary homes, *RESEC*, is influenced by the external image of the "pays", *IMAGE*, and by the distance to urban conglomerates, *DISVIL*. In addition, we consider the spread effect which makes a "pays" with a given number of secondary-home residents develop a positive image and attract more residents in turn (up to a saturation point where the process degenerates).
- Intensity of internal social relations: *RELINT* (intermediary variable)
  This, as well as *IMAGE*, refers to the quality of the geographical and sociological environment. It consists of the atmosphere created in social relations and takes socio-psychological factors into account. It holds a value through its influence upon scattered tourist activities, *TURDIF*, and upon the capacity for innovation, *APTINO*, as well as on migration, thus as a factor influencing *POPTOT*. The quality of these relations reflects the number of women between the ages 25 and 35, *FEMME*, secondary homes, *RESEC*, of the closing or maintaining of school classes and the relatively closed-in topography of the "pays".

Stock variables are written as differential equations:

$$\frac{\partial X}{\partial t} = \text{inputs - outputs}$$

where X is a stock variable.

Example: $$\frac{\partial POPTOT}{\partial t} = IMMIG - EMIG + VNAT$$

Given the intermediary variables and the exogenous variables (Figure 3), the model contains some 150 equations. The simulation language used is DYNAMO.

2.3. *Integrating Time and Dynamics*

The time unit taken as reference in the AMORAL model is the year. All interactions correspond to this reference. Actually, phenomena must be related in time in order to combine together. Building the model was rather easy on that point. Since human activities follow yearly rhythms, most data will automatically be expressed within the most common time units, for instance, economic results or demographic balance. Yet in the relation of Society to Nature, physical phenomena occur over much longer periods of time; therefore, the climatic phenomena or topographic elements appearing in the model have been considered as constants. Furthermore, it has been assumed that no phenomena acting on a smaller time scale will fundamentally affect the system's functioning. Thus, time, as well as space, has a homogeneous character. Dynamic changes rise from synchronic interactions referring to the same basic unit: the year.

Most often, when constructing this type of model, one can know very precisely the development of several if not all the variables considered individually. However, it is much less easy to evaluate the effects of their interactions when observed from the point of view of dynamic combinations, which gives a global as well as dynamic view of the whole process. From this perspective, model building is not a mere heaping up of phenomena; for, according to a fundamental principle of sociology we know that: "the whole is never equal to the sum of all parts".

3. SIMULATIONS AND SCENARIOS: RELATIONS BETWEEN PARTICULAR AND GENERAL ELEMENTS

3.1. *Simulations with the Reference-Model*

The model corresponds to the specifications of a typical "pays" in the Southern PreAlps. The simulation results illustrate the most frequent tendencies of that area. The initial values of the variables were chosen for a "pays" consisting of 2000 permanent residents, after having stored observations, official statistics and multivariate analysis. A simulation over a period of twenty years reveals the kind of changes in that "pays" with sufficient accuracy. Such a length of time gives the simulation a reasonable amount of reliability for evaluating the consequences of various planning decisions.

In order to follow the development of the chosen method, several simulation results are shown below. The results are only indicative.

3.1.1. *Development of the Stock Variables*

Without any form of planning intervention, the total permanent population (p) decreases regularly over the 20 simulated years (Figure 4). From 2,000 inhabitants it goes down to about 1800 persons. The number of women aged 25 to 35 (f) falls from 100 to 91.

In terms of jobs, the total number of non-agricultural workers engaged in a single form of employment (m) remains steady, at around 540 jobs, but increases from 27 to 30% proportionate to the total population. This increase is caused by on a slightly younger demographic profile and above all is counterbalanced by the decrease in the number of workers with multiple agricultural jobs (a): these drop from 50 to 29 individuals over the same length of time.

Among the economic activities, scattered touristic activities (*t*) increase gradually, but without any intervention, less than the number of secondary homes (*r*) - these increase by more than 50% over the 20 years. The number of available rented homes (*l*) remains relatively constant.

3.1.2. *Development of Intermediary Variables Selected as Socio-economic Indicators*

Four of the intermediary variables have been retained as most significant indicators regarding inner dynamic structures in the "pays" (Figure 5).

The demand for resident settlement (*d*) has a slightly growing influence over the years. This phenomenon contributes to a slightly younger demographic profile in spite of the total decline of the population.

In the same time, there is a slightly better capacity to innovate (*a*) and to develop internal relations (*r*). The image of the "pays" (*i*) whose effects depends partly on the demand for resident settlement (*d*) continues to climb slowly, independent of most local decisions.

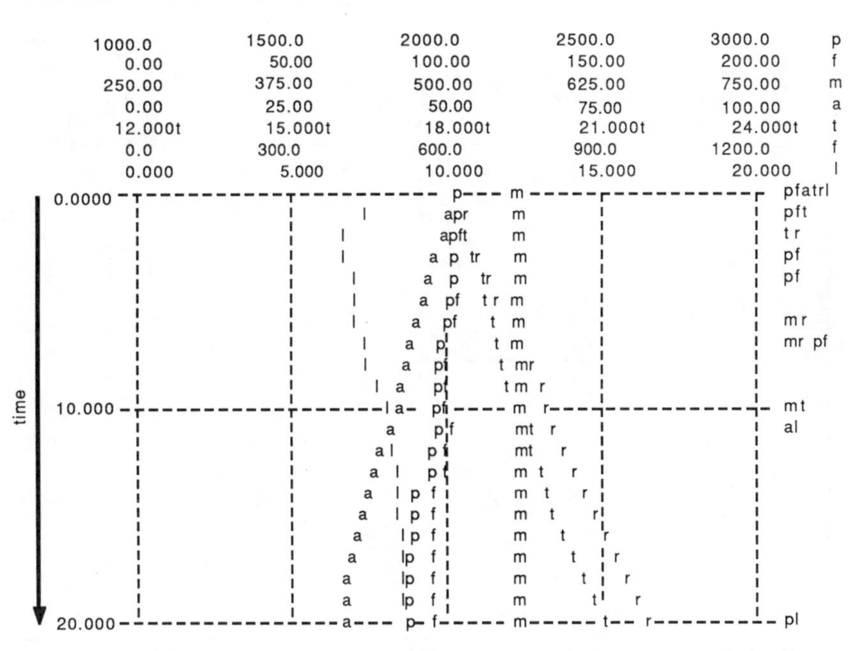

Fig. 4: Results of simulation: reference model of "pays" (without planning intervention).

3.2. *Model Adjustments and Scenarios*

The reference model corresponds to the scenario which has been constructed in order to understand the functioning of a typical "pays" in the Southern PreAlps. As the objective of the model consisted in helping decision-making in planning at specific locations, we could not confine our study to scenarios simulated on the basis of the reference model. In order to integrate the main spatial variations in the Southern PreAlps, three scenarios were built, each of them corresponding to the functioning of a different set of "pays".

These three "pays" sets have been classified with the aid of multivariate analysis (principal components analysis, agglomerative hierarchical clustering) using various demographic, economic and social criteria:

- "pays" centered on a market town functioning as a central place (more urban in character than most "pays" in the Southern PreAlps);
- "pays" where tourist activities are an incentive to economic growth;
- "pays" without any dynamic energy, because they lie far from the main traffic routes and offer few tourist attractions.

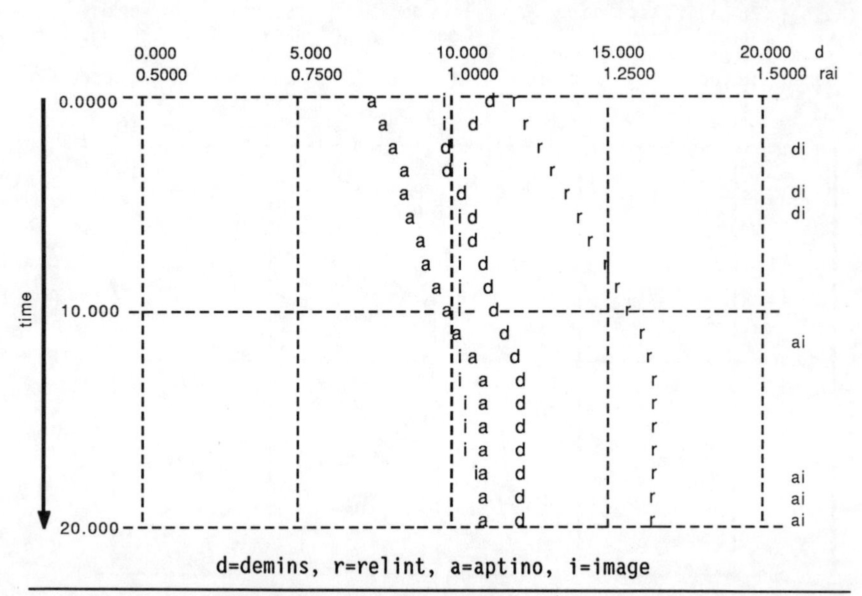

Fig. 5: Results of simulation: reference model of "pays" (without planning intervention).

We thus obtain three new models derived from the reference model (Figure 6) and adjusted to three types of "pays". Scenarios relating to specific planning projects were then simulated according to each of the three derived models. More specifically, they concern job improvement, demographic structure improvement, opening of access and tourist development.

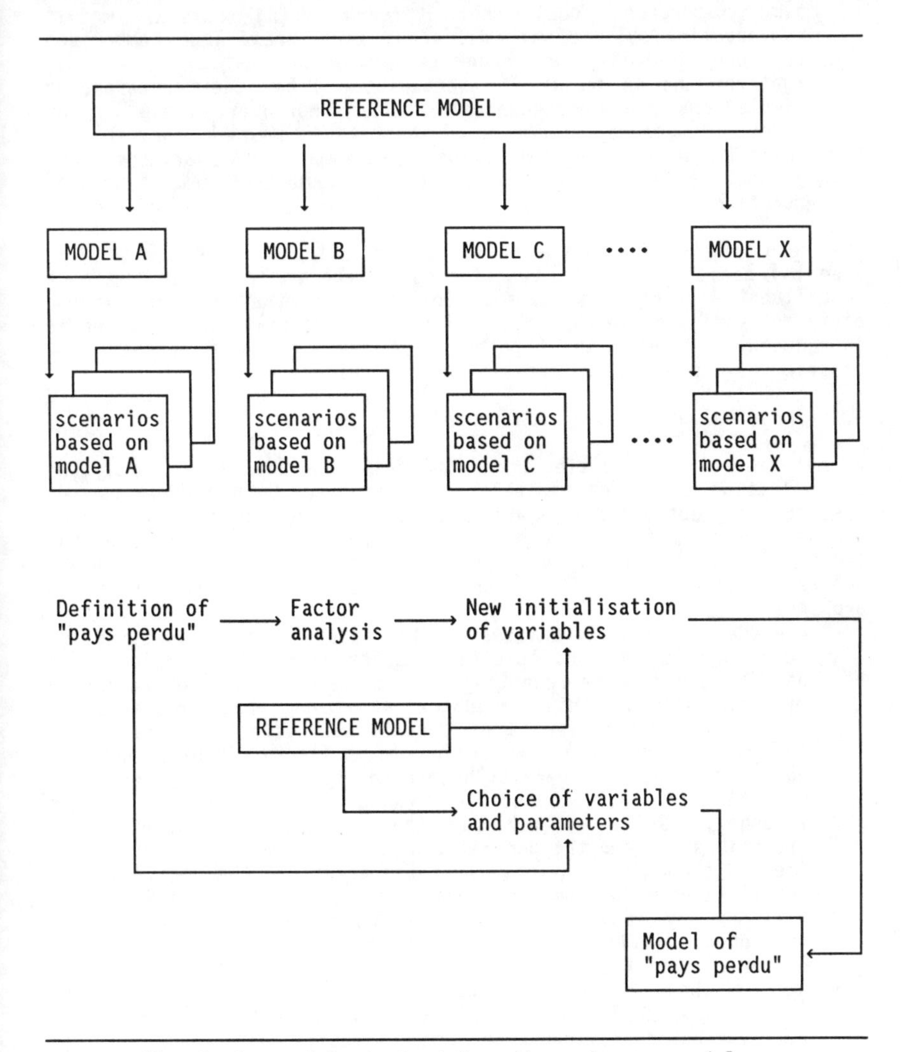

Fig. 6: New models derived from the reference model.

All these simulations were rather easy to build, as the AMORAL model is highly versatile due to the elements integrated into its structure:

- All the variables and the tabulated functions to be used in the construction of the various scenarios had been selected at the start of the modelling process.
- All simulations were made through successive derivations from the same reference model, not through using several models representing systems with different structures. Thus comparison from one simulation to another is possible.
- Whatever the nature of the scenarios may be, the corresponding simulations use few technical changes. Moreover, these do not affect the general structure of the reference model. They include changes in the initial values of some stock variables and parameters as well as changes in the values of some tabulated functions.

The resulting simulations clearly show the impacts, if present, of planning decisions on the functioning of the whole system and each constituent element. A systematic comparison of the reactions of the different variables in the model - first when no planning operation is introduced, then with the hypothesis of given planning operations - constitutes the basis for decision-making (Figure 7).

3.3. *Technical and Geographical Problems*

We met several difficulties when specifying the limits of the system and selecting the major variables used to describe its dynamics. We also met unexpected problems which appeared in the experimental period during the simulations. One is not always apt to measure the feasibility of an analogy between the mathematical equations and the geographic phenomena they express. We suggest three reasons for such problems:

- Are the modelling techniques and the nature of model construction able to integrate the specific aspects of spatial complexities? We may wonder if representing a given level of self-organization by means of a strong feedback structure does not tend to over-minimize the effects of certain sub-system interrelations. In other words, are the notion of self-organization and the notion of space homogeneity/heterogeneity as created by planning policies in fact compatible? Using G.-G. Granger's phrase (Granger, 1984), we may say that it poses the problem of "relations between the general and the particular".
- Are the phenomena convincingly integrated into the model? We may question the modelling technique, but the modelling actor just as well as any individual may make undiscernable mistakes. This widespread difficulty is even more serious here because of the number of variables considered.
- Do we have sufficient geographic information about the interactions to be modelled? In this case, we must confess we do not. Various social phenomena are not easily quantified and/or so poorly understood that we can hardly select a variable (or variables) to represent them. Moreover, the data are aggregated

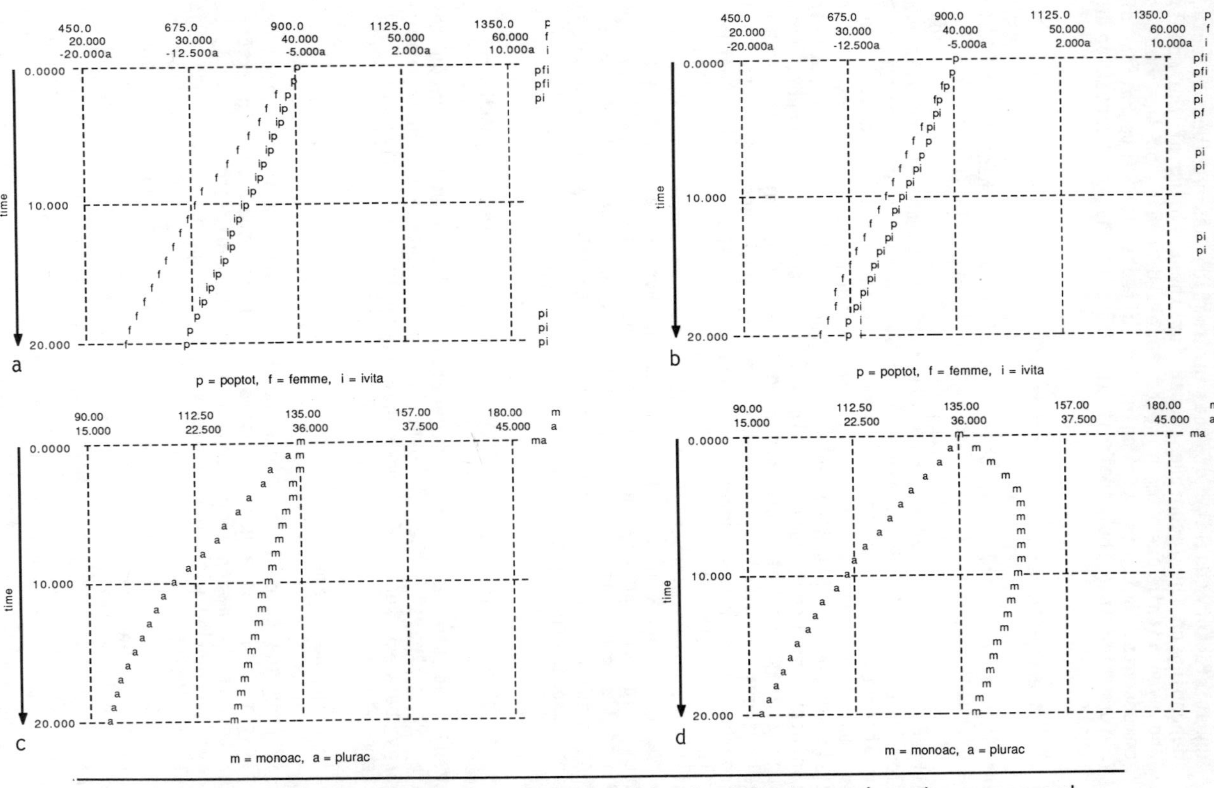

Fig. 7: Results of simulations: model of "pays" without any dynamic energy and scenario for creating jobs.

on spatial levels which are not adapted to the research. The unsatisfactory quality of the information has various effects, depending on the technique chosen for the building of the model, and even within one same type of model, depending on the various components. But, in any case, it affects the integration of information into the model; it also affects the interpretation of the results.

3.4. *Can the AMORAL Model be Transferred?*

Users should be warned about the risks of incorrect handling of this type of model which seems to be so easy and versatile to use.

Of course, the initial values of stock variables may be rapidly modified, some parameters may have altered values, and the results may follow by themselves. But these results will be profitable only if the user attempts to penetrate the inner thoughts of the modeller. And even more so, it is strongly recommended that the user (or decision-maker for planning policies) should take an active part in the process of model building. This being taken into account, even though it has not yet been subjected to experimentation, the AMORAL model seems to be transferable practically unchanged to all micro-regions whose structure is similar to that of the Southern PreAlps- to the Massif Central Southern areas for instance. Considering other micro-regions, one cannot be absolutely sure that the model will work adequately. Indeed, the hypotheses relating to the system's functioning may be inappropriate, and thus we fall outside the model's range of application. Yet, the process of building the AMORAL model can be transferred very easily to other areas regarded as having homogeneous functioning.

4. DIFFERENTIAL SPATIAL DYNAMICS IN THE SYSTEMS APPROACH

In many case studies, space homogeneity can be assumed: even on the scale of a "pays", the all-over dynamic exchanges result from a synergy between specific forces having their own individual logic patterns.

4.1. *Heterogeneous Spatial Units*

Considering the case of the Aigueblanche-Valmorel basin, the area can at first regarded as a unique system, which is, to a certain extent, accurate. It is a continous topographic area, divided by the Isere Valley, surrounded with mountain peaks, whose spatial organization is highly conditioned by the relief. It contains fields of activities and consequently has a socio-economic overlay. The various economic activities characterizing this area are: residual mountain agricultural activities (cultivation of fields, cattle breeding, orchards, forests), electro-metal works in Aigueblanche linked with hydroelectric plants, various touristic facilities (thermal spas in La Léchère, rural and wintersports activities in Les Avanchers, leisure and skiing activities at the new mountain resort in Valmorel). One can even simulate the likely development of these economic sectors and their actors, as well as the risks of stagnation due to the forester's desire to preserve the forests and the environmental lobbies wanting to preserve the sites.

Such a simplified verbal model which integrates the main interactions in this Northern Alps "pays" can be regarded as accurate. But, contrary to the conclusions drawn with the AMORAL model in the Southern PreAlps, the dynamics described here have no general significance for the area under consideration, as it is heterogenous. Some sectors live on industry, others on agriculture, yet others on tourism, etc. As evidenced by prior local intervention, the effects can be seen in isolation or aggregated (cattle breeding in summer, skiing in winter, or agricultural and industrial activities for that part of the rural population that also works in factories). These differential dynamic forces may be competitive or indifferent to each other, or even combine and reinforce their effects.

Each area is the territory of a social group using its own strategies. For the region as a whole each group may consist partly of a combination of secondary social groups (with multiple sources of employment in or outside agriculture) and not all strategies are determined locally. Part of the political and financial decision-making powers escape the local actors. The planning decisions become external. Some activities may even be totally in hands of outside interests, for instance, seasonal activities relating to tourism with outside labour and economic effects.

4.2. *The Significance of Heterogeneity of Space and Time*

The different dynamic forces in the "pays" are more or less similarly oriented and integrated. Some are almost autonomous and totally included in the sector, others work as functional links with the outside areas. This last category may even depend on the dynamics of contiguous "pays" functioning on the same scale (cf. the ski resort complex of the "Trois Vallées", or may just be a superficial effect of an activity functioning on a much larger scale (of regional, national or international importance). This is the case for all activities relating to the promotion of winter sports or to industrial strategies. This goes to show how necessary it is to model intervention taking place over several scales in space.

When building a model, many phenomena and their probable development can be considered individually. Some of them, however, remain uncertain over the period considered. They are mostly "natural catastrophes" such as landslides, avalanches and floods which may occur rather frequently and are intimately related to the land-use planning policies in the mountains. We cannot ignore them and they should be integrated in the model. But this can be accomplished only through the use of non-scheduled catastrophe scenarios. Other categories of data can be used only on finer time scales, for instance, snowfall scheduling. Such problems have not been solved yet, but may be in the future.

Anyway, within the limits of our research on heterogeneous spatial modelling, differentiated areas and their dynamics must be integrated in order to develop a generalized systems approach. It is less difficult to integrate phenomena with varying periods, as this problem can most often be solved by applying various types of tabulated functions. We are still left with the problem of the relation between space and time within the differential local dynamics, as they do not change according to the same logic at the same speed.

4.3. *Directions for Research*

One possibility for further research would consist in superimposing a grid over a given heterogeneous area, to reveal various dynamics, and then, with a matrix, building models of types that take their synergy into account.

Each one of those sub-models brings its own dynamics into the dynamics of the model. But the dynamics belonging to each element (agriculture, winter sports, thermalism, industry, housing development, environmental protection, etc.) should be attributed to determined spatial areas which can be read instantly. This spatial result may incur a change with every time step.

In fact, the dynamics of each element (or sub-model) follow their own inner curve while undergoing other changes besides those due to the relations with other sub-models. They are mutually transformed and also integrate outside influences (particularly when scenarios are in operation).

As a result of such multiple interactions, any zone which is characterized by a given type of land use may become spatially different in space. In reality, agricultural zones regress, urbanization spreads, tourist facilities may develop along their own dynamic lines and be affected by environmental protection, etc. One can imagine that by time $ti + 1$, a given zone will be modified; and by time $ti + 2$, that zone plus a second one will have changed, etc. This type of process has not yet been tested; it is still impossible to foretell the feasibility of the method. At present, it is just a direction for research.

It may be useful to note that the process is dynamic in itself. This is no cinematic model describing the successive images of a structure step by step through iteration, but is in fact a model where changes are integrated into the developing process.

The same approach has been used to build the CARPE model (Fournier, 1984) using the principles of systems dynamics: the urban zone of Carpentras (France) was divided into two interacting sub-systems, i.e. town center and surrounding district; their specific developments have been integrated into the dynamic environment of the whole region. The model has retained a reasonable size, in spite of the important increase of the number of stock and flow variables which result from this division. But the limits of this approach are apparent when the number of divisions increases. One may have to handle a gigantic model with many drawbacks: construction problems because of the large number of poorly understood relations, a considerable extent of expensive testing, risky interpretation of the simulation results.

Other approaches to spatial heterogeneity modelling are presently being attempted: Peter Allen's intra-urban model with its adaptation to French towns. At first glance, this model appears to be very suitable for dealing with heterogeneous areas, because the system is conceived to permit division into interrelated spatial compartments, each holding a specific attraction force. The latter notion is highly enlightening; it corresponds to a well-tried study of urban phenomena. Besides, in a recent article, R. Thom (Thom, 1984) showed how interesting the approach was; it consists of considering organization levels in terms of entities similar to those defined in the field of physics.

Presently, approaches to heterogeneous spatial modelling are converging. Yet, in order to be more efficient, this type of research necessitates a deeper look into geographic study and the processes of spatial complexities.

5. TOWARD A GEOGRAPHICAL THEORY: MODELLING PRINCIPLES FOR SPATIAL COMPLEXITY

When planning operations create a spatial system with the capacity to survive in its own environment, a general correspondence between geographic and systemic system principles is established (Figure 8). To demonstrate the role of the systems approach in explaining spatial complexity, we must elaborate on the correspondence by finding systemic equivalents to three groups of geographic notions: types of planning, appropriateness of various spatial organization forms, spatial dynamics.

5.1. *Planning Policies and Self-Organization of Spatial Systems*

Any human group attempting to answer its vital needs, such as surviving and reproducing, takes over and plans a more or less extended area of the earth's surface, which becomes its own territory. This area has a mode of functioning assigned by the group, which can be expressed in terms of spatial coherence. The group determines this kind of spatial homogeneity, as it generates and maintains a complex network of relations between its own members and between the various

BASIC VIEWS OF SYSTEM MODELLING		BASIC VIEWS OF SPATIAL COMPLEXITY
An action	⟶	spatial planning
for an aim	⟶	life, survival and reproduction of social groups
in an environment	⟶	dynamics of other territories

Fig. 8: Relations between the systemic model and representation of spatial complexity.

places inside its territory. Such spatial coherence equally characterizes the planned area: it changes with time, under the influence of given disturbing events, which help to identify the type of planning policies being implemented.

Planning can thus be regarded as "acting", in the systemic meaning of creating a spatial system. This system is identified by its organization, i.e. through a network of linked and enduring processes associated with the planning action, and endowed with specific properties, such as: functioning within time so as to achieve the aim set by the group, and evolving by being transformed. From the systemic point of view, those properties depend on the ability of the system to "memorize its own functioning, which thus conditions its own transformation space. The system informs the organization, which by transforming organizes the information which conditions the functioning of the autonomous system" (Le Moigne, 1984). This notion of system memory has a ready equivalent in spatial systems: indeed, contemporary geographers from nearly all countries have strongly confirmed the active part played by space.This part can be defined as a constraint, in the meaning applied in physics, i.e. the intensity of the inertia force exerted by planned space over future planning actions, the whole thing working as though the development was somehow conditioned by its past life (Figure 9). The notion of self-organization, rather than the notion of structure, seems to express more accurately the essential character of spatial systems.

5.2. *Hierarchic and Inter-twined Spatial Organization Levels*

To answer most of its needs, the human group also uses, throughout its territory, various selective processes which create a type of spatial differentiation: the territory tends to be organized into various sub-areas, functioning homogeneously because each of them is specialized to answer one specific need. This localizing process can be regarded as an self-organizing activity in the environment of the

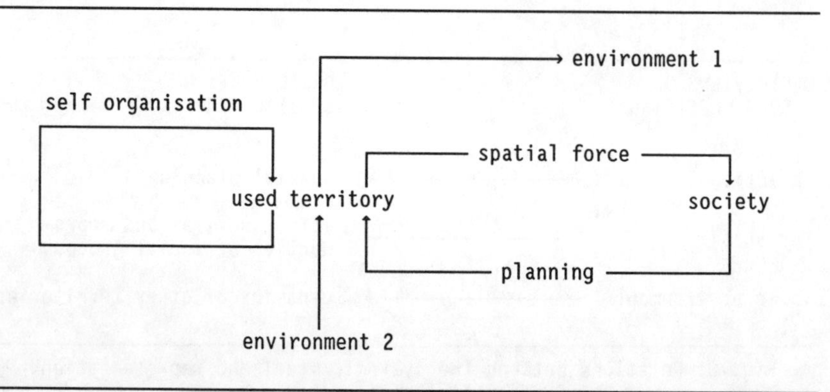

Fig. 9: Dynamics of a spatial system.

territory. So the variety inside the territory, even more than its spatial heterogeneity, results from the presence of a number of these homogeneous sub-areas which can be identified on the basis of their self-organization. Such sub-areas are not independent, but linked together as they contribute to the all-round functioning of the territory through their interactive network. To use systemic terms, we can say that the planning action creates several hierarchized spatial organization levels, because it generates, at the same time on different scales, spatial homogeneity and heterogeneity. These notions seem to be absolutely compatible with several essential systemic principles, such as:

- organization of a system with hierarchic levels;
- autonomous functioning of a level;
- circular causality between the various organization levels of the system.

Spatial heterogeneity is also caused by the existence of diversified social groups, with varying numbers of individuals, varying social organizations and varying needs. So these territories function on widely different spatial scales. The influence of the group on its own territory may be very diversified as well, according to the various components and needs, according to whether it is about to lose its social structure or whether it tries to assert and develop its relation in coherence with space.

Spatial heterogeneity can also be expressed in terms of appropriate and intertwining forms of different spatial organizations. These forms and the territorial limits can change and be transformed under various circumstances, at various speeds, and as the relations between a spatial system and its environment are subject to variations by themselves.

Spatial organization may show great variability; this seems compatible with essential systemic principles. In fact, there is no argument against considering different territories as different spatial systems with their own specific working principles.

5.3. *Spatial Systems and Time*

The preceding pages have been mostly dedicated to the morphology of spatial complexity, even if the functioning and transformation of space, which were repeatedly mentioned refer directly to time and dynamics.

It appears that systemics may give a globally suitable representation of spatial dynamics. By nature, systemics enables the introduction of spatial phenomena into irreversible time: the condition of a system by t time cannot be considered as a mere reproduction of the same system at $t-n$ time. Thus, systemics give a clear account of a development, i.e. of a transformation. Within a spatial system, this transformation can have different natures:

- It can be structural: through its own functioning, space produces a development of its components, which may entail variations or reversals of the trends; the above-mentioned notions of self-organization and memory of a spatial system are intimately related to space and time.

- It can be accidental: at any given moment, the social group may decide to institute a new planning policy which affects more or less numerous space components and consequently the functioning of the whole territory. Or again, an outside actor introducing a new action may also cause a change in the relations of a system to the environment, thus leading to a transformation of its functioning.

The forces influencing the development of spatial systems, whether they are structural or accidental, may be integrated into a mode of systemic representation.

Understanding spatial complexity presupposes taking into consideration those phenomena that play on varying spatial scales; similarly, the occurrence on varying time scales should be considered:

- A phenomenon may commence at different moments and follow different trajectories: overlapping of an intensity threshold or of a sequence duration. The activated phenomenon may take place at the same time as the activating phenomenon or be more or less delayed.
- The effects brought about by activating phenomena may have various aspects; at different times, one phenomenon may cause similar or differing actions; it may also have an impact on several others and have similar or differing effects.
- In a given number of geographic phenomena, time has a dialectic influence as a linear factor - a succession of events - or as a circular factor - the seasons in time, for instance.

In spite of all the diverging approaches as regards its epistemologic function, systemics cannot presently be ignored in geographic research. It contributes to the enrichment and specification of geographic studies, as it brings out the principles of system construction, these being considered as a means to express an understanding of complex phenomena. Systemics constitutes an approach to scientific knowledge which does not entail describing phenomena by separating their particular characteristics, but involves understanding them by identifying the network of relations and deterrents which links their components; it explains them by integrating them into an environment which affects their birth, their duration, their decline. How lucky are these geographers who are seduced by a fruitful hypothetic-deductive approach, and can, thanks to the systemic modelling key, consider exploring and understanding the dense universe of spatial interrelations and overlapping scales! What a good fortune also for traditional geography that has stubbornly refused all reduction of a detailed observation of landscapes and of men, and has always denied any interest in the construction of formal analytic models considered as too simplistic!

But these remarks should not be misleading. It does not imply a mere translation of old thoughts into new words; it is in fact a different conception of the world's representation. This was perceived very clearly by some geographers who then dropped this approach because it questioned their way of thinking. Indeed, the systemic model has its drawbacks, as any other representation. Geographic work then consists of looking for corresponding lines between the spatial complexity to be modelled and the essential characteristics of the system. So to get the best out of this approach, a dialectic force should run from the spatial complexity to the systemic model: the latter must allow an integration of notions relating to space, and also the fundamental notions of the systemic model should be transferable to the field of spatial complexity. Then, very drastic methods should be applied to the construction of formal simulation models, whose importance in geographic research is very specific and innovating.

In spite of the intellectual and technical problems which rise and also because of such problems, systemic modelling becomes a very powerful tool for contemporary geographic studies; it leads far beyond a mere enrichment of the specific knowledge of phenomena. It demands a new approach to space and engenders a deepening of the theories. This is a lasting advantage, whatever conception of systems is considered, whatever degree of formalization is used to represent spatial complexities, whatever interpretation of geography is selected.

The simulation models bring to geography a new and different direction for experimentation. They offer new insight into past phenomena and into "a theoretical exploring of various possible future events". If they are sensibly used and strongly built, some of them may give badly needed help in decision-making in regard to land-use planning.

Lastly, formal models, and more particularly simulation models, give an efficient means of detecting incoherence in a body of hypotheses, of revealing what components should be examined more soundly when studying a little-known phenomenon. In more general terms, systemic modelling offers a working tool for theoretical study because of the rigorous approach necessary in all the phases of model building. This last point should be stressed, as every one is acquainted with the weakness of geographic theory, and as systemics has not yet won recognition from the scientific community, for its function as well as for its epistemologic contents. These relations between systemics and geography should lead to a mutual enrichment.

U.F.R. de Géographie
Université de Grenoble I
Grenoble
France

BIBLIOGRAPHY

Groupe AMORAL: 1984a, 'Le modèle AMORAL, analyse systémique et modélisation régionale dans les Préalpes du Sud', Rapport de contrat D.A.T.A.R. - Université de Grenoble 1.

Groupe AMORAL: 1984b, 'Principes géographiques pour une modélisation dynamique d'un espace hétérogène, le bassin d'Aigueblanche-Valmorel, Savoie', Grenoble, Laboratoire Associé 344 du C.N.R.S.

D.A.T.A.R.: 1979, 'Schéma d'Orientation et d'Aménagement du Massif des Alpes du Sud'.

Fournier, S.: 1984, 'CARPE, un modèle spatialisé d'analyse du développement local adapté aux villes moyennes françaises', in Groupe Dupont (eds.), *Géopoint 84*, *Systèmes et localisations*, Avignon.

Granger, G-G.: 1984, 'Modèles qualitatifs, modèles quantitatifs dans la connaissance scientifique', in C.N.R.S. (eds.), *Querelles de modèles*, Cahiers Science Technologie Société 5.

Le Moigne, J-L.: 1984, 'Une localisation, des méthodes de modélisation systémique', in Groupe Dupont (eds.), *Géopoint 84*, *Systèmes et localisations*, Avignon.

Thom, R.: 1984, 'Modèles physiques et biologiques de la singularité', in C.N.R.S. (eds.), *Querelles de modèles*, Cahiers Science Technologie Société 5.

PAUL A. LONGLEY AND MICHAEL BATTY

MEASURING AND SIMULATING THE STRUCTURE AND FORM OF CARTOGRAPHIC LINES

1. INTRODUCTION

Cartographic lines are used to define boundaries between phenomena in a wide variety of spatial contexts. Such lines are abstractions whose particular geometrical character or form depends upon the scale at which they are defined, and thus there is an immediate task in seeking methods which enable consistent line generalisation to whatever scale. A more substantive task in cartographic line measurement involves interpreting the structure of the line in terms of the processes which have led to its form. Such lines are intrinsically irregular, reflecting the action of a variety of natural and man-made processes operating on the phenomena which they bound. Explanations of such irregularity, as well as the ability to consistently generalise such boundaries, thus becomes the central quest of cartographic line measurement.

This paper tackles this problem using the theory of fractals (Mandelbrot, 1967, 1982). Fractals are irregular geometrical objects whose form has hitherto defied conventional geometrical measurement, despite the widespread existence of phenomena which display such irregularity. Here we use the concept of fractal dimension to measure the irregularity of cartographic lines, emphasising the concept of self-similarity in phenomena at different scales. We argue that cartographic lines such as coastlines and urban boundaries have the same general form at whatever scale, and we detect this property of self-similarity using fractal measurement. This involves introducing a series of simulation methods to generalise lines at different scales, and thus we not only justify the concept of fractal description but also present methods that enable such fractal phenomena to be consistently simulated and hence predicted.

Although boundary definition is inherently subjective to a greater or lesser degree, we argue the virtues of generating quantitative indices of such boundaries as a means of both facilitating and directly contributing to other spatial analysis tasks. The anticipated proliferation of digitised data bases, coupled with the development and diffusion of computer systems of ever-increasing sophistication and power, is likely to routinise a complete range of different cartographic line-processing routines. In this paper we argue the virtues of preferring fractal-based methods in such a context. We describe how fractal dimensions may be determined empirically and then outline four alternative methods of measuring 'fractality' in empirical studies. These four methods are then applied to the digitised urban boundaries of the City of Cardiff, in the United Kingdom, in 1886, 1901, 1922 and 1949. We assess both the performance of each method and the implications of our empirical results for future research.

J. Hauer et al. (eds.), Urban Dynamics and Spatial Choice Behaviour, 269–292.

2. CARTOGRAPHIC LINES AND FRACTAL MEASUREMENT

Cartographic line generalisation involves procedures and processes of simplification (e.g. removing unwanted detail, smoothing of features, etc.), symbolisation (graphical coding of line character according to geographical and perceptual conventions), classification (aggregating and/or partitioning cartographic information into categories) and induction (the application of the creative logical assumptions which are made during generalisation); Buttenfield (1984) presents a full and detailed discussion. Whether carried out entirely by manual means or with the aid of automated techniques, it is thus clear that the depiction of cartographic information is the end result of a variety of codification conventions mediated by a human judgemental process. This is no less the case in the generation of computer-digitised data bases (whether by point, stream or fully-automated mode) than in traditional cartographic line-drawing (Jenks, 1981).

Nevertheless, attempts to measure boundaries and shapes have a fairly long history in natural science and have also been absorbed into the locational analysis tradition of human geography (see Haggett *et al*, 1977, pp. 309-312). Most of the earliest shape indices were based upon simple length, breadth and area relations, whilst the constraints associated with time-consuming manual measurement restricted assessment of line structure to simple indices of variation of selected points and monotonicity of line segments about a base line anchoring the end points of a line. These efforts were nevertheless well motivated since, as we shall demonstrate, such measurements can yield important quantitative indices for spatial analysis. First, and in the spirit of the spatial science paradigm, we feel that there are circumstances in which measurement can usefully help diagnose the processes which generate particular spatial forms, particularly if comparable measurements are taken of lines which measure a phenomenon in spatially and/or temporally diverse circumstances. Second, our wider research interests encompass the hitherto neglected task of presenting the results of synthetic urban land-use models as large-scale spatial simulations; measurement of edge irregularities and abstraction of a rule or rules which generate such irregularities provides an empirical basis to the partitioning of simulation space where no comprehensive areal data base is available for the mapping of model forecasts (Batty and Longley, 1986).

On a related point, just as the wide diffusion of computer graphics hardware and software has facilitated spatial simulation in a more routine manner, so too has this innovation eased the process of digitised data entry and analysis. The complementary procedures of data digitising and subsequent numerical processing at a single computer installation opens up possibilities for applying a wide range of techniques to diagnose line structure. Buttenfield (1984) develops a comprehensive classification and critique of line generalisation algorithms including various random and systematic point weeding routines, fitting of various mathematical functions, the epsilon neighbourhood concept (Perkal, 1966), and the use of both angular and band width tolerancing to dispense with successive points which fall outside a prespecified angular and/or band width threshold (Peucker, 1975). She concludes that the choice of method used can depend upon the often-conflicting emphases that different studies have placed upon geographical and perceptual accuracy.

In the present context we reiterate that most of these methods remain flawed in at least two fundamental ways. First, in a geographical sense, most are heavily reliant upon the *a priori* definition of the scale, starting point and ending point of constituent line features for the synthesis of total line structure. Second, and in a perceptual sense, many algorithms fail to preserve the qualitative visual character of a line when it is generalised (Muller, 1986). Seen in the context of the objectives of our own research, these are grave shortcomings: first, we cannot specify *a priori* those features which we expect to characterise the urban edge, and indeed their inductive generalisation remains one of the primary goals of the exercise; while our second goal of deriving visually acceptable space partitioning rules for computer-based land-use simulations dictates that our measurement parameters maintain perceptual accuracy in any of the lines which they are used to generate. We contend that use of *fractal* techniques represents a consistent and feasible route beyond this impasse, since (a) using very few parameters, the line can be measured as a *total entity* rather than as a piecemeal amalgam of constituent features, and (b) *visual character* is preserved by using the concept of *self-similarity* and by sensitive assessment of the *range of scales* over which this phenomenon holds.

The simplest way to convey the basis of fractal measurement is to investigate the well-known conundrum involving the indeterminancy of the length of a section of coastline. If one measures a length of coast using any very coarse-scale map, the resulting length is likely to be very much shorter than would have been the case had a much finer-scale map been used. The reason for this is that a much greater amount of edge detail is likely to be picked up at the higher level of resolution. Extension of this argument to measurements based upon a whole range of maps depicted at different resolutions leads to the notion that measured length will always increase at finer levels of resolution, and hence that length is scale dependent. Taken to a logical conclusion, of course, this implies that the 'real world' length of the coastline is infinite and indeterminate, although it will always be measureable for practical purposes. A second empirical characteristic of coastline measurement is that small-scale deviations and perturbations about the edge line frequently resemble larger-scale deviations in form. This phenomenon of fine scale irregular detail appearing to be a scaled version of that at preceding scales (*self-similarity*) can be discerned when measuring a wide range of natural and man-made objects.

The dual notions of scale-dependent length and self-similarity represent two concepts around which Mandelbrot (1967, 1982) formalises the theory of fractals. The term fractal as used here refers to the non-integral or fractional dimension of lines and surfaces which fill more space than, respectively, one-dimensional Euclidean lines and two-dimensional isotropic surfaces. In the present context, a sinuous coastline fills more space than a one-dimensional straight line but does not fill an entire two-dimensional plane and thus its dimension is seen to lie between one and two. As the coastline becomes more rugged, so the fractal dimension of the mapped edge increases within the bounds of one and two. The self-similarity of a perfect fractal is, in theory, formed by a single process which operates repeatedly at all scales, although in reality self-similar fractal shapes may arise as the consequence of the compound effect of a number of processes. The self-similar nature of such shapes may not, however, hold across all measureable scales.

The conventional wisdom of self-similarity fashioned by a single process clearly becomes increasingly strained as we make the transition from physical to social systems where, for example, urban edges clearly result from a wide range of social and physical processes. At a theoretical level, some spatial theories give cause to anticipate self-similarity, with central place theory being perhaps the best example (Arlinghaus, 1985); at a procedural level we anticipate at worst that the differences in the nature of self-similarity between social and physical systems are in degree rather than in kind. In short, at the outset of our project we had no reason to doubt the capacity of fractal techniques to abstract curve form, although, as our results in Section 4 will illustrate, we do not preclude the notion that edges fashioned by a multitude of processes may as a consequence be multi-fractal.

3. DEFINING AND MEASURING FRACTAL DIMENSION

The property of scale-dependent length may first be illustrated by considering the measurement of an irregular line between two fixed points. Defining a scale of resolution r_0, we see that n_0 chords are necessary to approximate the curve. Next, measurement is carried out based on a second scale of resolution r_1 which is half r_0 (i.e. $r_1 = r_0/2$), and n_1 chords are needed to approximate the line. If the line is fractal, halving the interval will always mean that *more* than twice as many steps will be needed to approximate the line, since increasing amounts of self-similar detail are registered. Thus if

$$\frac{r_0}{r_1} = 2, \text{ then } \frac{n_1}{n_0} > 2. \tag{1}$$

Denoting the perimeters of the two measured curves $P_0(= n_0 r_0)$ and $P_1(= n_1 r_1)$, it is also clear, using Equations (1), that $P_1 > P_0$. This is clearly demonstrated in Figure 1 which illustrates how a stretch of coastline can be approximated by approximately 11 chords (n_0) at scale r_0 and approximately 32 chords (n_1) at scale r_1. The relationship in Equation (1) can be generalised by assuming that the ratio of the number of chord sizes at *any* two scales is always in constant relation to the ratio of the lengths of the chords. Thus

$$\frac{n_1}{n_0} = \left[\frac{r_0}{r_1}\right]^D, \qquad 1 \leq D \leq 2 \tag{2}$$

where D is the fractal dimension common to the two measurement scales. The limits which bound this equation yield a Euclidean line ($D = 1$, i.e. halving the scale yields exactly twice the of chords) and a plane ($D = 2$, i.e. halving the scale gives four times the number of chords and the line encloses the space). Rearranging Equation (2) for prediction:

$$n_1 = (n_0 r_0^{\;D}) r_1^{\;-D} = a r_1^{\;-D} \tag{3}$$

where $n_0 r_0^{\;D}$ acts as the base constant a in predicting the number of chords n_1 from any interval size r_1 relative to this base.

In the case where a curve is measured at just two scales of resolution, D can be derived by rearranging Equation (2) thus:

$$D = \ell n(n_1/n_0) \;/\; \ell n(r_0/r_1)$$

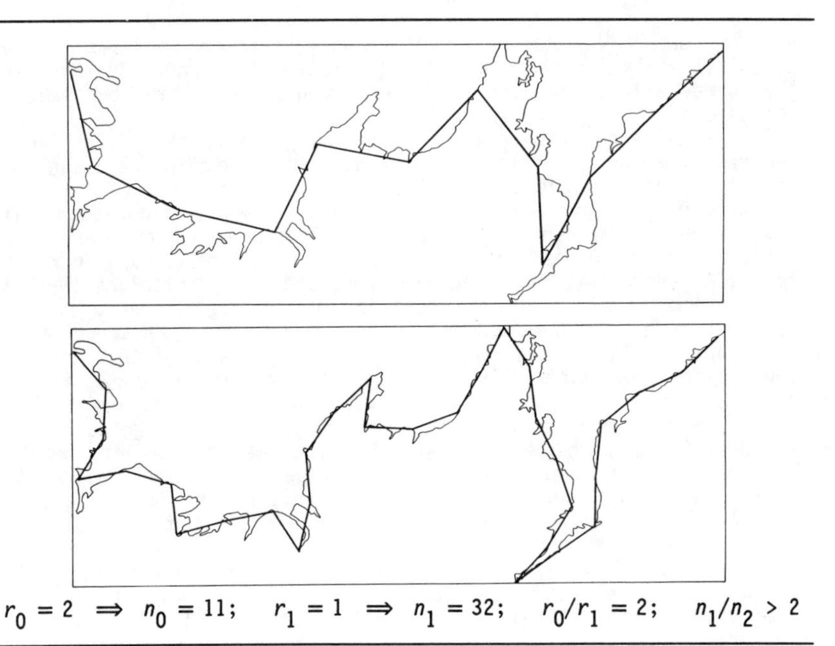

$r_0 = 2 \Rightarrow n_0 = 11$; $r_1 = 1 \Rightarrow n_1 = 32$; $r_0/r_1 = 2$; $n_1/n_2 > 2$

Fig. 1: Approximating a cartographic line at two adjacent order of magnitude scales.

However, in order to assess whether a line incorporates a fractal structure, it is necessary to demonstrate that D is constant over a number of scales. In this case, estimation of D proceeds by first rewriting Equation (3) as

$$n = ar^{-D} \tag{4}$$

where n is the number of chords associated with any r. Taking natural logarithms of Equation (4) yields

$$\ln n = \ln a - D \ln r \tag{5}$$

Assuming segment numbers (n) have been enumerated for a range of scales of resolution (r), D can be obtained using a standard simple regression procedure. A related equation is based upon the perimeter length P which is given using Equation (4) as

$$P = nr = ar^{(1-D)} \tag{6}$$

Taking natural logarithms:

$$\ln P = \ln a + (1-D)\ln r \tag{7}$$

Once again, D can be obtained using regression analysis. In our empirical analysis we use Equation (7) rather than Equation (5), since the former allows the most appropriate means of checking the range of scales used.

A wide range of methods of determining the fractal dimensions of curves have been developed in a range of disciplines including ecology, particle science and geology. In our empirical case study of the historical development of Cardiff, we apply algorithms which assess the scale dependency of curve length based on four rather different sets of assumptions; other methods employ area-perimeter relations (Woronow, 1981; Kent and Wong, 1982) or variograms (Mark and Aronson, 1984). Each of the methods involve specifying a range of scales which reflect orders of magnitude in a near geometric relationship. The reason for this is that estimation of the log-linear regression (Equation (7)) is likely to yield more reliable small-sample results when the observations (scales) are more or less equally spaced and hence are approximately equally weighted. Various rules-of-thumb have been invoked in previous studies in order to define a range over which scale changes are likely to detect meaningful changes in curve length. Our previous research (Batty and Longley, 1987) suggests an appropriate lower limit to this range is the mean length of the chords linking the successive digitised points (or the use of every point in the case of the equipaced polygon method; see below) and an upper limit is a scale which approximates the curve with no less than eight chords. In our empirical measurement, we usually interpolate 30 changes in scale within these limits in accordance with a geometric criterion. The rudiments of each of the methods which enable curves to be approximated at successive scales are as follows:

3.1. *The Structured Walk Method*

This is the original method used by Richardson (1961) to measure the length of frontiers and coastlines, and originally involved manually walking a pair of dividers along a mapped boundary; scale-dependent lengths could be obtained by altering the divider span between walks. This principle has been automated using an algorithm developed by Shelby *et al* (1982) which involves using different prespecified chord lengths to approximate the boundary of an object consisting of the segments between digitised co-ordinates. Points are interpolated onto the base level curve at the given chord length intervals using standard trigonometry. The structured walk is started at any point on the curve and proceeds in both directions to the curve's endpoints; as a consequence, the last steps are a fraction of the fixed step size (Kaye, 1978). In our own analysis we used 30 changes in scale, where each scale was related to the lowest scale r_0 by $r_k = G^k r_0$, $k = 0,1,\ldots,30$. r_0 was computed by first calculating the distances between each adjacent pair of (x,y) co-ordinates i and $i+1$:

$$d_{i,i+1} = [(x_i - x_{i+1})^2 + (y_i - y_{i+1})^2]^{1/2}, \qquad i=1,\ldots,N-1 \tag{8}$$

Next, the perimeter of the digitised base level curve was computed as

$$P = \sum_{i=1}^{N-1} d_{i,i+1} \tag{9}$$

The lowest scale r_0 was then taken to be the average chord length, which from Equation (9) is given as

$$r_0 = P/(N-1) \tag{10}$$

At any given scale specified by chord length r, after a starting point (x_p,y_p) and a direction towards the curve end has been chosen, the distance $d_{p,i}$ from the start point to the next co-ordinate pair (x_i,y_i) is computed using Equation (8). If this distance is less than the chord length r, the next co-ordinate pair (x_{i+1},y_{i+1}) is selected, the distance $d_{p,i+1}$ computed and the test against chord length r is made again. This process continues until the distance $d_{p,i+k} < r$ and when this is achieved, a new point x_{p+1},y_{p+1} is interpolated on the line segment $[x_{i+k-1},y_{i+k-1} : x_{i+k},y_{i+k}]$ using the Shelby *et al* algorithm referred to above. In this way, chords of exactly length r are computed which span the curve until the end point is approached. If the distance between the last interpolated point and the end point is less than r as it will generally be, a fraction of the chord length r is computed and the interpolation is thus closed on the end point.

This process is operated from any starting point in both directions along the curve to its end points, and the curve's entire perimeter and chords are computed by adding both sets of computations.

3.2. *The Equipaced Polygon Method*

This method has been suggested by Kaye and Clarke (1985) and involves using the chords associated with the original digitised base level curve, rather than the new points interpolated onto the curve as used in the structured walk method. As a consequence, chords are not generally equal in length when the equipaced polygon method is used. Perimeter lengths are computed by taking the sequence of chord lengths between adjacent co-ordinates, then between co-ordinates which are spaced at more than one pair of co-ordinates apart, and so on. In order to give more equal weighting in the log-log regressions, the sequences are constructed geometrically using adjacent co-ordinates, then every 2nd, every 4th, every 8th co-ordinate pair and so on are used. The chord length r in Equation (3) is then given by the average chord length measured in each of these sequences (and thus the lowest scale is given by the average chord length r_0 as in Equation (10) above). Use of the original digitised co-ordinates to anchor each of the scale sequences means that, first, measured perimeter length is likely to depend upon the choice of the co-ordinate used to institute the structured walk and, second, that the spacing of digitised points (itself a function of digitising mode and/or operator behaviour) is likely to affect results in a broader sense.

From a given starting point which is always a digitised point on the base level curve (x_i,y_i), a direction is established and a chord constructed to a digitised point (x_{i+k},y_{i+k}) which is k steps away from the starting point: k is thus an index of scale. The distance $d_{i,i+k}$ is computed using Equation (8), and then the reset chord involving the point (x_{i+2k},y_{i+2k}) is constructed from (x_{i+k},y_{i+k}). This process continues until the endpoint is approached and when the step length k is less than or equal to the remaining number of chords at the base level, the curve is closed on the endpoint. Computations in both directions along the curve are added to determine the entire perimeter and total number of chords.

3.3. *The Hybrid Walk Method*

This method has been suggested by Clark (1986) as a compromise between the structured walk and equipaced polygon procedures. Like the structured walk method, it is based upon specification of a series of chord lengths rather than point sequences and involves the same geometric chord length series and lowest scale of resolution (r_0) as the structured walk method. However, it is more akin to the equipaced polygon method in that no new points are interpolated onto the digitised base level curve. Instead, the chord is either extended or contracted to coincide with the nearest digitised base point; this point is then used as the origin from which the next chord length is sought; and so on.

The method proceeds in exactly the same manner as the structured walk detailed above, but when a point (x_{i+k},y_{i+k}) is reached where $d_{p,i+k} < r$, a new point is not interpolated using the Shelby *et al* algorithm, but point (x_{i+k},y_{i+k}) is chosen if $|d_{p,i+k} - r| \leq |d_{p,i+k-1} - r|$. If not, then the point (x_{i+k-1},y_{i+k-1}) is chosen, that point being closer to the chord length r than the first point. Closure of the curve and computation of its perimeter and chord number is achieved as in the structured walk method.

3.4. *The Cell Count Method*

This method has been suggested as a computationally inexpensive first approximation by a number of authors (Goodchild, 1980; Dearnley, 1985; Morse *et al*, 1985). In concept it is akin to imposing a regular lattice upon the digitised map, the squares of which are based upon a similar sequence of scales to those used in the other methods. Rather than computing lengths using detailed trigonometry, the cell count algorithm simply searches out and counts all of the cells that the base level curve passes through. This process is then repeated at the full range of desired resolutions. Although strictly speaking each cell scale should itself be defined with respect to the start and end points of the base level curve, convenience and the desire to maintain clear comparability led to our using the same 31 scales adopted in the structured and hybrid walk methods.

The algorithm for the cell count method is computationally very straightforward. From a given starting point (x_p,y_p) with a chosen cell size r and direction of traverse, the next co-ordinate point (x_i,y_i) in the given direction is chosen. A test is made to see if $|x_i-x_p| \geq r$ or $|y_i-y_p| \geq r$. If one of these conditions holds, a new point is established where the co-ordinate in question is updated in the direction of greatest increase. That is, if $|x_i-x_p| \geq |y_i-y_p|$, $x_{p+1} = x_p+r$ and $y_{p+1} = y_p$ while if the converse holds, $x_{p+1} = x_p$ and $y_{p+1} = y_p+r$. If the first test is not met, that is if the increase in either direction is less than the grid size r, then a new co-ordinate point (x_{i+1},y_{i+1}) is chosen, and the same operations occur. Each time the direction is updated, a cell has been crossed and is thus counted. When the end point of the curve is approached, the cell curve is not closed on this point but simply finishes when the cell in which the end point exists has been identified. Computations in both directions along the base curve are added to produce the composite results in the usual manner.

Clearly an artifact of all of these methods is that both the number of chords and the perimeter lengths will depend upon the chosen starting point along the curve. Some previous researchers have sought to reduce the arbitrariness of this variation by starting the measurements at a small number of different points and then averaging the results (e.g. Kaye *et al*, 1985; Kent and Wong, 1982). In view of the exclusive dedication of a MicroVAX 2 computer to this stage of our project we were able to pursue this approach to its logical end, and our empirical results reported in the next section are the averages obtained by beginning measurements at every single one of our digitised points. In the following section we will explore some of the benefits which accrue as the result of this averaging.

4. FRACTAL MEASUREMENT OF THE CARDIFF URBAN EDGE

Our digitised data base attempts to depict the urban edge of Cardiff, in 1886, 1901, 1922 and 1949, using available 1:10560 scale Ordnance Survey (OS) maps for the earliest three time slices and a 1:25000 OS map for 1949. These particular dates were chosen because they depict the period of Cardiff's fastest population growth rate and most rapid industrial change (i.e. 1860 until World War I), which then led into the beginnings of the contemporary suburbanisation process (see Daunton, 1977, for an exposition of Cardiff's growth between 1870 and 1914). Development of the South Wales Coalfield, the configuration of landownership, attitudes of landowners, and the attendant focussing of the regional railway system upon Cardiff all conspired to develop the docks area as one of Britain's foremost coal-exporting ports by the late 19th century. The period 1872-1914 also saw the development of the tramway system which opened up tracts of suburban land for more spacious housing development. However, the City's heavy over-reliance upon coal coupled with the failure of the coal export trade to generate any local industrial linkages led to the demise of the docks during the inter-war period, and thus the early 20th Century remains the heyday of Cardiff's industrial history. Subsequent development of Cardiff as the Welsh capital has, however, led to a growth in administrative functions in the Post-World War II period.

The data were captured using the point-mode digitising package MicroPLOT (Bracken, 1985) which met the specific requirements of our project. From the outset we were aware that definition of the 'urban' edge is inherently subjective. Whilst we were able to readily devise practical rules-of-thumb consistent with a functional conception of the urban edge (e.g. allotments were classified as being of inherently

TABLE I
Summary statistics of the digitised data sets

Data set	No.of points	Mean distance between points	St. dev. of distance	1st[1] quartile	3rd[1] quartile
1886	2458	12.48	4.60	9.22	15.03
1901	2757	11.02	4.32	8.54	12.73
1922	4615	10.49	3.53	8.06	12.21
1949	1558	47.20	37.99	23.67	52.94
1949 (M)[2]	1925	38.23	36.50	13.42	52.33

[1] cut-off points.
[2] corrected for inconsistencies: see text.

'urban' character), it was much more difficult to discern the land vacancy and dereliction which relieves land of its 'urban' function (e.g. desertion of land adjoining significant tracts of railway sidings which became disused during the evolution of Cardiff's urban form). Once defined, digitising the urban boundary in point mode proved to be a more tractable process although as always, data capture is reliant upon operator precision, and is thus likely to include errors and to reflect operator learning behaviour to a limited extent. Efforts were made to record points at very frequent intervals and to digitise erratic line features particularly intensively. Intermediate processing was used to scale the 1949 1:25000 digitised data up to the same level of resolution as the other three series. Table I illustrates that the 1949 series is both the coarsest and most erratic series, facts which are given visual expression in Figure 2 which shows the shape of Cardiff in 1949 from the digitised point data only.

When the four maps were plotted, it became clear that at least two interpretive inconsistencies existed between the 1922 and 1949 maps, since the 1922 boundary projected through the 1949 envelope in two places. This was remedied by substituting 1922 values into the

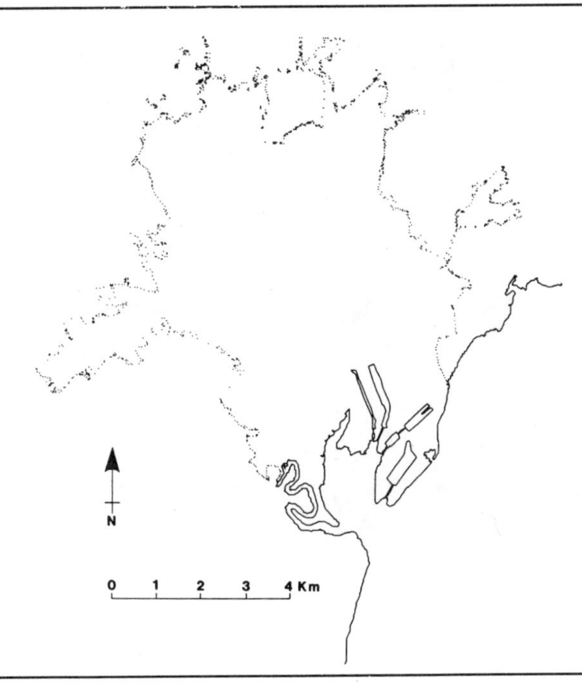

Fig. 2: The density of point digitisation forming the 1949 Cardiff urban edge.

1949 files along segments for which a digitising inconsistency was visually apparent; the summary statistics of this merged file are contained in the 1949 (M) row of Table I, and this modified file was used in the fractal measurements which we report below. In Figure 3, we show all four urban edges and this provides an immediate visual impression of the growth of Cardiff during this period as well as a clue to changes in the irregularity of its urban boundary.

We described in Section 3 how scales may be specified in a geometric ratio in order to derive equally weighted observations in log-linear regression analysis. These points may also be displayed during analysis on a log-log scatterplot as a means of permitting visual inspection of any trend amongst the points. This presents a simple yet clear means of assessing the appropriateness of the standard fractal measurement hypothesis that the fractal dimension of a curve is constant across a range of scales. In Section 2 we made the point that the spatial form of social systems is never the result of a

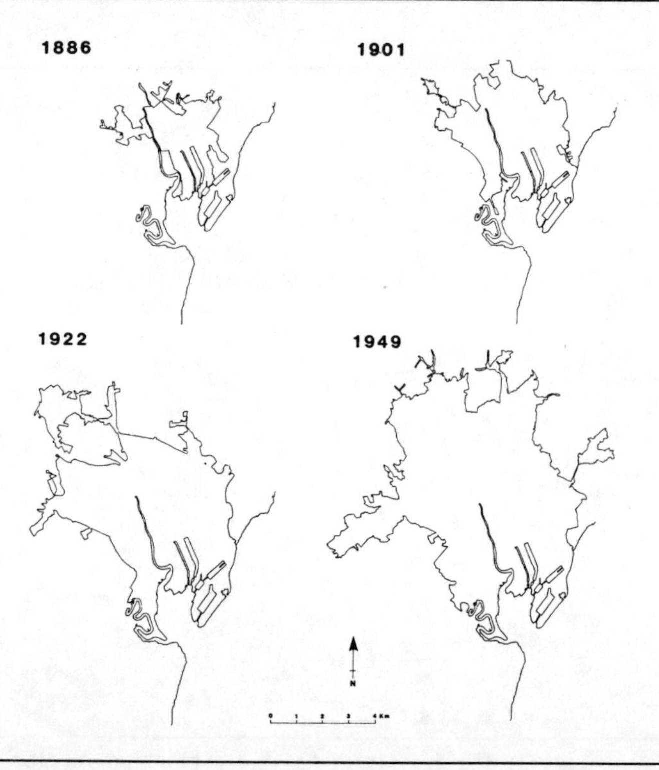

Fig. 3: Urban edges of Cardiff in 1886, 1901, 1922, and 1949.

single process and that, as a consequence, we might anticipate our urban edge being multi-fractal; use of log-log scatterplots provides a means of diagnosing ill-fitting points and of informing the choice of a more appropriate global functional form. Our use of a different (smaller-scale) Ordnance Survey series to digitise the 1949 map raises some doubt as to the strict comparability of results for this time period with results from the other three; inspection of scatterplots thus helps fulfil a second role of assessing the generality of any single functional form across the differing ranges of scales within the four-data series. Taken together, our various empirical explorations have two goals: at a technical level, to evaluate four

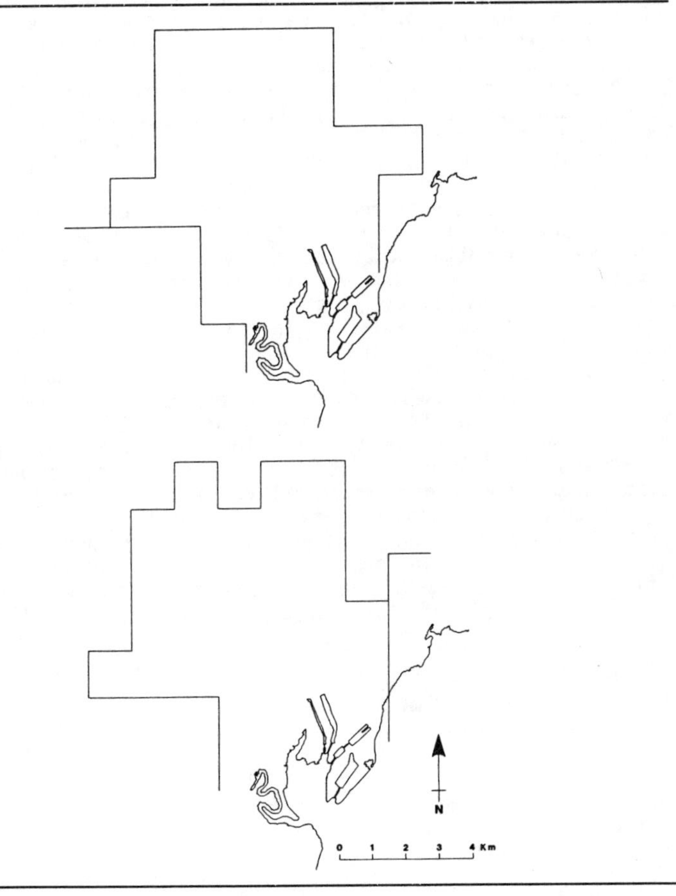

Fig. 4: Variations in shape approximation using different starting points for the cell count method.

different fractal measurement methods using data which have been captured from base maps at different scales; and, at a substantive level, to measure the fractal dimension of Cardiff over a long period of time in the hope that these results might be used to inform urban theory and/or to articulate it through the medium of computer-based displays.

The first stage involved applying each of the four fractal measurement techniques to our four time slice data series. 31 scales, consisting of the first scale based on r_0, the average chord length, and approximately 30 successive order of magnitude changes, provided the observations based on the criteria described in Section 3. Although choice of starting point makes little or no difference to the results when the base curve is being traversed in very small increments, this is not the case for large step increments which only crudely approximate the curve and yield a measurement which is highly sensitive to the lengths of the large residual steps which are left at either end of the curve. Figure 4 illustrates how different are the fractal measurements at a fairly coarse resolution using the cell count method, depending upon whether the chosen starting point is co-ordinate 456 (upper diagram) or co-ordinate 1234 (lower diagram). In order to make our results asymptotically acceptable, we began successive measurements for each method at every single point on each of the four base curves; final observations for each method, year and scale were then obtained by taking appropriate averages. The total CPU times (using a MicroVAX 2 machine) for this entire process are recorded in Table II, whilst the scatterplots of observations obtained using the structured walk, equipaced polygon, hybrid walk and cell count methods are depicted in Figures 5, 6, 7 and 8 respectively. These results, coupled with our own reflections upon the different measurement techniques, led us to suggest four important criteria which should guide the adoption of a particular measurement method in further work. First, we felt that deep fissures in the base curves (e.g. those representing suburban communities connected to the main urban area by ribbon development) could distort measurements made by some methods more than those made using others. The equipaced polygon method was particularly susceptible to such phenomena, since the measured curve could suddenly dislocate when the point weeding criteria missed some fissured points for the first time. Figure 9 shows an example of this susceptibility at two adjacent scales.

TABLE II
CPU times associated with the different methods (days:hours:minutes)

	1886	1901	1922	1949 (M)
Structured Walk	15:23	19:11	2:07:10	7:49
Equipaced Polygon	0:36	0:45	2:09	0:20
Hybrid Walk	12:34	15:52	1:21:56	6:52
Cell Count	3:04	3:51	11:30	1:37

Second, some methods were more vulnerable than others to the way in which the base map had been digitised. The structured walk method precisely interpolated points on the base curve using trigonometry, whilst the cell count method simply tracks the path of the base curve across a regular grid; as a consequence, neither is influenced by the number and spacing of base curve anchor co-ordinates to any great degree. This is not the case for either the hybrid walk or equipaced polygon methods in which the chord lengths are constrained by the digitising criteria used to create the base map. Third, we felt that measurements obtained using the structured and hybrid walk methods were both sensitive to changes in scale; by contrast the log-log scatterplots obtained using the cell count method flatten towards the coarser scales at which the method is less capable of representing base curve shape. Finally, Table II shows the immense range in CPU times required for the repeated simulations, ranging in the case of the 1922 curve from approximately two hours to over two and a quarter days. The equipaced polygon method clearly has the lowest resource demands, with the cell count method coming a fairly strong second in this respect. Nevertheless, after taking our previous three comments into account we would continue to advocate use of the structured walk method in the measurement process. In circumstances of more acute

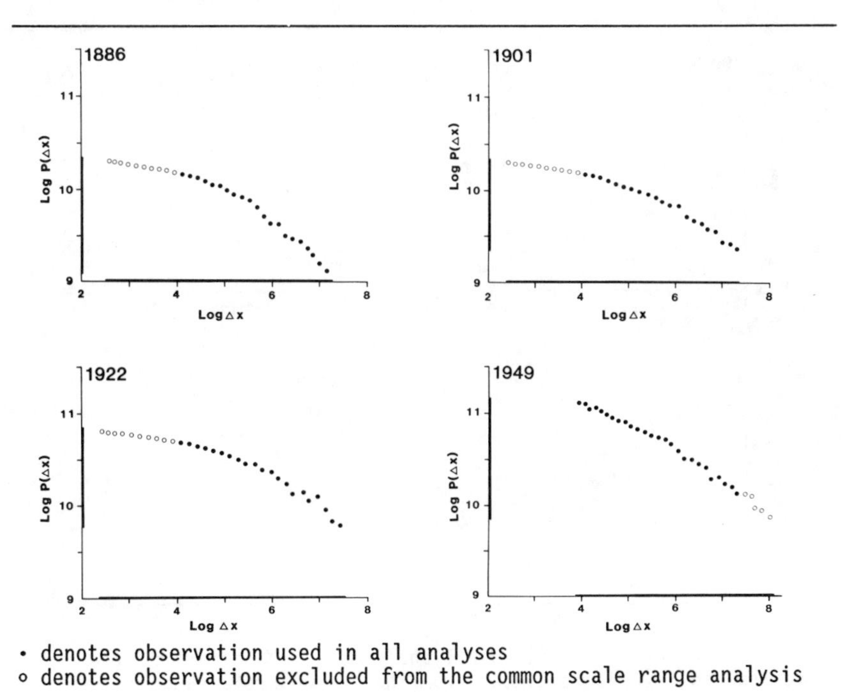

• denotes observation used in all analyses
o denotes observation excluded from the common scale range analysis

Fig. 5: Variations in perimeter length over many scales: the structured walk method.

resource constraint, the low variability of measurements made from different starting points using the structured walk method means that many fewer starting points could be used in future.

Most empirical fractal measurement studies assume that the standard log-linear functional form (Equation (7)) should be adopted in regression analysis. This is consistent with the notion that the self-similar detail about a curve is fashioned by a single process at a variety of scales. Regression analysis using this form and all of the points depicted for the four methods in Figure 5 to 8 yielded the results shown in Table III. The R^2 values are within the range 0.87-0.93 for the first three time periods for the structured walk, equipaced polygon and hybrid walk methods; these might be considered to be on the low side in view of the sample size and the nature of the regression analysis. Only the results of the less reliable cell method broach an acceptable threshold of 0.95. The R^2 values are, however, much better for the 1949 data, which were generated by scaling up digitised data from a smaller scale map and hence were not suitable for generating observations at the finer levels of resolution.

Visual inspection of Figures 5 to 8 suggests that a curvilinear functional form might be more appropriate: if allowance is made for

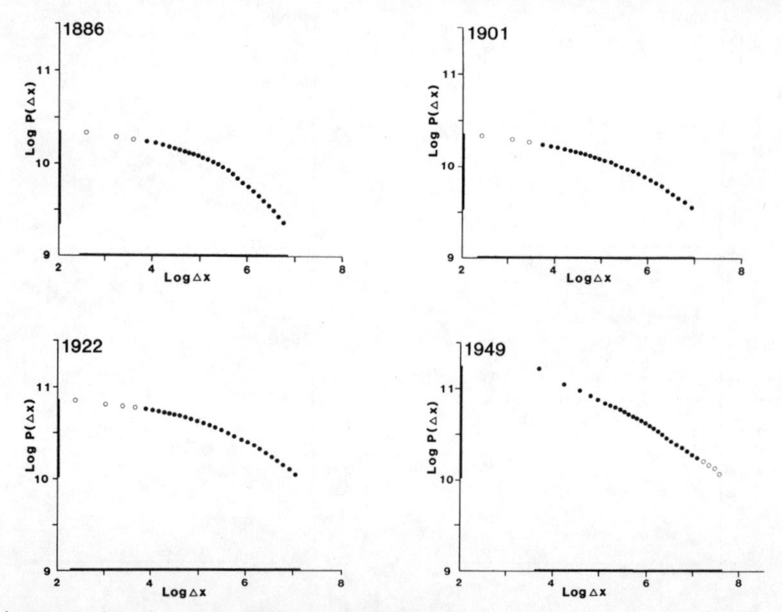

• denotes observation used in all analyses
o denotes observation excluded from the common scale range analysis

Fig. 6: Variations in perimeter length over many scales: the equipaced polygon method.

the measurement difficulties associated with the cell method at coarse scales, it becomes impossible to identify clear breaks in the slopes of plots and so fitting several linear functions to the points would be arbitrary. We thus postulate that the fractal dimension itself is a function of scale and that the scaling coefficient ($1-D$) is given by

$$(1-D) = b_1 + b_2 r \qquad (11)$$

Substituting this into Equation (7) gives

$$\ell n\, P = \ell n\, a - b_1\, \ell n\, r - b_2\, r\, \ell n\, r \qquad (12)$$

As the scale $r \to 0$, $D \to 1 + b_1$. Thus the term $b_2\, r\, \ell n\, r$ in Equation (12) acts as a dispersion factor which increases the fractal dimension at increased scales. If $b_2 = 0$, then this factor which introduces the non-linearity into the plots is redundant and Equation (12) collapses back into Equation (7). In essence, the model postulates increasing fractal dimension with scale.

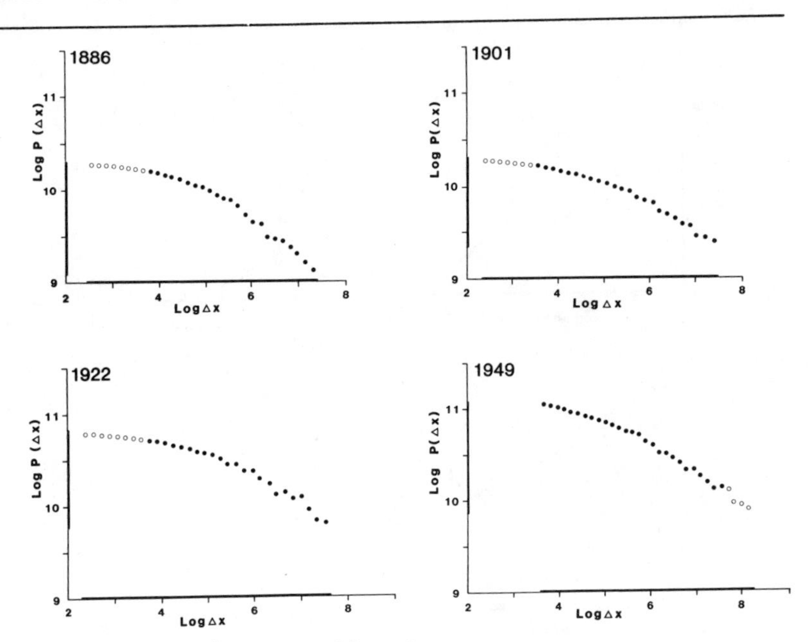

• denotes observation used in all analyses
o denotes observation excluded from the common scale range analysis

Fig. 7: Variations in perimeter length over many scales: the hybrid walk method.

The results of regressions based on Equation (12) and using measurements using all observations from the four methods are shown in Table IV. These show a dramatic improvement in the R^2 values for the 1886, 1901 and 1922 time slices, and suggest that this more complicated model is still the most parsimonious means of accounting for the variation in the data. A further specification based upon

$$(1-D) = b_1 + b_2 \ln r$$

was also investigated, although results were significantly inferior to the model specified in Equation (11) and are thus not worthy of reporting.

As we have already suggested, effective comparison across the four time slices is hindered by the fact that the 1949 data were digitised at a coarser interval. This means that measurements of the fractal dimension at fine scales are not meaningful for this series, although the much greater areal extent of Cardiff in 1949 means that measurements at very coarse scales remain quite reliable. As an attempt to generate more direct comparability between our four time

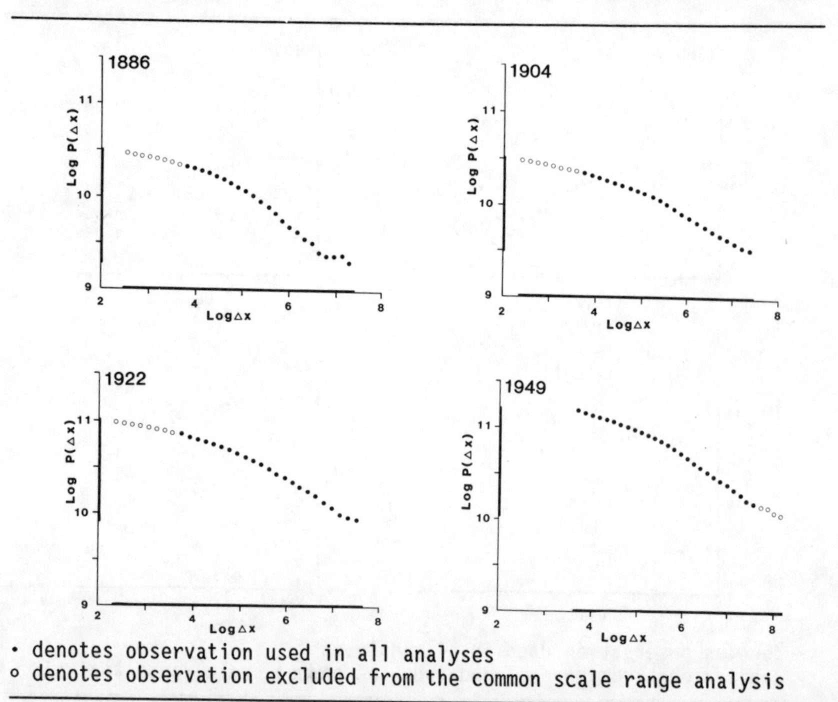

Fig. 8: Variations in perimeter length over many scales: the cell count method.

slices the following procedure was adopted: for each of the four methods, the minimum log chord size was taken to be the smallest value used in the 1949 series, and thus all smaller values on the 1886, 1901 and 1922 files were deleted; the maximum log chord size was then taken to be the largest value used in any of the first three (1886, 1901 and 1922) series, and the larger values on the 1949 file were also deleted; regression analyses based on Equations (7) and (12) were then used to derive the results presented in Tables V and VI respectively. The points which were used in these analyses are those marked by solid dots in Figures 5 to 8, the full sample used previously to form Tables III and IV being based on the range of points including both solid and open dots.

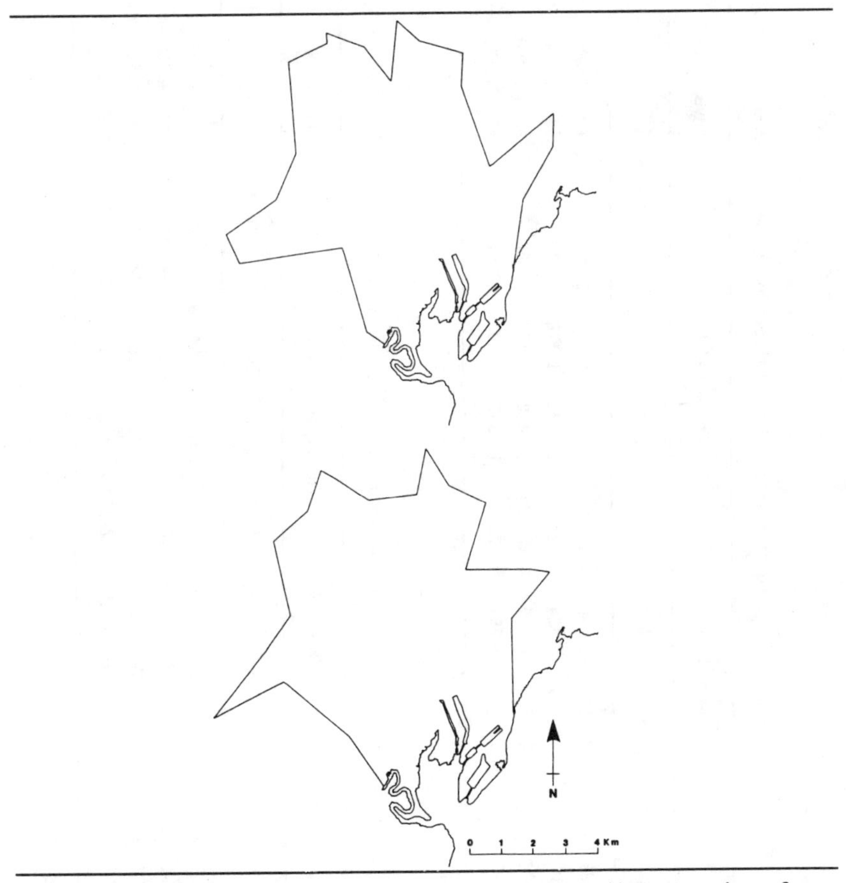

Fig. 9: Sudden changes in approximation at two adjacent order of magnitude scales using the equipaced polygon method.

Table III

Scaling constants and fractal dimensions from Equation (7): full sample

	Structured Walk			Equipaced Polygon			Hybrid Walk			Cell Count		
Date	$\ln a$	D	R^2	$\ln a$	D	R^2	$\ln a$	D	R^2	$\ln a$	D	R^2
1886	11.080	1.239	0.914	11.176	1.236	0.875	11.119	1.248	0.913	11.326	1.267	0.953
1901	10.866	1.184	0.927	10.923	1.178	0.917	10.895	1.190	0.929	11.079	1.200	0.967
1922	11.393	1.186	0.907	11.420	1.172	0.902	11.412	1.190	0.906	11.617	1.209	0.957
1949	12.150	1.267	0.975	12.342	1.293	0.992	12.416	1.308	0.989	12.288	1.274	0.985

Table IV

Scaling constants and fractal dimensions from Equation (12): full sample

	Structured Walk				Equipaced Polygon				Hybrid Walk				Cell Count			
Date	$\ln a$	d^*	$b_2 \times 10^{-5}$	R^2	$\ln a$	d^*	$b_2 \times 10^{-5}$	R^2	$\ln a$	d^*	$b_2 \times 10^{-5}$	R^2	$\ln a$	d^*	$b_2 \times 10^{-5}$	R^2
1886	10.719	1.141	5.865	0.983	10.589	1.086	11.200	0.995	10.715	1.137	7.256	0.987	11.109	1.207	3.525	0.973
1901	10.622	1.117	3.947	0.985	10.594	1.094	5.920	0.994	10.633	1.117	4.560	0.990	10.919	1.156	2.592	0.989
1922	11.114	1.109	3.901	0.984	11.078	1.085	5.187	0.992	11.111	1.106	4.567	0.989	11.426	1.156	2.686	0.988
1949	11.883	1.211	1.202	0.991	12.132	1.250	1.211	0.998	12.197	1.262	1.001	0.996	12.144	1.244	0.646	0.990

* $d = 1 + b_1$ (when $r = 0$)

Table V

Scaling constants and fractal dimensions from Equation (7): common scale range sample

Date	Structured Walk			Equipaced Polygon			Hybrid Walk			Cell Count		
	$\ell n\ a$	D	R^2	$\ell n\ a$	D	R^2	$\ell n\ a$	D	R^2	$\ell n\ a$	D	R^2
1886	11.513	1.312	0.964	11.505	1.294	0.937	11.702	1.345	0.968	11.685	1.327	0.983
1901	11.162	1.234	0.964	11.120	1.213	0.963	11.284	1.255	0.970	11.303	1.238	0.988
1922	11.729	1.241	0.950	11.661	1.213	0.959	11.863	1.263	0.953	11.890	1.254	0.985
1949	12.061	1.250	0.972	12.287	1.283	0.991	12.340	1.293	0.989	12.260	1.269	0.978

Table VI

Scaling constants and fractal dimensions from Equation (12): common scale range sample

Date	Structured Walk				Equipaced Polygon				Hybrid Walk				Cell Count			
	$\ell n\ a$	d^*	$b_2 \times 10^{-5}$	R^2	$\ell n\ a$	d^*	$b_2 \times 10^{-5}$	R^2	$\ell n\ a$	d^*	$b_2 \times 10^{-5}$	R^2	$\ell n\ a$	d^*	$b_2 \times 10^{-5}$	R^2
1886	11.071	1.213	3.866	0.990	10.741	1.119	9.784	0.997	11.148	1.225	4.605	0.991	11.599	1.307	0.076	0.984
1901	10.842	1.162	2.740	0.990	10.735	1.124	4.709	0.998	10.898	1.171	3.038	0.993	11.144	1.202	1.359	0.994
1922	11.331	1.153	2.909	0.988	11.247	1.120	4.031	0.998	11.360	1.156	3.370	0.990	11.684	1.208	1.507	0.994
1949	11.781	1.187	2.129	0.991	12.066	1.235	1.935	1.00	12.091	1.238	2.021	0.996	11.978	1.206	2.141	0.994

* $d = 1 + b_1$ (when $r = 0$)

The results obtained using all observations and the standard (Equation (7)) model indicate that fractal dimension appears to decrease during the period 1886-1922, but increases for the 1949 time slice. The goodness-of-fit (R^2) statistics for the standard model remain generally on the low side, and the curvilinear (Equation (12)) model is necessary to make these fits more acceptable. The curvilinear model reveals the same trends in fractal dimensions, and further indicates a consistent decrease in the effect of scale (b_2) over time. When the smaller common scale range samples are used the same trends are apparent (with the exception of the anomalous results for the cell count estimates of the b_2 parameters). The main difference is that the exclusion of the smaller scale intervals has the effect of increasing most of the R^2 statistics for the simple model above the 0.95 benchmark; exclusion of these observations thus has the effect of truncating the curvilinear trend which is apparent in Figures 5 to 8.

5. CONCLUSIONS

As stated in the introduction, one of the aims of our measurement exercise was to devise rules for partitioning space in our computer simulations. In this sense our results provide a plausible generating rule for replicating urban boundaries in similar contexts in which the precise path of any required boundary is unknown (see Batty and Longley, 1986, for an example of this approach). Although this 'black-box' role of the fractal measurements provided an important motivation to this research, the results also go some way towards informing urban theory about the evolution of Cardiff's spatial form.

The phenomenon of first decreasing and then increasing fractal dimension suggests that Cardiff's urban edge became increasingly regular as it first reached, and was then constrained by the railway network which was centred upon the docks area; by 1949 the pressure of urban development had bridged these barriers and (aided by the development of non-rail-based transit) had spilled into the beginnings of the major suburban developments which characterise Cardiff's contemporary form. This steady release of land further relaxed the impact of the rail network upon the macro-scale shape of Cardiff, as is manifest in the decreasing magnitude of the scale (b_2) parameter, and leads to the more linear form of our 1949 scatterplots (Figures 5 to 8). Taken together, we anticipate that the urban form is becoming more irregular as tentacles of development reach out along road-based transport routes. These conclusions should nevertheless be tempered with two qualifications: first, 1949 really only represents the beginning of this process and some later time slices should be examined to attempt a strengthening of this hypothesis; and, second, the coarser scale at which the 1949 data were digitised curtails the range of scales over which this hypothesis can be reliably evaluated. That said, our limited evidence does suggest that the irregularity of urban form may become increasingly replicable by a standard log-linear relationship between perimeter and scale as time progresses.

Of further and more general import is the finding that the three data series which extend to the finer scales clearly exhibit multifractal structures. This leads to the clear implication that the fractal dimensions of these urban boundaries are a function of scale. It remains for further research in diverse spatial and temporal contexts to assess the generality of this finding, and to identify its correspondence with the particular social and economic processes which exist or have existed in these different contexts. As a final conclusion, it is also clear that we have established methods that not only enable substantive explanations of spatial geometries but also enable consistent predictions and generalisation of a variety of boundaries and shapes. We would argue that this synthesis of the substantive and methodological is an essential requirement to progress in the area of cartographic generalisation.

Department of Town Planning
University of Wales Institute of Science and Technology
Cardiff
United Kingdom

BIBLIOGRAPHY

Arlinghaus, S. L.: 1985, 'Fractals take a central place', *Geografiska Annaler* **67B**, 83-88.

Batty, M. and P. A. Longley: 1986, 'The fractal simulation of urban structure', *Environment and Planning* **A 18**, 1143-1179.

Batty, M. and P. A. Longley: 1987, 'Fractal-based description of urban form', *Environment and Planning* **B 14**, 123-134.

Bracken, I.: 1985, 'Computer-aided cartography with microcomputers: a practical guide to MICROPLOT', Working Paper 90, Department of Town Planning UWIST, Cardiff, UK.

Buttenfield, B.: 1984, 'Line structure in graphic and cartographic space', PhD Thesis, University of Washington and University Microfilms International.

Clark, N. N.: 1986, 'Three techniques for implementing digital fractal analysis of particle shape', *Powder Technology* **46**, 45-52.

Daunton, M. J.: 1977, *Coal Metropolis: Cardiff 1870-1914*, Leicester University Press, Leicester.

Dearnley, R.: 1985, 'Effects of resolution on the measurement of grain 'size'', *Mineralogical Magazine* **49**, 539-546.

Goodchild, M. F.: 1980, 'Fractals and the accuracy of geographical measures', *Mathematical Geology* **12**, 85-98.

Haggett, P., A. D. Cliff, and A. Frey: 1977, *Locational Analysis in Human Geography*, Edward Arnold, London.

Jenks, G. F.: 1981, 'Lines, computers and human frailties', *Annals of the Association of American Geographers* **71**, 1-10.

Kaye, B. H.: 1978, 'Specification of the ruggedness and/or texture of a fineparticle profile by its fractal dimension', *Powder Technology* **21**, 1-16.

Kaye, B. H. and G. G. Clark: 1985, 'Fractal Description of Extraterrestrial Fineparticles', Department of Physics, Laurentian University, Sudbury, Ontario, Canada.

Kaye, B. H., J. E. Leblanc, and P. Abbot: 1985, 'Fractal description of the structure of fresh and eroded aluminium shot fineparticles', *Particle Characterization* **2**, 56-61.
Kent, C. and J. Wong: 1982, 'An index of littoral zone complexity and its measurement', *Canadian Journal of Fisheries and Aquatic Sciences* **39**, 847-853.
Mandelbrot, B. B.: 1967, 'How long is the coast of Britain? statistical self-similarity and fractional dimension', *Science* **156**, 636-638.
Mandelbrot, B. B.: 1982, *The Fractal Geometry of Nature*, W. H. Freeman and Company, New York.
Mark, D. M. and P. B. Aronson: 1984, 'Scale-dependent fractal dimensions of topographic surfaces: an empirical investigation, with applications in geomorphology and computer mapping', *Mathematical Geology* **16**, 671-683.
Morse, D. R., J. H. Lawton, M. M. Dodson, and M. H. Williamson: 1985, 'Fractal dimension of vegetation and the distribution of arthropod body lengths', *Nature* **314**, 731-733.
Muller, J-C.: 1986, 'Fractal and automated line generalisation', Paper presented to the Canadian Cartographic Association, Simon Fraser University, British Columbia.
Perkal, J.: 1966, (Trans. R. Jackowski) 'On the length of empirical curves', Michigan Inter-University Community of Mathematical Geographers, Discussion Paper no. 10.
Peucker, T. K.: 1975, 'A theory of the cartographic line', *Proceedings*, Auto-Carto II, 508-518, Reston, Virginia.
Richardson, L. F.: 1961, 'The problem of contiguity: an appendix to statistics of deadly quarrels', *General Systems Yearbook* **6**, 139-187.
Shelby, M. C., H. Moellering, and N. Lam: 1982, 'Measuring the fractal dimensions of empirical cartographic curves', *Auto-Carto* **5**, 481-490.
Woronow, A.: 1981, 'Morphometric consistency with the Hausdorff-Besicovich dimension', *Mathematical Geology* **13**, 201-216.

INDEX OF NAMES

INDEX OF SUBJECTS